ELECTROCHEMISTRY

ELECTROCHEMISTRY

PHILIP H. RIEGER

Brown University

PRENTICE-HALL, INC.
Englewood Cliffs, N. J. 07632

Library of Congress Cataloging-in-Publication Data

RIEGER, PHILIP HENRI, date
    Electrochemistry.

    Includes bibliographies and index.
    1. Electrochemistry. I. Title.
QD553.R53   1987   541.3'7   87-2229
ISBN   0-13-248907-4

Cover design: Edsal Enterprises
Manufacturing buyer: Harry Baisley

© 1987 by Prentice-Hall, Inc.,
A Division of Simon & Schuster,
Englewood Cliffs, New Jersey 07632

Printed in the United States of America

10   9   8   7   6   5   4   3   2   1

ISBN 0-13-248907-4   01

Prentice-Hall International (UK) Limited, *London*
Prentice-Hall of Australia Pty. Limited, *Sydney*
Prentice-Hall Canada Inc., *Toronto*
Prentice-Hall Hispanoamericana, S.A., *Mexico*
Prentice-Hall of India Private Limited, *New Delhi*
Prentice-Hall of Japan, Inc., *Tokyo*
Prentice-Hall of Southeast Asia Pte. Ltd., *Singapore*
Editora Prentice-Hall do Brasil, Ltda., *Rio de Janeiro*
Whitehall Books Limited, *Wellington, New Zealand*

*Dedicated to the memory of those who
inspired my interest in electrochemistry:*

Arthur F. Scott

William H. Reinmuth

Donald E. Smith

CONTENTS

CHAPTER 2. THE ELECTRIFIED INTERFACE

PREFACE

It has been fashionable to describe electrochemistry as a discipline which interfaces with many other sciences. A perusal of the table of contents will affirm that view. Electrochemistry finds applications in all branchs of chemistry as well as in biology, biochemistry, and engineering; electrochemistry gives us batteries and fuel cells, electroplating and electrosynthesis, and a host of industrial and technological applications which are barely touched on in this book. However, I will maintain that electrochemistry is really a branch of physical chemistry. Electrochemistry grew out of the same tradition which gave physics the study of electricity and magnetism. The reputed founders of physical chemistry—Arrhenius, Ostwald, and van't Hoff—made many of their contributions in areas which would now be regarded as electrochemistry. With the post–World War II capture of physical chemistry by chemical physicists, electrochemists have tended to retreat into analytical chemistry, thus defining themselves out of a great tradition. G. N. Lewis defined physical chemistry as "the study of that which is interesting." I hope that the readers of this book will find that electrochemistry qualifies.

While I have tried to touch on all the important areas of electrochemistry, there are some which have had short shrift. For example, there is virtually nothing on spectroelectrochemistry or the use of dedicated microcomputers in electrochemical instrumentation, and there is rather little on ion–selective electrodes and chemically modified electrodes. The selection of topics has been far harder than I anticipated, a reflection of my ignorance of some important areas when I started. On the other hand, there may be a few topics which may appear to have received too much attention. I confess that my interest in electrochemistry is primarily in mechanistic studies, particularly with organometallic systems. This orientation may be all too apparent for some readers.

Since this is a textbook with the aim of introducing electrochemistry to the previously uninitiated, breadth has been sought at the expense of depth. I have tried, however, to provide numerous entries into the review literature so that a particular topic of interest can be followed up with a minimum of effort. References in the text are of four types. Some are primarily of historical interest; when I have traced ideas to their origins, I have tried to give the original reference, fully aware that only a science history buff is likely to read them but equally aware that such references can be hard to find. A second class of references is to specific results from the recent literature, and a third class leads to the review literature. These references are collected at the end of each chapter. A fourth class of references includes the books and monographs which are collected in a classified Bibliography, Appendix 1.

SI units have been employed throughout the book. References to older units are given in footnotes where appropriate. In most cases, the use of SI units eliminates unit conversion problems and greatly simplifies numerical calculations. The major remaining source of units ambiguity comes from concentrations. When a concentration is used as an approximation to an activity, molar units (mol L^{-1}) must be used to conform to the customary standard state. But when a concentration acts as a mechanical variable, $e.g.$, in a diffusion problem, the SI unit, mol m^{-3}, should be used. The mol m^{-3} concentration unit is equivalent to mmol L^{-1} and, in a sense, is a more practical concentration scale since voltammetric experiments often employ substrate concentrations in the millimolar range.

This book has had a long, slow evolution. It began as chapters on electrochemistry and transport processes in a textbook on biophysical chemistry, a joint project with J. H. Gibbs and J. M. Steim which unfortunately came to naught. There is a slightly biological flavor in parts of Chapters 1 and 2 which reflects this origin. It then evolved into a set of lecture notes for a short course in electrochemistry for seniors and graduate students at Brown; the emphasis in the course was organometallic applications and again that flavor remains, particularly in Chapters 4 and 5. Chapters and sections were added over the years until about 200 pages had accumulated. In the spring of 1983, the chemistry editor at Prentice–Hall, Elizabeth Perry, expressed interest in the book and the lecture notes started on a painful journey toward a book. Progress was slow until the summer of 1984, by which time Chapters 1–3 had attained nearly their present form, and Ms. Perry's successor, Nancy Forsyth, formalized the nascent book with a contract. Chapters 4 and 5 grew during 1984 and early 1985, and Chapter 6 was largely written during the spring of 1985 while I enjoyed a sabbatical leave at the University of Vermont. Ms. Forsyth was succeeded at Prentice–Hall by Curtis Yehnert in 1985 and by Dan Joraanstad in 1986, and they shepherded the book into final form.

Many people have helped along the way. I am particularly grateful to my wife, Anne L. Rieger, for her patience in listening to my problems and for her encouragement in times of discouragement.

David Gosser has listened to my ideas and offered many helpful suggestions; the cyclic voltammogram simulations of Chapters 4 and 5 are his work. I have benefitted from the advice and encouragement of many of my colleagues at Brown, most particularly Joe Steim, John Edwards, Dwight Sweigart, and Ed Mason. In addition to providing a stimulating atmosphere at the University of Vermont, Bill Geiger offered some timely advice on electroanalytical chemistry. I am particularly grateful to James Anderson of the University of Georgia, to Arthur Diaz of IBM, San Jose, to Harry Finklea of Virginia Polytechnic Institute, and to Franklin Schultz of Florida Atlantic University for their careful reading of the manuscript and numerous helpful suggestions; the book has been much improved through their efforts. I am also grateful to Catherine Schwab, who contributed several problems; to Nancy Stone, whose careful reading turned up some errors; to Bertha Hansen, Carol White, and Patricia Fay, who typed early versions of Chapters 1–3; and to Barbara Tellier, who helped with the assembly of the figures.

Since late 1984, I have been linked to Brown's mainframe computer, and Chapters 4–6 have been written and edited directly at a terminal. The book has been produced using the IBM Waterloo SCRIPT word–processing system and a Xerox 9700 laser printer equipped with Century Schoolbook roman, italic, bold, bold italic, greek, and mathematics fonts. The help of Richard Damon and Allen Renear, of the Brown Computer Center, and Virginia Huebner, production editor at Prentice–Hall, has been essential to the completion of a book which at least approachs professional standards. All the figures have been produced using TOPDRAW software (SLAC Computation Group, Stanford, CA) and a Calcomp plotter. TOPDRAW was never intended for line drawings and some of the figures turned out to be *tours de force*. Multiple drafts were often required to get a figure just right; the patience of the Computer Center staff has been appreciated.

Philip H. Rieger

December 1986

1

ELECTROCHEMICAL POTENTIALS

1.1 INTRODUCTION

Origins of Electrode Potentials

When a piece of metal is immersed in an electrolyte solution, an electric potential difference is developed between the metal and the solution. This phenomenon is not unique to a metal and electrolyte; in general whenever two dissimilar conducting phases are brought into contact, an electric potential is developed across the interface. In order to understand this effect, let us consider first the related case of two dissimilar metals in contact.

When individual atoms condense to form a solid, the various atomic orbital energy levels broaden and merge, generally forming two bands of allowed energy levels. The band of levels corresponding to the bonding molecular orbitals in a small molecule is called the *valence band* and usually is completely filled. The band of levels corresponding to nonbonding molecular orbitals is called the *conduction band*. This band is partially filled in a metal and is responsible for the electrical conductivity. As shown in Figure 1.1, electrons fill the conduction band up to an energy called the *Fermi level*. The energy of the Fermi level, relative to the zero defined by ionization, depends on the atomic orbital

1

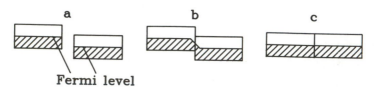

Fermi level

Figure 1.1 The conduction bands of two dissimilar metals (a) when the metals are not in contact; (b) at the instant of contact; and (c) at equilibrium.

energies of the metal and on the number of electrons occupying the band and thus varies from one metal to another. Now consider two dissimilar metals. Since the Fermi levels are at different energies, when the two metals are brought into contact, electrons flow from the metal with the higher Fermi level into the metal with the lower Fermi level. This electron transfer results in a separation of charge and an electric potential difference across the phase boundary. The effect of the electric potential difference is to raise the energy of the conduction band of the second metal and to lower the energy of the conduction band of the first until the Fermi levels are equal in energy; when the Fermi levels are equal, no further electron transfer takes place. In other words, the intrinsically lower energy of the second metal's conduction band is exactly compensated by the electrical work required to move an electron from the first metal to the second against the electric potential difference.

A very similar process occurs when a metal, say a piece of copper, is placed in a solution of copper sulfate. Some of the copper ions may deposit on the copper metal, accepting electrons from the metal conduction band and leaving the metal with a small positive charge and the solution with a small negative charge. With a more active metal, it may be the other way around: a few atoms leave the metal surface as ions, giving the metal a small negative charge and the solution a small positive charge. The direction of charge transfer depends on the metal, but in general charge separation occurs and an electric potential difference is developed between the metal and the solution.

When two dissimilar electrolyte solutions are brought into contact, there is generally a charge separation at the phase boundary owing to the different rates of diffusion of the various ions. The resulting electric potential difference, called a *liquid junction potential*, is discussed in Section 3.4.

In general, whenever two conducting phases are brought into contact, an interphase electric potential difference will develop. The exploitation of this phenomenon is one of the subjects of electrochemistry.

Galvanic Cells

Consider the electrochemical cell shown in Figure 1.2. A piece of zinc metal is immersed in a solution of $ZnSO_4$ and a piece of copper metal is immersed in a solution of $CuSO_4$. The two solutions make contact with one another through a fritted glass disk (to prevent mixing), and the two pieces of metal are attached to a voltmeter

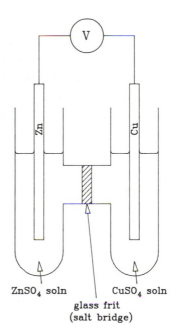

ZnSO$_4$ soln   CuSO$_4$ soln
                    glass frit
Figure 1.2 The Daniell cell.   (salt bridge)

through copper wires. The voltmeter tells us that a potential is developed, but what is its origin? There are altogether four sources of potential: (1) the copper–zinc junction where the voltmeter lead is attached to the zinc electrode; (2) the zinc–solution interface; (3) the junction between the two solutions; and (4) the solution–copper interface. The measured voltage is the sum of all four interphase potentials.

In the discussion which follows, we shall neglect potentials which arise from junctions between two dissimilar metals or two dissimilar solutions. This is not to say that such junctions introduce negligible potentials; however, our interest lies primarily in the metal–solution interface and solid or liquid junction potentials make more or less constant additive contributions to the measured potentials of electrochemical cells. In careful work, it is necessary to take explicit account of solid and liquid junction potentials.

The electrochemical cell we have been discussing was invented by in 1836 by Daniell.[‡] It was one of many such cells developed to supply electrical energy before electrical generators were available. Such cells are called *galvanic cells,* remembering Galvani,[#] who in 1791 accidentally discovered that static electricity could cause a convulsion in a frog's leg; he then found that a static generator was unnecessary for the effect, that two dissimilar metals (and an electrolyte solution) could also result in the same kinds of muscle contractions. Galvani thought of the frog's leg as an integral part of the experiment, but in 1800 Volta[§] showed that the generation of electricity had nothing to do with the frog and was able to construct a battery (the *voltaic pile*) from alternating plates of silver and zinc separated by cloth soaked in salt solution. This invention was quickly taken up by many other scientists, most immediately by Nicholson and Carlisle,[¶] who

---

[‡]   John F. Daniell (1790–1845) was Professor of Chemistry at King's College, London. Daniell was a prolific inventor of scientific apparatus but is best known for the electrochemical cell which bears his name.

[#]   Luigi Galvani (1737–1798) was a physiologist at the University of Bologna. Beginning about 1780, Galvani became interested in "animal electricity" and conducted all kinds of experiments looking for electrical effects in living systems.

[§]   Alessandro G. A. A. Volta (1745–1827) was Professor of Physics at the University of Pavia. Volta had worked on problems in electrostatics, meteorology, and pneumatics until Galvani's discovery attracted his attention.

[¶]   William Nicholson (1753–1815) started his career as an East India Company civil servant, was then a salesman for Wedgwood pottery in Holland, an aspiring novelist, a teacher of mathematics, a

discovered that water is decomposed to hydrogen and oxygen by electrolysis. The Nicholson–Carlisle experiment, published only a few weeks after Volta's pile was announced in England, caused a sensation in scientific circles throughout Europe. Among many scientists who followed this lead was Davy,[‡] who, in 1806, used the voltaic pile as a source of electricity to isolate metallic sodium and potassium from the hydroxides. Davy's assistant, Faraday,[#] went on in the next decades to lay the foundations of the science of electrochemistry.[§]

A galvanic cell is a means of converting chemical energy into electrical energy, and, as such is of enormous technical importance. To begin with, however, our interest in galvanic cells will be less in their technical consequences than in the kinds of chemical information we can obtain from them. We shall find that the electric potential produced by a cell is proportional to the Gibbs free energy change of the cell reaction; electrochemical cell potentials thus are an important source of chemical thermodynamic data. We will also examine some of the many important analytical applications of electrochemical cells.

Electrochemical cells can operate to convert chemical energy to electrical energy. Alternatively, current may be passed through a cell from an external source to effect a chemical transformation (as in the

---

physics textbook writer and translator, a civil engineer, patent agent, and inventor of scientific apparatus. He founded the *Journal of Natural Philosophy, Chemistry, and the Arts* in 1797, which he published monthly until 1813. Sir Anthony Carlisle (1768–1840) was a socially prominent surgeon who dabbled in physics and chemistry on the side.

[‡] Sir Humphry Davy (1778–1829) was Professor of Chemistry at the Royal Institution. Davy was an empiricist who never accepted Dalton's atomic theory and spent most of his career looking for defects in Lavoisier's theories, but in the process he made some very important discoveries in chemistry.

[#] Although Michael Faraday (1791–1867) began his career as Davy's assistant at the Royal Institution, he soon made an independent reputation for his important discoveries in organic chemistry, electricity and magnetism, and in electrochemistry. Although his electrochemical work was seemingly an extension of Davy's electrolysis experiments, in fact Faraday was asking much more fundamental questions. Faraday is responsible (with the classicist

experiments of Nicholson, Carlisle, and Davy); such cells are called *electrolysis cells.* Some confusion is possible if we attempt to discuss both types at once, and we confine our attention in this chapter to galvanic cells. We will return to electrolysis cells in Chapter 4.

Since the most obvious feature of a galvanic cell is its ability to convert chemical energy to electric energy, we shall begin our study by investigating the thermodynamic role of electrical work. In Section 1.3, we discuss some applications of data obtained from electrochemical cells. We turn to some experimental details in Sections 1.4–1.6 and conclude this chapter with introductions to analytical potentiometry in Section 1.7 and to batteries and fuel cells in Section 1.8.

1.2 ELECTROCHEMICAL CELL THERMODYNAMICS

Electrical Work

The first law of thermodynamics may be stated as

$$\Delta U = q + w \tag{1.1}$$

where ΔU is the change in the internal energy of the system, q is the heat absorbed by the system, and w is the work done on the system. In elementary thermodynamics, we usually deal only with mechanical work, for example, the work done when a gas is compressed under the influence of pressure ($dw = -P dV$) or the expansion of a surface area under the influence of surface tension ($dw = \gamma dA$). However, other kinds of work are possible and here we are especially interested in electrical work, the work done when an electrical charge is moved through an electric potential difference.

Consider a system which undergoes a reversible process at constant temperature and pressure in which both mechanical (P–V) work and electrical work are done:

William Whewell) for many of the terms still used in electrochemistry, such as *electrode, cathode, anode, electrolysis, anion,* and *cation.*

§   The early history of electrochemistry is brilliantly expounded in Ostwald's 1896 book, now available in English translation (*C1*).

$$w = -P\Delta V + w_{elec}$$

Since, for a reversible process at constant temperature, $q = T\Delta S$, eq (1.1) becomes

$$\Delta U_{T,P} = T\Delta S - P\Delta V + w_{elec} \qquad (1.2)$$

At constant pressure, the system's enthalpy change is

$$\Delta H_P = \Delta U_P + P\Delta V \qquad (1.3)$$

and at constant temperature, the Gibbs free energy change is

$$\Delta G_T = \Delta H_T - T\Delta S \qquad (1.4)$$

Combining eqs (1.2)–(1.4), we have

$$\Delta G_{T,P} = w_{elec} \qquad (1.5)$$

This quite remarkable result immediately demonstrates the utility of electrochemical measurements: We have a direct method for the determination of changes in the Gibbs free energy without recourse to measuring equilibrium constants or enthalpy and entropy changes. Now let us see how electrical work is related to the experimentally measurable parameters which characterize an electrochemical system. Consider an electrochemical cell (the thermodynamic system) which has two terminals across which there is an electric potential difference, E.[‡] The two terminals are connected by wires to an external load (the

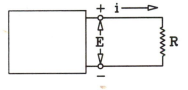

Figure 1.3 Electrochemical cell doing work on an external resistance.

[‡]   In Chapter 2, where we will be dealing with electric potential in a slightly different context, we will use the symbol Φ for potential. Here, we follow tradition and denote the potential difference produced by an electrochemical cell by the symbol E, which comes from the archaic term *electromotive force*. The electromotive force or emf is synonymous with potential difference or voltage.

surroundings), represented by a resistance R. When a charge Q is moved through a potential difference E, the work done on the surroundings is

$$w_{elec} = EQ$$

The charge passed in the circuit is the product of the number of charge carriers and the charge per charge carrier. If we assume that the charge carriers are electrons, then

$$Q = \text{(number of moles of electrons)} \times \text{(charge/mole)}$$

or

$$Q = nF$$

where F is the Faraday constant, the charge on one mole of electrons, 96,484.6 coulombs (C), and n is the number of moles of electrons transferred. Thus the work done by the system on the resistor (the resistor's energy is raised) is simply nFE. However, according to the sign convention of eq (1.1), work done on the system is positive so that the electrical work is negative if the system transfers energy to the surroundings,

$$w_{elec} = -nFE \qquad\qquad (1.6)$$

Substituting eq (1.6) into eq (1.5), we obtain the change in Gibbs free energy of the system,

$$\Delta G_{T,P} = -nFE \qquad\qquad (1.7)$$

If E is measured in volts (V), F in C mol^{-1}, and n is the number of moles of electrons per mole of reaction (mol mol^{-1}), then ΔG will have the units of joules per mole (J mol^{-1}) since 1 J = 1 V–C. This quite remarkable result immediately demonstrates the utility of electrochemical measurements: We have a direct method for the determination of changes in the Gibbs free energy without recourse to measuring equilibrium constants or enthalpy and entropy changes.

According to the second law of thermodynamics, a spontaneous process at constant temperature and pressure results in a decrease in Gibbs free energy. Thus a positive potential is expected when the cell reaction is spontaneous. There is room for ambiguity here since we could equally well measure the potential in either direction. However, we recall the convention for the sign of ΔG for a chemical reaction: if the chemical reaction is spontaneous, *i.e.*, proceeds from left to right *as written*, we say that ΔG is negative. We need a convention for the sign of E which is consistent with that for ΔG.

Electrochemical Cell Conventions

In developing the required conventions, let us consider as a specific example the Weston cell[‡] shown in Figure 1.4. It is customary, in discussing electrochemical cells, to use a shorthand notation to represent the cell rather than drawing a picture of the experimental

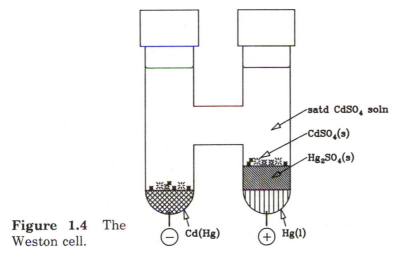

Figure 1.4 The Weston cell.

apparatus. The shorthand representation uses vertical lines to represent phase boundaries and starts from left to right, noting the composition of each phase in the system. Thus, the Weston cell may be represented as:

$$Cd(12.5\% \text{ amalgam})\,|\,CdSO_4(s)\,|\,CdSO_4(aq,satd)\,|\,Hg_2SO_4(s)\,|\,Hg(l)$$

[‡] The Weston cell was developed in 1893 by Edward Weston (1850–1936), an inventor and manufacturer of precision electrical measuring instruments. The cell has long been used as a standard potential source since it produces a very reproducible potential, 1.0180 V at 25°C.

We now agree by convention that, if the right–hand electrode is positive with respect to the left–hand electrode, we will say that the cell potential is positive.

Look now at the chemical processes going on at the two electrodes. Consistent with the convention of reading from left to right, we say that at the left–hand electrode, the process is

$$Cd(Hg) \rightarrow Cd^{2+}(aq) + 2 e^-$$ (1.8)

and, at the right–hand electrode,

$$Hg_2SO_4(s) + 2 e^- \rightarrow 2 Hg(l) + SO_4^{2-}(aq)$$ (1.9)

The overall cell reaction then is the sum of these two "half–cell" reactions:

$$Cd(Hg) + Hg_2SO_4(s) \rightarrow Cd^{2+}(aq) + SO_4^{2-}(aq) + 2 Hg(l)$$ (1.10)

According to convention, the free energy change for the cell reaction is negative if the reaction proceeds spontaneously to the right and, according to eq (1.7), the cell potential should then be positive, *i.e.*, the right–hand electrode (Hg) should be positive with respect to the left–hand electrode (Cd).

Let us see if this is consistent. If the Hg electrode is positive, then conventional (positive) current should flow in the external circuit from + to – (from Hg to Cd) and electron (negative) current in the opposite direction. Thus electrons should enter the cell at the Hg electrode, converting Hg_2SO_4 to Hg and SO_4^{2-} [as in eq (1.9)], and leave the cell at the Cd electrode, converting Cd to Cd^{2+} [as in eq (1.8)] and this is consistent with the overall cell reaction proceeding from left to right as in eq (1.10).

The cell convention can be summarized as follows: For an electrochemical cell *as written*, a positive potential at the right–hand electrode with respect to the left–hand electrode is equivalent to a negative ΔG for the corresponding cell reaction. Conventional positive current flows from right to left in the external circuit, from left to right in the cell. Negative electron current flows from left to right in the external circuit, from right to left in the cell. The left–hand (negative) electrode is called the *anode* and the electrode process is an oxidation (removal of electrons); the right–hand (positive) electrode is called the *cathode* and the electrode process a reduction (addition of electrons).[‡]

Activities and Activity Coefficients

Consider a general chemical reaction

$$a\,A + \beta\,B \rightleftharpoons \gamma\,C + \delta\,D \qquad (1.11)$$

According to chemical thermodynamics, the Gibbs free energy change when the reaction proceeds to the right is

$$\Delta G = \Delta G^\circ + RT\,\ln\frac{a_C{}^\gamma a_D{}^\delta}{a_A{}^a a_B{}^\beta} \qquad (1.12)$$

where R is the gas constant, T the absolute temperature, and, for example, a_C is the activity of species C. At equilibrium, $\Delta G = 0$, and eq (1.12) reduces to the familiar relation

$$\Delta G^\circ = -\,RT\,\ln K_{eq} \qquad (1.13)$$

where

$$K_{eq} = \frac{a_C{}^\gamma a_D{}^\delta}{a_A{}^a a_B{}^\beta} \qquad (1.14)$$

In the derivation of eq (1.12), the activities were introduced to account for nonstandard states of the species. Thus for an ideal gas with standard state partial pressure $P^\circ = 1$ bar, the activity is $a = P/P^\circ$; for a component of an ideal solution with standard state concentration $C^\circ = 1$ mol L^{-1} (1 M), the activity is C/C°. Pure solids or liquids are already in standard states, so that their activities are unity. The solvent in an ideal solution is usually assumed to be essentially the pure liquid with unit activity.

[‡]   The identification of the cathode with the reduction process and the anode with the oxidation process is common to both galvanic and electrolysis cells and is a better definition to remember than the electrode polarity, which is different in the two kinds of cells.

In order to preserve the form of eqs (1.12), (1.13), and (1.14) for non–ideal solutions or mixtures of nonideal gases, so–called *activity coefficients* are introduced which account for the departure from ideality. Thus for a solute in a real solution, we write

$$a = \gamma C/C^\circ \tag{1.15}$$

where γ is the unitless activity coefficient, C is the concentration, and C° is the standard state concentration, 1 M.[‡] Since $C^\circ = 1$ M, activities are numerically equal to γC and we will normally leave C° out of expressions. We must remember, however, that activities, whether they are approximated by molar concentrations or by partial pressures or corrected for nonideality, are unitless. Thus equilibrium constants and the arguments of logarithms in expressions such as eq (1.12) are also unitless.

In this chapter, we will usually assume ideal behavior and ignore activity coefficients. We will return to the problem of non–ideality in Section 2.6.

The Nernst Equation

If we take eq (1.11) to be the overall reaction of an electrochemical cell, then eqs (1.7) and (1.12) may be combined to give

$$-nFE = \Delta G^\circ + RT \ln \frac{a_C^\gamma a_D^\delta}{a_A^\alpha a_B^\beta} \tag{1.16}$$

If we introduce the standard potential, defined by

$$E^\circ = -\Delta G^\circ/nF \tag{1.17}$$

and rearrange eq (1.16) slightly, we have the equation first derived by Nernst[#] in 1889:

[‡] We will use the 1 M standard state in this book, but another common choice is 1 molal, 1 mole solute per kilogram of solvent. Although activity coefficients are unitless, they do depend on the choice of the standard state (see Section 2.6).

[#] Hermann Walther Nernst (1864–1941) was Ostwald's assistant at

$$E = E° - \frac{RT}{nF} \ln \frac{a_C{}^{\gamma} a_D{}^{\delta}}{a_A{}^{a} a_B{}^{\beta}} \tag{1.18}$$

The Nernst equation relates the potential generated by an electrochemical cell to the activities of the chemical species involved in the cell reaction and to the standard potential, $E°$.

In using the Nernst equation or its predecessors, eqs (1.7) and (1.17), we must remember the significance of the parameter n. We usually use ΔG in units of energy *per mole* for a chemical equation *as written*. The parameter n refers to the number of moles of electrons transferred through the external circuit for the reaction as written, taking the stoichiometric coefficients as the number of moles of reactants and products. In order to determine n, the cell reaction must be broken down into the processes going on at the electrodes (the half-cell reactions) as was done in discussing the Weston cell.

If the Nernst equation is to be used for a cell operating at 25°C, it is sometimes convenient to insert the values of the temperature and the Faraday and gas constants and convert to common (base 10) logs. The Nernst equation then is

$$E = E° - \frac{0.0592}{n} \log \frac{a_C{}^{\gamma} a_D{}^{\delta}}{a_A{}^{a} a_B{}^{\beta}} \tag{1.19}$$

Let us now apply the Nernst equation to the Weston cell. Identifying the appropriate species for substitution into eq (1.18) we obtain (with n = 2 moles of electrons per mole of Cd)

$$E = E° - \frac{RT}{2F} \ln \frac{a_{Cd^{2+}} a_{SO_4^{2-}} a_{Hg}}{a_{Cd} a_{Hg_2SO_4}} \tag{1.20}$$

In the cell, Hg and Hg_2SO_4 are pure materials and so should have

---

the University of Leipzig and later Professor of Physical Chemistry at the Universities of Göttingen and Berlin. Nernst made many important contributions to thermodynamics (he discovered the Third Law) and to solution physical chemistry.

unit activities. The solution is saturated in $CdSO_4$, so that the Cd^{2+} and $SO_4{}^{2-}$ ion activities should be constant as long as the temperature remains constant. Finally, the Cd activity should be constant provided that the concentration of cadmium in the amalgam (mercury solution) remains constant. If sufficient current is drawn from the cell, enough of the cadmium could be oxidized to change this concentration appreciably, but if only small currents are drawn, the potential is seen to be relatively insensitive to changes such as evaporation of the solvent. The potential of the Weston cell, 1.0180 V at 25°C, is easily reproducible and has a relatively small temperature coefficient. The cell has been widely used as a standard potential source.

Half-Cell Potentials

As we have already seen, it is possible to think of the operation of a cell in terms of the reactions taking place at the two electrodes separately. Indeed we must know the half-cell reactions in order to determine n, the number of moles of electrons transferred per mole of reaction. There are advantages in discussing the properties of individual half-cells since (1) each half-cell involves a separate reaction which we would like to understand in isolation, and (2) in classifying and tabulating results, there will be a great saving in time and space if we can consider each half-cell individually rather than having to deal with all the cells which can be constructed using every possible combination of all the available half-cells.

Consider again the Weston cell, and think of the potential an electron sees in its trip through the cell. The potential is constant within the metallic phase of the electrodes and in the bulk of the electrolyte solution and changes from one constant value to another over a few molecular diameters at the phase boundaries.[‡] The profile of the potential then must look something like the sketch shown in Figure 1.5.

The Nernst equation for the Weston cell, eq (1.20), gives us the cell potential as a function of the standard potential and of the activities of all participants in the cell reaction. We would like to break eq (1.20) into two parts which represent the electrode–solution potential drops shown in Figure 1.5. It is customary in referring to

---

[‡] The details of this variation of potential with distance are considered in Chapter 2.

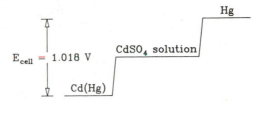

Figure 1.5
Hypothetical electric potential profile in the Weston cell.

these so–called "half–cell potentials" to speak of the potential of the solution relative to the electrode. This is equivalent to referring always to the electrode process as a reduction. When, in the actual cell reaction, the electrode process is an oxidation, the contribution to the cell potential will then be the negative of the corresponding reduction potential. Thus in the case of the Weston cell, we can write for the cell potential

$$E_{cell} = -E_{Cd^{2+}/Cd} + E_{Hg_2SO_4/Hg} \qquad (1.21)$$

Breaking the standard cell potential into two components in the same way, the half–cell potentials defined by eq (1.21) can be written in the form of the Nernst equation such that substitution in eq (1.21) gives eq (1.20). Thus

$$E_{Cd^{2+}/Cd} = E^{\circ}{}_{Cd^{2+}/Cd} - \frac{RT}{2F} \ln \frac{a_{Cd}}{a_{Cd^{2+}}}$$

$$E_{Hg_2SO_4/Hg} = E^{\circ}{}_{Hg_2SO_4/Hg} - \frac{RT}{2F} \ln \frac{a_{Hg}\, a_{SO_4{}^{2-}}}{a_{Hg_2SO_4}}$$

Unfortunately, there is no way of measuring directly the potential difference between an electrode and a solution, so that single electrode potentials cannot be uniquely defined.[‡] However, since the quantity is not measurable we are free to assign an arbitrary standard potential to one half–cell which will then be used as a standard reference. This

---

[‡] We will see in Section 2.5 that some properties (the electron–solution interfacial tension and the electrode–solution capacitance) depend on the electrode–solution potential difference and so could provide an indirect means of establishing the potential zero.

in effect establishes a potential zero against which all other half–cell potentials may be measured.

By universal agreement among chemists, the hydrogen electrode was chosen as the standard reference half–cell for aqueous solution electrochemistry.[‡] The hydrogen electrode, shown in Figure 1.6,

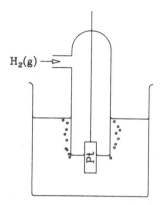

$H_2(g)$ →

Figure 1.6 The hydrogen electrode.

consists of a platinum electrode coated with finely divided platinum (platinum black) over which hydrogen gas is bubbled. The half–cell reaction

$$2\ H^+(aq) + 2\ e^- \rightarrow H_2(g)$$

is assigned the standard potential of 0.000 V. Thus the potential of the cell

$$Pt\,|\,H_2(g, a = 1)\,|\,H^+(a = 1),\ Zn^{2+}(a = 1)\,|\,Zn$$

is equal to the standard potential of the Zn^{2+}/Zn half–cell. By measuring the potentials of many such cells containing the hydrogen electrode, chemists have built up extensive tables of half–cell potentials. A selection of such data, taken from the recent compilation by Bard, Parsons, and Jordan *(H12)*, is given in Appendix Table A.4.

[‡]   See Section 4.2 for discussion of the problem of reference potentials for nonaqueous solutions.

1.3 SOME USES OF STANDARD POTENTIALS

With the data of Table A.4 or more extensive collections of half–cell potentials *(H1–H12)*, we should be able to predict the potentials of a large number of electrochemical cells. Since the standard potential of a cell is related to the standard Gibbs free energy change for the cell reaction by eq (1.17), we can also use standard cell potential data to compute $\Delta G°$, predict the direction of spontaneity, or calculate the equilibrium constant *(D6)*. In this section we will work through several examples of such calculations.

Many half–cell reactions in aqueous solutions involve H^+, OH^-, weak acids, or weak bases, so that the half–cell potential is a function of pH. While this dependence is predictable using the Nernst equation, it is often inconvenient to take explicit account of pH effects and a variety of techniques have been developed to simplify qualitative applications of half–cell potential data. An understanding of these methods is particularly important in biochemical applications. Several graphical presentations of half–cell potential data have been developed in attempts to make it easier to obtain qualitative predictions of spontaneity for redox reactions.

Potentials, Free Energies, and Equilibrium Constants

Building a hypothetical cell from two half–cells is straightforward on paper. (The construction of an actual working cell often can present insurmountable problems.) Consider two half–cells

$$A + n\,e^- \rightarrow B \qquad\qquad E°_{A/B}$$

$$C + m\,e^- \rightarrow D \qquad\qquad E°_{C/D}$$

The half–cell reactions are combined to give the cell reaction

$$m\,A + n\,D \rightarrow m\,B + n\,C$$

where we have multiplied the first equation by m, the second by n, and subtracted. To see that the cell potential is just

$$E°_{cell} = E°_{A/B} - E°_{C/D} \tag{1.22}$$

we need a few lines of proof. When we add or subtract chemical equations, we similarly add or subtract changes in thermodynamic state functions such as U, H, S, or G. Thus in this case

$$\Delta G°_{cell} = m \Delta G°_{A/B} - n \Delta G°_{C/D}$$

$$\Delta G°_{cell} = m (-nFE°_{A/B}) - n (-mFE°_{C/D})$$

$$\Delta G°_{cell} = -mnF(E°_{A/B} - E°_{C/D}) = -(mn)FE°_{cell}$$

where mn is the number of electrons transferred per mole of reaction as written.

Why can we combine half–cell potentials directly without taking account of stoichiometric coefficients, whereas ΔG's must be properly adjusted before combination? There is a significant difference between $E°$ and $\Delta G°$: G is an extensive property of the system so that when we change the number of moles we are discussing, we must adjust G; E, on the other hand, is an intensive variable—it is independent of the size of the system—related to ΔG by the extensive quantity nF.

Example 1.1 Compute $E°$ and $\Delta G°$ for the cell

$$Pt(s) \mid I^-(aq), I_2(aq) \parallel Fe^{2+}(aq), Fe^{3+}(aq) \mid Pt(s)$$

(the double vertical line refers to a salt bridge used to separate the two solutions).

Referring to Table A.4, we find the following data:

$$Fe^{3+} + e^- \rightarrow Fe^{2+} \qquad E°_{Fe^{3+}/Fe^{2+}} = 0.771 \text{ V}$$

$$I_2 + 2 e^- \rightarrow 2 I^- \qquad E°_{I_2/I^-} = 0.536 \text{ V}$$

To obtain the overall cell reaction, we multiply the first equation by 2 and subtract the second:

$$2 Fe^{3+} + 2 I^- \rightarrow 2 Fe^{2+} + I_2$$

$$E°_{cell} = E°_{Fe^{3+}/Fe^{2+}} - E°_{I_2/I^-} = 0.235 \text{ V}$$

$$\Delta G° = -2FE°$$

$$\Delta G° = -45.3 \times 10^3 \text{ J mol}^{-1}$$

Thus the oxidation of iodide ion by ferric ion is spontaneous under standard conditions. The same standard free energy change should apply for the reaction under nonelectrochemical conditions. Thus we can use the standard free energy change computed from the cell potential to calculate the equilibrium constant for the reaction of Fe^{3+} with I^- to give Fe^{2+} and I_2.

$$\ln K = -\Delta G^\circ / RT$$

$$K = 8.6 \times 10^7 = \frac{[Fe^{2+}]^2[I_2]}{[Fe^{3+}]^2[I^-]^2}$$

Because K is large, it might be difficult to determine directly by measurement of all the constituent concentrations, but we were able to compute it relatively easily from electrochemical data. Indeed, most of the very large or very small equilibrium constants we encounter in aqueous solution chemistry have their origins in electrochemical cell potential measurements.

Example 1.2 Given the half–cell potentials

$$Ag^+(aq) + e^- \rightarrow Ag(s) \qquad\qquad E^\circ = 0.7991 \text{ V}$$

$$AgBr(s) + e^- \rightarrow Ag(s) + Br^-(aq) \qquad E^\circ = 0.0711 \text{ V}$$

compute the solubility–product constant for silver bromide.

Subtraction of the first half–cell reaction from the second gives the AgBr solubility equilibrium

$$AgBr(s) \rightleftharpoons Ag^+(aq) + Br^-(aq)$$

The standard potential of the hypothetical cell with this reaction is

$$E_{cell} = E^\circ{}_{AgBr/Ag} - E^\circ{}_{Ag^+/Ag} = -0.7280 \text{ V}$$

$$\Delta G^\circ = -nFE^\circ = 70.24 \text{ kJ mol}^{-1}$$

$$K = 4.95 \times 10^{-13} = a_{Ag^+}a_{Br^-}$$

While potentials of half–cells are simply subtracted to compute the potential of a cell, there are seemingly similar calculations where this approach leads to the wrong answer. Consider the half–cell processes

$$A + n\,e^- \rightarrow B \qquad E^\circ_{A/B}$$

$$B + m\,e^- \rightarrow C \qquad E^\circ_{B/C}$$

$$A + (n+m)\,e^- \rightarrow C \qquad E^\circ_{A/C}$$

Clearly the last reaction is the sum of the first two, but $E^\circ_{A/C}$ is *NOT* the sum of $E^\circ_{A/B}$ and $E^\circ_{B/C}$. It must be true that

$$\Delta G^\circ_{A/C} = \Delta G^\circ_{A/B} + \Delta G^\circ_{B/C}$$

Converting to standard potentials, we have

$$-(n+m)FE^\circ_{A/C} = -nFE^\circ_{A/B} - mFE^\circ_{B/C}$$

or

$$E^\circ_{A/C} = \frac{nE^\circ_{A/B} + mE^\circ_{B/C}}{n + m} \qquad (1.23)$$

so that when half–cell potentials are combined to produce a new half–cell, the potentials are not additive. The best advice for calculations involving half–cell potentials is: **When in doubt, convert to free energies before doing the calculation.** Since the desired result is often a free energy change or an equilibrium constant, this strategy usually involves no more computations and is much less likely to lead to errors.

Example 1.3 Compute the standard half–cell potential for the reduction of Fe^{3+} to $Fe(s)$ given $E^\circ_{Fe^{3+}/Fe^{2+}}$ and $E^\circ_{Fe^{2+}/Fe}$.

The half–cell potentials from Table A.4 are

$$Fe^{3+} + e^- \rightarrow Fe^{2+} \qquad E^\circ = 0.771 \text{ V}$$

$$Fe^{2+} + 2\,e^- \rightarrow Fe(s) \qquad E^\circ = -0.44 \text{ V}$$

These half–cell potentials are combined using eq (1.23) to obtain the half–cell potential for the three–electron reduction of Fe^{3+}:

$$Fe^{3+} + 3\ e^- \rightarrow Fe(s)$$

$$E^\circ_{Fe^{3+}/Fe} = \tfrac{1}{3}\ (E^\circ_{Fe^{3+}/Fe^{2+}} + 2E^\circ_{Fe^{2+}/Fe})$$

$$E^\circ_{Fe^{3+}/Fe} = \tfrac{1}{3}\ (0.771 - 0.88) = -0.04\ V$$

Formal Potentials

Standard potentials refer to standard states, which for solution species are the hypothetical 1 M ideal solutions. Very dilute solutions can be assumed ideal and calculations using standard potentials are then reasonably accurate without activity coefficient corrections. For electrolyte concentrations less than about 0.01 M, activity coefficients can be computed reasonably accurately using Debye–Hückel theory (Section 2.6), but for more concentrated solutions, empirical activity coefficients are required.

One way around the problem of activity coefficients is through so–called *formal potentials*. A formal half–cell potential is defined as the potential of the half–cell when the *concentration* quotient of the Nernst equation equals 1. Consider the Fe(III)/Fe(II) couple. The Nernst equation gives

$$E = E^\circ_{Fe^{3+}/Fe^{2+}} - \frac{RT}{F}\ \ln \frac{a_{Fe^{2+}}}{a_{Fe^{3+}}}$$

or

$$E = E^\circ_{Fe^{3+}/Fe^{2+}} - \frac{RT}{F}\ \ln \frac{\gamma_{Fe^{2+}}}{\gamma_{Fe^{2+}}} - \frac{RT}{F}\ \ln \frac{[Fe^{2+}]}{[Fe^{3+}]}$$

Thus when the concentrations of Fe^{3+} and Fe^{2+} are equal, the last term on the right–hand side is zero and the formal half–cell potential is

$$E^{\circ\prime} = E^{\circ} - \frac{RT}{F} \ln \frac{\gamma_{Fe^{2+}}}{\gamma_{Fe^{3+}}}$$

As we will see in Section 2.6, activity coefficients depend primarily on the total electrolyte concentration (ionic strength) of the solution, so that in a solution where the ionic strength is determined mostly by a high concentration of an inert electrolyte, the activity coefficients are nearly constant. Molar concentrations can then be used, together with formal potentials appropriate to the medium, in calculations with the Nernst equation. A representative sample of formal potentials for 1 M $HClO_4$, 1 M HCl, and 1 M H_2SO_4 solutions is given in Table A.5.

The formal potential of the Fe(III)/Fe(II) couple in 1 M $HClO_4$ is $E^{\circ\prime} = 0.732$ V, significantly different from the standard potential of 0.771 V, suggesting that the activity coefficient ratio is about 0.22 in this medium. In a medium with coordinating anions, e.g., aqueous HCl, Fe(II), and Fe(III) form a variety of complexes. In order to compute the half-cell potential from the standard potential, we would have to know not only the activity coefficients but the formation constants of all the complexes present. The formal potential of the Fe(III)/Fe(II) couple in 1 M HCl, $E^{\circ\prime} = 0.700$ V, thus differs from the standard potential both because of activity coefficient effects and because of chloro complex formation. As long as the medium is constant, however, the relative importance of the various complexes is constant and the formal potential can be used as an empirical parameter to compute the overall Fe(II)/Fe(III) concentration ratio.

Latimer Diagrams

When we are interested in the redox chemistry of an element, a tabulation of half-cell potential data such as that given in Table A.4 can be difficult to assimilate at a glance. A lot of information is given and it is not organized to give a qualitative understanding of a redox system. Nitrogen, for example, exists in compounds having nitrogen oxidation states ranging from –3 Figure 1.7 for the aqueous nitrogen system. In the Latimer diagram, any (NH_3) to +5 (NO_3^-) and all intermediate oxidation states are represented. One way of dealing with complex systems like this is to use a simplified diagram introduced by Latimer[‡] (H1) and usually referred to with his name.

[‡] Wendell M. Latimer (1893–1955) was a student of G. N. Lewis and

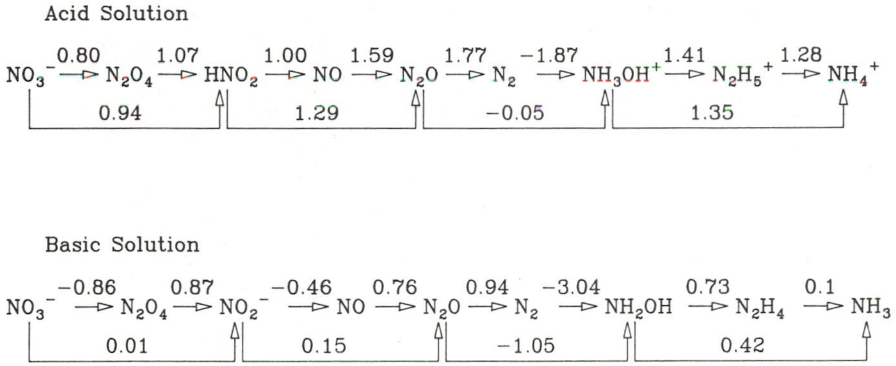

Acid Solution

$$NO_3^- \xrightarrow{0.80} N_2O_4 \xrightarrow{1.07} HNO_2 \xrightarrow{1.00} NO \xrightarrow{1.59} N_2O \xrightarrow{1.77} N_2 \xrightarrow{-1.87} NH_3OH^+ \xrightarrow{1.41} N_2H_5^+ \xrightarrow{1.28} NH_4^+$$

0.94   1.29   -0.05   1.35

Basic Solution

$$NO_3^- \xrightarrow{-0.86} N_2O_4 \xrightarrow{0.87} NO_2^- \xrightarrow{-0.46} NO \xrightarrow{0.76} N_2O \xrightarrow{0.94} N_2 \xrightarrow{-3.04} NH_2OH \xrightarrow{0.73} N_2H_4 \xrightarrow{0.1} NH_3$$

0.01   0.15   -1.05   0.42

Figure 1.7 Latimer diagram showing the half–cell potentials for the various nitrogen redox couples in acidic and basic aqueous solutions. Hydrogen or hydroxide ions or water required to balance the half–cell reactions have been omitted for clarity.

An example of a Latimer diagram is shown in H^+, OH^-, or H_2O required to balance the half–cell reaction is omitted for clarity. Thus, if we wish to use the half–cell potential for the NO_3^-/HNO_2 couple, for example, we must first balance the equation

$$NO_3^- + 3\,H^+ + 2\,e^- \rightarrow HNO_2 + H_2O \qquad E° = 0.94\ V$$

Since the pK_a of nitrous acid is 3.3, the N(III) species in neutral or basic solution is NO_2^-. Figure 1.7 also includes a Latimer diagram for nitrogen species in basic aqueous solution. Here the two–electron reduction of nitrate is

$$NO_3^- + H_2O + 2\,e^- \rightarrow NO_2^- + 2\,OH^- \qquad E° = 0.01\ V$$

As usual, the standard potential refers to 1 M standard state concentrations (activities) for all solution species including OH^-; thus $E° = 0.01$ V corresponds to pH 14. The two half–cell reactions for the reduction of nitrate are related and one potential can be computed from the other given the pK_a of HNO_2 and pK_w (the ionization constant of water).

---

later a Professor of Physical Chemistry at the University of California (Berkeley). Latimer's contributions were primarily in applications of thermodynamics to chemistry.

Example 1.4 Compute the half–cell potential for the reduction of NO_3^- to NO_2^- in basic solution given the potential for the reduction in acid solution, 0.94 V, and the ionization constants of nitrous acid and water, $pK_a = 3.3$, $pK_w = 14.00$.

The desired half–cell reaction is the sum of the following:

$$NO_3^- + 3\,H^+ + 2\,e^- \rightarrow HNO_2 + H_2O$$

$$\Delta G^\circ = -2FE^\circ_{NO_3^-/HNO_2}$$

$$HNO_2 \rightarrow H^+ + NO_2^-$$

$$\Delta G^\circ = 2.303\,RT\,pK_a$$

$$2\,H_2O \rightarrow 2\,H^+ + 2\,OH^-$$

$$\Delta G^\circ = 2 \times 2.303\,RT\,pK_w$$

The standard free energy change for the desired half–cell then is

$$\Delta G^\circ = -2F(0.94) + 2.303\,RT\,(3.3 + 28.00)$$

$$\Delta G^\circ = -2700\ \text{J mol}^{-1}\text{K}^{-1}$$

$$E^\circ = -\Delta G^\circ/2F = 0.01\ \text{V}$$

Free Energy – Oxidation State Diagrams

While Latimer diagrams compress a great deal of information into a relatively small space, they are expressed in potentials which, as we have seen, are not simply additive in sequential processes. Some simplification is possible if the potentials are converted to free energy changes relative to a common reference point. If we use the zero oxidation state (the element itself) as the reference point, then half–cell potentials can be converted to a kind of free energy of formation where the species of interest is formed from the element in its most stable form at 25°C and from electrons, hydrogen or hydroxide ions, water, or other solution species. Thus we could form HNO_2 from $N_2(g)$ by the following half–cell process:

$$\tfrac{1}{2} N_2(g) + 2 H_2O \rightarrow HNO_2 + 3 H^+ + 3 e^-$$

The free energy change for this process can be obtained from half–cell potential data such as those given in Figure 1.7. Since the reductions of HNO_2 to NO, NO to N_2O, and N_2O to N_2 all involve one electron per nitrogen atom, the potentials can be added directly to obtain the (negative) overall free energy change for the three–electron reduction; changing the sign gives $\Delta G°$ for the oxidation:

$$\Delta G° = F(1.77 + 1.59 + 1.00) = 4.36\, F$$

$\Delta G°$ is usually expressed in kJ mol^{-1}, but for the present purpose it is easier to think of $\Delta G°/F$, which is equivalent to putting $\Delta G°$ in units of electron–volts (1 eV = 96.485 kJ mol^{-1}).

Similar calculations for the free energies of the other nitrogen oxidation states can be done using the data of Figure 1.7. These free energies are plotted *vs.* nitrogen oxidation number in Figure 1.8. Free energy – oxidation state diagrams were introduced by Frost *(1)* and often are called Frost diagrams. These diagrams were popularized in England by Ebsworth *(2)* and sometimes are referred to as Ebsworth diagrams.

A free energy – oxidation state diagram contains all the information of a Latimer diagram but in a form which is more easily comprehensible for qualitative purposes. The slope of a line segment connecting any two points, $\Delta G°/\Delta n$, is just the potential for the reduction half–cell connecting the two species. Thus we can see qualitatively, for example, that $E°$ is positive for the reduction of NO_3^- to NH_4^+ in acid solution but is negative for the reduction of NO_3^- to NH_3 in basic solution. Species corresponding to minima, *e.g.*, $N_2(g)$, NH_4^+ (acid solution), NO_2^-, and NO_3^- (basic solution), are expected to be thermodynamically stable since pathways (at least from nearby species) are energetically downhill to these points. Conversely, points that lie at maxima are expected to be unstable since disproportionation to higher and lower oxidation states will lower the free energy of the system. Thus, for example, the disproportionation of N_2O_4 to NO_2^- and NO_3^- is highly exoergic in basic solution. The point for N_2O_4 in acid solution is not a maximum, but does lie above the line connecting HNO_2 and NO_3^-; thus disproportionation is spontaneous in acid solution as well.

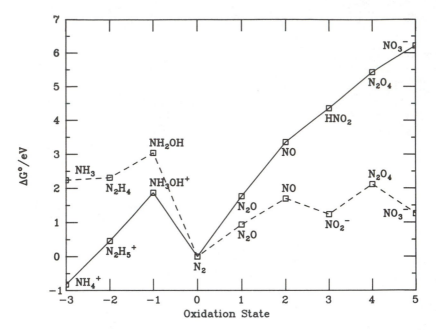

Figure 1.8 Free energy – oxidation state diagram for the nitrogen oxidation states in acid (solid lines) and basic (dashed lines) solutions.

The Biochemical Standard State

Oxidation–reduction reactions play important roles in biochemistry and half–cell potential data are often used in thermodynamic calculations. The usual standard reduction potentials tabulated in chemistry books or used in Latimer diagrams or free energy – oxidation state diagrams refer of course to 1 M standard states and thus to pH 0 or pH 14, depending on whether we choose to balance the half–cell reactions with H^+ or OH^- ions. Because life rarely occurs in strongly acidic or strongly basic solutions, however, neither of these choices is convenient for biochemical purposes, and biochemists usually redefine the standard states of H^+ and OH^- as pH 7, i.e., $[H^+]° = [OH^-]° = 10^{-7}$ M.

Biochemical standard free energy changes, standard potentials, and equilibrium constants are usually distinguished from the corresponding chemical standard quantities by writing $\Delta G'$, E', or K' in place of $\Delta G°$, $E°$, or K. A collection of biochemical standard potentials for some half–cell reactions of biological interest is given in Table A.6.

With the standard states of H^+ and OH^- defined as 10^{-7} M, we must be careful to recognize that the activities of these species are no longer even approximately equal to their molar concentrations. The activity of a solute i is defined (neglecting nonideality) as

$$a_i = C_i/C_i{}^\circ$$

so that when $C_i{}^\circ = 1$ M, $a_i = C_i$. But when $C_i{}^\circ = 10^{-7}$ M, $a_i = 10^7$ C_i. Consider the Nernst equation for the hydrogen electrode:

$$E = E^\circ - \frac{RT}{F} \ln \frac{(a_{H_2})^{\frac{1}{2}}}{a_{H^+}}$$

With $H_2(g)$ at unit activity, this expression can be written

$$E = E^\circ + \frac{RT}{F} \ln(a_{H^+})$$

Using the chemical standard state and E° = 0.00 V, we have

$$E = \frac{RT}{F} \ln[H^+]$$

With the biochemical standard state, the corresponding equation is

$$E = E' + \frac{RT}{F} \ln(10^7[H^+])$$

For equal $[H^+]$, the two equations must give the same potential; subtracting one equation from the other thus gives

$$E' = -0.414 \text{ V}$$

at 25°C. By a similar calculation, it is easy to show for a half–cell reaction

$$A + m\,H^+ + n\,e^- \rightarrow B$$

that E' and E° are related by

$$E' = E° - \frac{mRT}{nF} \ln(10^7) \tag{1.24}$$

or, at 25°C,

$$E' = E° - 0.414 \, m/n$$

Similar relations can be derived between $\Delta G'$ and $\Delta G°$ and between K' and K (see Problems). For an equilibrium reaction

$$A + m \, H^+ \rightleftharpoons B$$

we find that

$$K' = 10^{-7m} \, K \tag{1.25}$$

$$\Delta G' = \Delta G° + mRT \ln(10^7) \tag{1.26}$$

or, at 25°C,

$$\Delta G'/kJ \, mol^{-1} = \Delta G° + 40.0 \, m$$

Conversion back and forth between chemical and biochemical standard states is straightforward as long as the reactions involve H^+ or OH^- in clearly defined roles. Unfortunately, it is not always clear exactly how many H^+ or OH^- ions are involved in a half–cell reaction. The problem is that many species of biochemical interest are polyelectrolytes having weak acid functionalities with pK_a's near 7. For example, the pK_a of $H_2PO_4^-$ is 7.21, so that at pH 7, $[HPO_4^{2-}]/[H_2PO_4^-] = 0.62$. When phosphate is involved in a half–cell reaction, it is often bound (esterified) to give a species with a pK_a similar but not identical to that of $H_2PO_4^-$. Thus the exact number of protons involved in the electrode process is a complex function of pH. Biochemists often simply ignore this problem in qualitative discussions, referring to the mixture of $H_2PO_4^-$ and HPO_4^{2-} as inorganic phosphate, P_i.

Example 1.5 Calculate the biochemical standard free energy change and equilibrium constant and the chemical standard free energy change and equilibrium constant for the reaction of

3–phosphoglyceraldehyde with nicotinamide adenine dinucleotide (NAD^+) and inorganic phosphate, P_i, to give 1,3–diphospho–glycerate.

The half–cell potentials from Table A.6 are (writing R = $-CHOHCH_2OPO_3^{2-}$)

$$RCO_2PO_3^{2-} + 2\,H^+ + 2\,e^- \rightarrow RCHO + HPO_4^{2-}$$

$$E' = -0.286\ V$$

$$NAD^+ + H^+ + 2\,e^- \rightarrow NADH \qquad E' = -0.320\ V$$

The cell reaction is

$$RCHO + HPO_4^{2-} + NAD^+ \rightarrow RCO_2PO_3^{2-} + H^+ + NADH$$

and the standard cell potential at pH 7 is

$$E' = (-0.320) - (-0.286) = -0.034\ V$$

The free energy change and equilibrium constant at pH 7 are

$$\Delta G' = -nFE'$$

$$\Delta G' = +6560\ J\ mol^{-1} \qquad (6.56\ kJ\ mol^{-1})$$

$$K = \exp(-\Delta G'/RT)$$

so that

$$K = 0.071 = \frac{a_{RCO_2PO_3^{2-}}\,a_{H^+}\,a_{NADH}}{a_{RCHO}\,a_{HPO_4^{2-}}\,a_{NAD^+}}$$

or

$$K' = \frac{[RCO_2PO_3^{2-}](10^7[H^+])[NADH]}{[RCHO][HPO_4^{2-}]}$$

Equations (1.25) and (1.26) give

$$K = 7.1 \times 10^{-9}$$

$$\Delta G° = 46.6 \text{ kJ mol}^{-1}$$

This result shows one reason for the use of a special standard state for H^+ and OH^- in biochemistry. The reaction of NAD^+ with 3–phosphoglyceraldehyde appears to be almost hopelessly endoergic under chemical standard conditions. The values of $\Delta G°$ and K, while correct, are misleading since at pH 7 the equilibrium constant is not so very small and significant amounts of product are expected, particularly if the phosphate concentration is high.

Potential – pH Diagrams

Chemists interested in reactions at pH 0 or pH 14 or biochemists willing to stay at or near pH 7 are well served by tables of half–cell potentials, Latimer diagrams, or free energy – oxidation state diagrams. For systems at other pH values, the Nernst equation gives us a way of correcting potentials or free energies. Consider again the general half–cell process

$$A + m\, H^+ + n\, e^- \rightarrow B$$

with potential

$$E = E° - \frac{RT}{nF} \ln \frac{a_B}{a_A (a_{H^+})^m}$$

When the A and B activities are equal, we have

$$E = E° - 2.303 \frac{mRT}{nF} \text{ pH}$$

Thus a half–cell potential is expected to be linear in pH with a slope of $-59.2\ (m/n)$ mV per pH unit at 25°C. Electrode processes involving a weak acid or weak base have potential – pH variations which show a change in slope at pH $= pK_a$. For example, the reduction of N(V) to N(III) in acid solution is

$$NO_3^- + 3\, H^+ + 2\, e^- \rightarrow HNO_2 + H_2O$$

so that the E *vs.* pH slope is -89 mV pH^{-1}. In neutral or basic solution, the process is

$$NO_3^- + 2H^+ + 2e^- \rightarrow NO_2^- + H_2O$$

or

$$NO_3^- + H_2O + 2e^- \rightarrow NO_2^- + 2OH^-$$

so that $dE/dpH = -59$ mV pH^{-1}. The N(V)/N(III) half–cell potential is plotted *vs.* pH in Figure 1.9. Plots of the half–cell potentials of the N(III)/N(0) and N(0)/N(–III) couples are also shown in Figure 1.9, together with the O_2/H_2O and H_2O/H_2 couples.

The significance of the O_2/H_2O and H_2O/H_2 couples is that these define the limits of thermodynamic stability of an aqueous solution. Any couple with a half–cell potential greater than that of the O_2/H_2O couple is in principle capable of oxidizing water. Similarly, any couple with a potential less than (more negative than) that of the H_2O/H_2 couple is in principle capable of reducing water. As it happens, there are many species which appear to be perfectly stable in aqueous solution which are part of couples with half–cell potentials greater than that of the O_2/H_2O couple or less than that of the H_2O/H_2 couple. This is a reflection, however, of the intrinsically slow O_2/H_2O and H_2O/H_2 reactions rather than a failure of thermodynamics. The reactions do not occur because there is a very large activation barrier, not because there is no overall driving force.

Predominance Area Diagrams

We see in Figure 1.9 that nitrous acid and the nitrite ion are in principle capable of oxidizing water and are thus unstable in aqueous solution. Although this instability is not manifested because the reactions are slow, it raises the related question: What is the thermodynamically most stable form of nitrogen at a given pH and electrode potential? Referring to the free energy – oxidation state diagram, Figure 1.8, we see that the points for N(–I) and N(–II) both lie above the line connecting N(0) and N(–III) and therefore these species are unstable with respect to disproportionation. Similarly, the points for N(I), N(II), N(III), and N(IV) all lie above the N(0) – N(V) line, so that these species are also unstable. The existence of any of these species in aqueous solution therefore reflects kinetic stability rather than thermodynamic stability. In fact, only NO_3^-, NH_3, NH_4^+, and N_2 are thermodynamically stable in aqueous solution.

Furthermore, according to thermodynamics, NO_3^- is unstable in the presence of NH_4^+ or NH_3. One way of expressing the conclusions regarding thermodynamic stability is by means of a *predominance area diagram*, commonly called a *Pourbaix diagram* after the Belgian physical chemist who introduced their use *(3,H5)*.

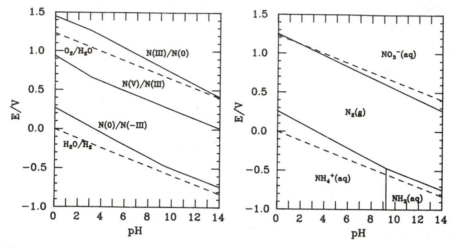

Figure 1.9 Potential – pH diagram showing the variation in half–cell potentials with pH for the N(III)/N(0), N(V)/N(III), and N(0)/N(–III) couples. Note changes in slope of the first two lines at pH 3.14 (the pK_a of HNO_2) and of the third line at pH 9.24 (the pK_a of NH_4^+). Also shown (dashed lines) are the potentials of the O_2/H_2O and H_2O/H_2 couples.

Figure 1.10 Predominance area diagram (Pourbaix diagram) for nitrogen in aqueous solution. The labeled regions correspond to areas of thermodynamic stability for the indicated species. The dashed lines showing the O_2/H_2O and H_2O/H_2 couples indicate the area of water stability.

A Pourbaix diagram is a potential – pH plot, similar to that of Figure 1.9, in which regions of thermodynamic stability are labelled. Lines separating regions represent the potential and pH at which the various thermodynamically stable species are in equilibrium at unit activity. Pourbaix diagrams usually also include the region of water stability. The Pourbaix diagram for nitrogen is shown in Figure 1.10. We see that in most of the potential – pH area of water stability the most stable form of nitrogen is $N_2(g)$. At low potential and low pH, NH_4^+ is most stable and at low potential and high pH NH_3 is most stable. Nitrate ion is stable in a narrow high–potential region above pH 2. Below pH 2, NO_3^- is in principle capable of oxidizing water, though the reaction is so slow that we normally do not worry about the decomposition of nitric acid solutions.

Pourbaix diagrams for kinetically stable systems like the oxidation states of nitrogen are not particularly useful. Most of the interesting chemistry involves species which are thermodynamically unstable and therefore do not appear on the diagram. For more labile redox systems, on the other hand, where thermodynamic stability is more significant, a Pourbaix diagram can be very useful in visualizing the possibilities for aqueous chemistry.

1.4 MEASUREMENT OF CELL POTENTIALS

In order to obtain cell potentials which have thermodynamic significance, the potential must be measured under reversible conditions. In thermodynamics, a reversible process is one which can be reversed in direction by an infinitesimal change in the conditions of the surroundings. For example, the direction of a reversible chemical reaction can be changed by an infinitesimal increase in the product concentration. Similarly, an electrochemical cell is reversible if the direction of current flow in the external circuit can be reversed by an infinitesimal change in one of the concentrations. Alternatively, we can think in terms of an *operational* definition of electrochemical cell reversibility: An electrochemical cell is regarded as reversible if a small amount of current can be passed in either direction without appreciably affecting the measured potential.

The words "small" and "appreciable" are ambiguous; the meaning depends upon the context of the experiment. A cell which appears reversible when measured with a device with a high input resistance (and thus small current) may appear quite irreversible when measured by a voltmeter with a low input resistance. There are a number of causes of irreversible behavior in electrochemical cells which are discussed in more detail in Chapter 5. Irreversible behavior usually results from an electrode process having slow reaction kinetics, either in the actual electron transfer process, in a coupled chemical reaction, or in the delivery of reactants to the electrode. For the purposes of discussion of cell potential measurements, however, we can think of an electrochemical cell as having an internal resistance which limits the current which can be delivered. Strictly speaking, in any reasonable model the resistance would be nonohmic, *i.e.*, current would not be linear in potential. We can, however, use a simple ohmic model to show one of the consequences of irreversibility.

Consider an electrochemical cell with an internal resistance R_{int} connected to a voltage–measuring device having an input resistance

Figure 1.11 Circuit showing the effects of cell resistance and leakage resistance on potential measure-ments.

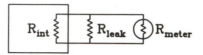

R_{meter} as shown in Figure 1.11. If we could draw zero current, the voltage across the cell terminals would be the true cell potential E_{cell}. In practice, however, we draw a current

$$i = E_{cell}/(R_{int} + R_{meter})$$

and actually measure a voltage

$$E_{meas} = iR_{meter}$$

Combining these expressions, we get

$$E_{meas} = \frac{E_{cell}R_{meter}}{R_{int} + R_{meter}}$$

In order for the measured voltage to equal the cell potential, we apparently require that $R_{meter} \gg R_{int}$. Our perception of the reversibility of a cell thus depends on the measuring instrument.

For cells having a high internal resistance ($>10^6$ Ω) another practical problem may arise even if a voltmeter is available with a high input resistance. In very high resistance circuits, leakage of current between the meter terminals becomes critically important. Dust, oily films, or even a fingerprint on an insulator can provide a current path having a resistance of $10^7 - 10^9$ Ω. This resistance in parallel with the meter (as shown in Figure 1.11) results in an effective meter resistance equal to the parallel combination of the true meter resistance and the leakage resistance,

$$1/R_{eff} = 1/R_{meter} + 1/R_{leak}$$

which may be several orders of magnitude less than the true meter resistance and even comparable with the cell internal resistance.

The problem of leakage resistance is generally encountered for any cell which has a high internal resistance and thus tends toward irreversibility. In particular the glass electrode commonly used for pH measurements has a very high resistance and is prone to just such leakage problems. Careful experimental technique with attention to clean leads and contacts is essential to accuracy.

Potentiometers

The classical method of potential measurement makes use of a null–detecting potentiometer circuit. Such a circuit, shown in Figure 1.12, involves a linear resistance slidewire R calibrated in volts, across which is connected a battery; the current through the slidewire is adjustable with a rheostat R'. With a known potential source E_s (often a Weston standard cell) connected to the circuit, the slidewire is set at the potential of the standard cell and the rheostat is adjusted to give zero–current reading on the galvanometer G. The potential drop between A and the slidewire tap is then equal to E_s, and, if the slidewire is linear in resistance, the voltage between A and any other point on the slidewire can be determined. The switch is then thrown to the unknown cell, E_x, and the slidewire adjusted to zero the galvanometer. The unknown potential can then be read from the

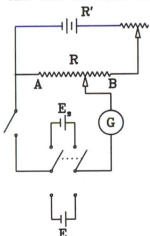

Figure 1.12 A potentiometer circuit.

slidewire calibration. Reversible cell potentials measured with a potentiometer are usually accurate to ± 0.1 mV and with care can be even better. The apparatus is relatively inexpensive, but the method is slow, cumbersome, and requires some technical skill. Furthermore, although in principle the potential is read under zero current conditions, some current must be drawn in order to find the zero. The current is small, typically on the order of 10^{-8} A with a sensitive galvanometer, but may be large enough to cause serious errors for a cell with high internal resistance.

Electrometers

With the development of electronic circuitry beginning in the 1930's, d.c. amplifiers with very high input impedances became available. Such devices, called electrometers, are particularly well suited to measurement of the potentials of cells with high internal resistance.

A number of different designs have been used for electrometer circuits, but one which is particularly common in electrochemical instrumentation is the so-called voltage follower shown in Figure 1.13. A voltage follower employs an *operational amplifier*, indicated by the

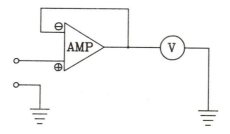

Figure   1.13   An operational amplifier voltage follower.

triangle in the figure. An operational amplifier has two inputs and the output voltage is proportional to the difference between the two input voltages with very high gain ($> 10^4$). In a voltage follower, the output is connected to the negative (inverting) input. Suppose that the output voltage is slightly greater than that at the positive input. The difference between the inputs will be amplified, driving the output voltage down. Conversely, if the output voltage is low, it will be driven up. The output is stable, of course, when the output and input are exactly equal. Since the input impedance is high ($> 10^{10}$ Ω), a very small current is drawn from the cell. Since the output impedance is low (ca. 10 Ω), a voltage–measuring device such as a meter or digital voltmeter can be driven with negligible voltage drop across the output impedance of the voltage follower. Direct–reading meters typically are limited to an accuracy on the order of 0.1% of full scale (± 1 mV for a full–scale reading of 1 V), but the accuracy can be improved considerably by using an electrometer in combination with a potentiometer circuit.

1.5 REFERENCE AND INDICATOR ELECTRODES

The standard half–cell potentials of all aqueous redox couples are given with respect to the hydrogen electrode. The hydrogen electrode is thus the standard reference electrode. Unfortunately, however, the hydrogen electrode is awkward and inconvenient to use, requiring hydrogen gas and a specially prepared platinum electrode. The platinum surface is easily poisoned and other electrode processes compete with the H^+/H_2 couple in determining the electrode potential. For these reasons, other electrodes are more commonly used as secondary references (F1).

Reference Electrodes

A practical reference electrode should be easily and reproducibly prepared and maintained, relatively inexpensive, stable over time, and usable under a wide variety of conditions. Two electrodes are particularly commonly used, meeting these requirements quite well.

The calomel[‡] electrode, shown in Figure 1.14, and represented in shorthand notation as

$$Cl^-(aq) \,|\, Hg_2Cl_2(s) \,|\, Hg(l)$$

is frequently used as a secondary reference. The half–cell reaction is

$$Hg_2Cl_2(s) + 2\,e^- \rightarrow 2\,Cl^-(aq) + 2\,Hg(l) \qquad E^\circ = +0.2682 \text{ V}$$

The standard potential, of course, refers to unit activity for all species,

---

[‡] Calomel is another name for mercurous chloride, Hg_2Cl_2.

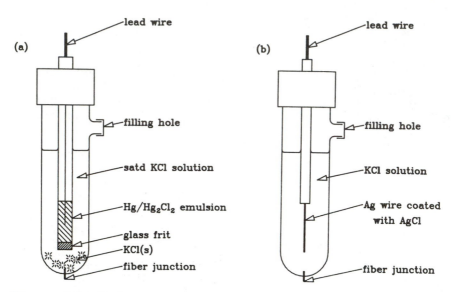

Figure 1.14 Reference electrodes. (a) Saturated calomel electrode. (b) Silver–silver chloride electrode.

including chloride ion. In practice it is usually more convenient to use saturated KCl solution as the electrolyte with a few crystals of solid KCl present in the electrode to maintain saturation.[‡] In this way the chloride ion concentration is held constant without having to prepare solutions of exactly known concentration. The potential of the saturated calomel electrode (abbreviated s.c.e.) equipped with a saturated KCl salt bridge is $+0.244$ V at 25°C. Because the solubility of KCl is temperature dependent, there is a significant variation in potential with temperature, -0.67 mV K^{-1}. When this is a problem, 0.1 M KCl can be used as the electrolyte; the potential and temperature coefficient then are 0.336 V and -0.08 V K^{-1}, respectively.

Another common reference electrode is the silver–silver chloride electrode, also shown in Figure 1.14 and represented by

---

[‡] As we will see in Section 3.4, the use of a saturated KCl solution minimizes the potential of the liquid junction connecting the reference electrode to the rest of the cell.

$$Cl^-(aq) \,|\, AgCl(s) \,|\, Ag(s),$$

The half–cell reaction is

$$AgCl(s) + e^- \to Ag(s) + Cl^-(aq) \qquad E° = +0.2223 \text{ V}$$

The Ag/AgCl electrode is usually used with 3.5 M KCl solution and has a formal half–cell potential of $+0.205$ V and a temperature coefficient of -0.73 mV K^{-1}. The Ag/AgCl electrode is operationally similar to the calomel electrode but is more rugged; AgCl adheres very well to metallic silver and there is no liquid mercury or Hg_2Cl_2 paste to deal with. The Ag/AgCl electrode is easily miniaturized and is thus convenient for many biological applications.

Indicator Electrodes

If one of the electrodes of an electrochemical cell is a reference electrode, the other is called a *working* or *indicator* electrode. The latter designation implies that this electrode responds to (indicates) some specific electrode half–reaction. At this point, it is appropriate to ask what determines the potential of an electrode. Consider a solution containing $FeSO_4$ and H_2SO_4 in contact with an iron wire electrode. If we make a cell by adding a reference electrode, there are at least three electrode processes which might occur at the iron wire:

$$Fe^{2+} + 2\,e^- \to Fe(s) \qquad\qquad E° = -0.44 \text{ V}$$

$$SO_4{}^{2-} + 4\,H^+ + 2\,e^- \to SO_2(aq) + 2\,H_2O \qquad E° = 0.16 \text{ V}$$

$$H^+ + e^- \to \tfrac{1}{2}\,H_2(g) \qquad\qquad E° = 0.00 \text{ V}$$

What then is the actual potential of the iron wire electrode? We first note that in order to have a finite reversible potential at an electrode, all the participating species must be present in finite concentration so that appreciable current can be drawn in either direction. Thus, as we have defined the system with no sulfur dioxide or gaseous hydrogen present, only the first electrode reaction qualifies and we might guess that the potential should be determined by the Fe^{2+}/Fe couple. This is the right answer but for the wrong reason. Suppose that we bubbled some hydrogen gas over the iron wire electrode or added a little sodium sulfite—what then? There is another way to look at this system: according to the half–cell potentials, the iron wire should be oxidized spontaneously by either H^+ ($\Delta G° = -85$ kJ mol^{-1}) or by $SO_4{}^{2-}$ ($\Delta G° = -116$ kJ mol^{-1}). That neither reaction occurs to any appreciable extent is because the reactions are very slow. Just as these homogeneous reactions are very slow, the electrochemical reduction of SO_4^{2-} or H^+ at an iron electrode is slow. If one electrode reaction is much faster than other possible reactions, only the fastest one will contribute to the potential—the slower electrode processes will appear irreversible under the conditions where we can measure the potential due to the faster process. If two or more processes have significant rates then we expect a homogeneous reaction to occur until the electrode potentials of the various half–cell reactions are equal. Thus in the example, we expect the potential of the iron wire electrode will be determined by the Fe^{2+}/Fe couple with

$$E = E° - \frac{RT}{2F} \ln \frac{1}{a_{Fe^{2+}}}$$

or, at 25°C,

$$E/V = -0.440 - 0.0296 \; pFe$$

In other words, the iron wire electrode acts in this case as an indicator of the Fe^{2+} activity.

Although it is difficult to predict the rates of electrode processes without additional information, there are a few useful generalizations. First, electrode processes which involve gases are usually very slow unless a surface is present which catalyzes the reaction. Thus a surface consisting of finely divided platinum (platinum black) catalyzes the reduction of H^+ to H_2, but this process is slow on most other surfaces. Second, as a general rule, simple electron transfers which do not involve chemical bond breaking, e.g., $Fe^{3+} + e^- \rightarrow Fe^{2+}$, are usually fast compared with reactions which involve substantial reorganization of molecular structure, e.g., the reduction of sulfate to sulfite.

1.6 ION–SELECTIVE ELECTRODES

The potential of a Ag/AgCl electrode depends on [Cl$^-$] and so we might think of the Ag/AgCl electrode as an indicator electrode, the potential of which is a measure of ln[Cl$^-$], or pCl. We could easily set up an electrochemical cell to take advantage of this property. Suppose that we have two Ag/AgCl electrodes, one a standard reference electrode with KCl solution at unit activity, the other in contact with a test solution having an unknown Cl^- activity. The cell can be represented schematically by

$$Ag \mid AgCl(s) \mid KCl(aq,a=1) \parallel Cl^-(aq) \mid AgCl(s) \mid Ag$$

At the left–hand electrode, the half–cell reaction is

$$Ag(s) + Cl^-(aq,a=1) \rightarrow AgCl(s) + e^-$$

and at the right–hand electrode, the process is

$$AgCl(s) + e^- \rightarrow Ag(s) + Cl^-(aq)$$

so that the overall cell reaction is

$$Cl^-(aq,a=1) \rightarrow Cl^-(aq)$$

In other words the cell "reaction" is simply the dilution of KCl. The potential of the cell is given by the Nernst equation:

$$E = E^\circ - (RT/F) \ln(a/1)$$

or, since $E^\circ = 0$, the cell potential at 25°C is

$$E/V = +0.0592 \ pCl$$

This arrangement would work well as a technique for determination of unknown Cl^- activities. However, the Ag/AgCl electrode is not particularly selective. If will respond, for example, to any anion which forms an insoluble silver salt (*e.g.*, Br^- or SCN^-).

Glass Membrane Electrodes

It would be nice to have an indicator electrode which would respond to only one specific ion. This could be accomplished in principle if we could design a membrane which is permeable to only one species. A membrane with perfect selectivity is yet to be found, but there are quite a number of devices which come close.[‡] There are several approaches to membrane design, and we will not discuss them

---

[‡] See books by Koryta *(A9)*, Vesely, Weiss, and Stulik *(D11)*, and by Koryta and Stulik *(D14)* for further details.

all. The oldest[‡] membrane electrode is the glass electrode, which has been used to measure pH since 1919 *(4)* but properly understood only relatively recently. The electrode, shown in Figure 1.15, consists of a glass tube, the end of which is a glass membrane about 0.1 mm thick (and therefore very fragile!). Inside the tube is a Ag/AgCl electrode and 1 M HCl solution. The glass electrode is used by dipping it into a test solution and completing the electrochemical cell with a reference electrode.

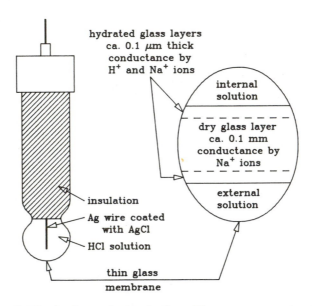

Figure 1.15 A glass electrode for pH measurements showing a schematic view of the glass membrane.

The glass used in the membrane is a mixture of sodium and calcium silicates—Na_2SiO_3 and $CaSiO_3$—and silicon dioxide, SiO_2. The silicon atoms tend to be four–coordinate, so that the glass is an

---

[‡] The first membrane potential was discovered by nature eons ago when animals first developed nervous systems. A nerve cell wall can be activated to pass Na^+ ions and so develop a membrane potential which triggers a response in an adjacent cell. Chemists were slower in appreciating the potential of such a device, but we are catching up fast.

extensively cross–linked polymer of SiO_4 units with electrostatically bound Na^+ and Ca^{2+} ions. The glass is weakly conductive, with the charge carried primarily by the Na^+ ions. (The Ca^{2+} ions are much less mobile than Na^+ and contribute little to the conductance.) The glass is also quite hygroscopic and takes up a significant amount of water in a surface layer perhaps as much as 0.1 μm deep. In the hydrated layers (one on either side of the membrane) there is equilibrium between H^+ and Na^+ electrostatically bound to anionic sites in the glass and in solution.

$$H^+(aq) + Na^+(gl) \rightleftarrows H^+(gl) + Na^+(aq) \qquad (1.27)$$

If the concentration of $H^+(aq)$ is low, this equilibrium shifts to the left; Na^+ from the interior of the glass tends to migrate into the hydrated region to maintain electrical neutrality. Hydrogen ions on the other side of the glass penetrate a little deeper into the glass to replace the Na^+ ions that have migrated. This combination of ion migrations gives sufficient electric current that the potential is measurable with a high–impedance voltmeter. Since the H^+ ions are intrinsically smaller and faster moving than Na^+, most of the current in the hydrated region is carried by H^+ and the glass electrode behaves as if it were permeable to H^+ and thus acts as an indicator electrode sensitive to pH.

In a solution with low $[H^+]$ and high $[Na^+]$, NaOH solutions for example, the Na^+ concentration in the hydrated layer of the glass may be much greater than the H^+ concentration, and Na^+ ions then carry a significant fraction of the current; the potential developed across the membrane is then smaller than might have been expected. For this reason, glass electrodes do not give a linear pH response at very high pH, particularly when the alkali metal ion concentration is high. Glass electrodes respond with virtually perfect selectivity to hydrogen ions over the pH range 0–11. Above pH 11, response to alkali ions becomes important with some glasses and such electrodes become unusable above pH 12.

We will see in Section 3.4 that the potential across a glass membrane can be written

$$E_{membrane} = \text{constant} + (RT/F) \ln[a_H + k_{H,Na}a_{Na}] \qquad (1.28)$$

where a_H and a_{Na} are the activities of $H^+(aq)$ and $Na^+(aq)$ and $k_{H,Na}$ is called the *potentiometric selectivity coefficient*. The selectivity coefficient depends on the equilibrium constant of the reaction of eq

(1.27) and on the relative mobilities of the H^+ and Na^+ ions in the hydrated glass. These properties depend on the composition and structure of the glass and can be controlled to some extent. Thus some glass electrodes now available make use of glasses where lithium and barium replace sodium and calcium, giving a membrane which is much less sensitive to sodium ions, allowing measurements up to pH 14. Corrections for Na^+ ions may still be required, however, above pH 12.

Because of the importance of the hydration of the glass surface, glass electrodes must be conditioned before use by soaking in water or an aqueous buffer solution (*i.e.*, the glass surface must be hydrated. The glass surface is dehydrated on prolonged exposure to nonaqueous solvents or to aqueous solutions of very high ionic strength. Nonetheless, glass electrodes can be used (*e.g.*, to follow acid–base titrations) in alcohols or polar aprotic solvents such as acetonitrile or dimethyl sulfoxide provided that the exposure is of relatively short duration.

Glasses containing about 20% Al_2O_3 are significantly more sensitive to alkali metal ions and can be used to produce electrodes which are somewhat selective to specific ions. Thus a $Na_2O - Al_2O_3 - SiO_2$ glass can be fabricated which is selective for sodium ions (*e.g.*, $k_{Na,K} = 0.001$). Although these electrodes retain their sensitivity to hydrogen ion ($k_{Na,H} = 100$), they are useful in neutral or alkaline solutions over the pNa range 0–6.

Other Solid Membrane Electrodes

Rapid progress has been made in recent years in the design of ion–selective electrodes which employ an insoluble inorganic salt as a membrane. For example, a lanthanum fluoride crystal, doped with a little EuF_2 to provide vacancies at anionic sites, behaves like a membrane permeable to F^- ions. The only significant interference is from OH^- ion. F^- and OH^- are almost exactly the same size, but other anions are too large to fit into the F^- sites in the crystal. The LaF_3 crystal, together with a solution of KF and KCl and a Ag/AgCl electrode, gives a fluoride ion–selective electrode usable in the pF range 0–6.

Other solid membrane electrodes make use of pressed pellets of insoluble salts such as Ag_2S. In this case, Ag^+ ions are somewhat mobile in the solid, so that a Ag_2S membrane can be used in a Ag^+-selective electrode. Other metals which form insoluble sulfides could in principle replace a Ag^+ ion at the surface of the membrane, but in practice only Hg^{2+} is a serious interference. Since Ag^+ can move in either direction through the membrane, either away from a source of Ag^+ or toward a sink—a source of S^{2-}—the Ag_2S membrane electrode can be used either for measurement of pAg or pS in the range 0 to 7. Other solid membrane electrodes are commercially available for Cl^-, Br^-, I^-, CN^-, SCN^-, NH_4^+, Cu^{2+}, Cd^{2+}, and Pb^{2+}.

Liquid Membrane Electrodes

By replacing the glass or inorganic crystal with a thin layer of a water–immiscible liquid ion exchanger, another type of ion–selective electrode may be constructed. For example, by using the calcium salt of an organophosphoric acid in an organic solvent as the liquid ion exchanger and contacting the ion–exchange solution with the aqueous test solution through a thin porous membrane, a membrane system permeable to Ca^{2+} ions (and to some degree to other divalent ions) is obtained. On the inner side of the membrane is a Ag/AgCl electrode with $CaCl_2$ aqueous electrolyte. Thus a membrane potential proportional to $\ln([Ca^{2+}]_{out}/[Ca^{2+}]_{in})$ is developed. Liquid membrane electrodes are commercially available for Ca^{2+}, K^+, Cl^-, NO_3^-, ClO_4^-, and BF_4^-.

Other Types of Selective Indicator Electrodes

By surrounding the glass membrane of an ordinary glass electrode with a dilute aqueous solution of $NaHCO_3$ which is separated from the test solution by a membrane permeable to CO_2, an electrode responsive to dissolved CO_2 is obtained. This technique, where an ion-selective electrode is converted to respond to some other species by interposition of a reaction system involving ordinary chemical reactions, is extendable to a wide variety of applications. Electrodes are commercially available for dissolved CO_2, NH_3, and NO_2.

Another example of the above approach is in the construction of an electrode sensitive to urea. The enzyme urease is immobilized in polyacrylamide, which forms a thin layer on a cation–sensing glass electrode which then responds to the NH_4^+ produced as a product of the hydrolysis of urea. Similarly, by coating a glass electrode with immobilized L–amino acid oxidase, an electrode is obtained which responds to L–amino acids in solution.

Clearly, the field of ion–selective electrodes is large, and, more significantly, is growing rapidly.[‡] There are literally thousands of potential applications in chemistry, biochemistry, and biology which await an interested investigator.

1.7 CHEMICAL ANALYSIS BY POTENTIOMETRY

There are a number of analytical methods based on measurements of electrochemical cell potentials *(7,8,D2,D7,D10,D11,D13,D14)*. For convenience these can be divided into two groups: those which determine concentration (or activity) directly from the measured potential of an electrochemical cell; and those in which the potential of a cell is used to determine the equivalence point in a titration. Both types of methods have some important advantages which will become apparent as we discuss them.

Direct Methods

The measurement of pH using a glass electrode is by far the most common of all electroanalytical techniques *(9,D7,D10)* and is also characteristic of direct determinations using cell potentials. A typical electrode arrangement for pH measurements is as follows:

$$Ag(s) \mid AgCl(s) \mid KCl(1\ M) \parallel H^+(aq) \mid glass \mid HCl(1\ M) \mid AgCl(s) \mid Ag(s)$$

i.e., a solution in which a Ag/AgCl reference electrode and a glass electrode are immersed. The half–cell reactions,

$$Ag(s) + Cl^-(1\ M) \rightarrow AgCl(s) + e^-$$

$$AgCl(s) + e^- \rightarrow Ag(s) + Cl^-(1\ M)$$

lead to no net change so that the overall cell reaction is the nominal transfer of H^+ from the test solution to the 1 M solution inside the glass membrane.

$$H^+(aq) \rightarrow H^+(aq, 1\ M)$$

[‡] For a good review, see Murray *(5)*. The field is reviewed every two years in *Analytical Chemistry, e.g.*, by Arnold and Solsky in 1986 *(6)*.

The free energy change for this process is

$$\Delta G = RT \ln(1/a_H)$$

so that the cell potential is

$$E = \frac{RT}{F} \ln a_H{}^+$$

or, at 25°C,

$$E/V = -0.0592 \text{ pH}$$

Thus the cell potential is directly proportional to the pH of the test solution. In practice, this may not be exactly true because of a liquid junction potential at the salt bridge linking the reference electrode to the test solution and a small potential intrinsic to the glass electrode. Thus buffers of known pH are used to calibrate the pH meter so that the pH is determined relative to a standard rather than absolutely.

Most of the commercial ion–selective electrodes can be used directly to determine pX = $-\log a_X$, just as the glass electrode is used to measure pH. Two general procedures are used commonly.

(1) **Calibration curve.** The potential of a given cell can be measured with standard solutions and a potential *vs.* concentration curve constructed. The concentration of an unknown may then be read from the calibration curve given the potential of the unknown. When this method is used, it is important that the unknown solution have the same ionic strength as that of the standards so that the activity coefficients remain constant. The concentrations of other species, such as complexing agents, should also be identical.

(2) **Standard addition method.** By first measuring the potential of an unknown solution and then adding a known amount of the substance detected, the incremental response of the voltmeter in effect calibrates the scale. The advantages of this procedure are that the concentration is determined without the tedium of a calibration curve, and more important, calibration is obtained on the same solution under the same measurement conditions, possibly avoiding some systematic errors. On the other hand, if the calibration curve were nonlinear, the known addition method would introduce a large uncorrectable error.

As a rule, the calibration curve method is preferable if many analyses are to be done. If only a few analyses are to be performed, the known addition method is usually faster. In either case, the method should have been carefully investigated to ferret out systematic errors.

Example 1.6   100 mL of a solution containing an unknown concentration of fluoride was analyzed using a fluoride ion–selective electrode and a Ag/AgCl reference electrode. The cell potential was –97 mV at 25°C. After addition of 10 mL of 2.00×10^{-3} M F^- standard solution, the potential was –70 mV. What was the original concentration of fluoride? If the potentials are accurate to ± 0.5 mV, what is the uncertainty in the determination?

The Nernst equation gives the cell potential

$$E = E^\circ - \frac{RT}{F} \ln \frac{1}{a_{F^-}} - E_{ref}$$

so that

$$-0.0970 \pm 0.0005 = E^\circ - 0.0592 \, pF_0 - E_{ref}$$

$$-0.0700 \pm 0.0005 = E^\circ - 0.0592 \, pF_1 - E_{ref}$$

Subtracting, we get[‡]

$$pF_0 - pF_1 = (0.0270 \pm 0.0007)/0.0592 = 0.456 \pm 0.012$$

$$pF_0 - pF_1 = \log(a_1/a_0)$$

Assuming that we can replace the activities by molar concentrations, we have

$$[F^-]_1/[F^-]_0 = \text{antilog}(0.456) = 2.86 \pm 0.08$$

The difference, $[F^-] - [F^-]_0$, corresponds to the number of moles of

---

[‡]   In addition or subtraction, $z = x \pm y$, the errors propagate as $s_z^2 = s_x^2 + s_y^2$.

F$^-$ added.

$$[F^-]_1 - [F^-]_0 = (2.00 \times 10^{-3} \text{ mol L}^{-1})(10 \text{ mL}/110 \text{ mL})$$

$$[F^-]_1 - [F^-]_0 = 0.181 \times 10^{-3} \text{ mol L}^{-1}$$

Substituting $[F^-]_1 = (2.86 \pm 0.08)[F^-]_0$, we obtain

$$[F^-]_0 (2.86 \pm 0.08 - 1) = 0.181 \times 10^{-3}$$

or[‡]

$$[F^-]_0 = (9.7 \pm 0.4) \times 10^{-5} \text{ M}$$

Notice that in the example above, the fluoride ion concentration was determined to approximately 4% expected accuracy. Such accuracy is characteristic of direct methods. For concentrations of chemical species above 10^{-4} M, there are usually methods (such as the titration methods discussed below) which are capable of considerably greater accuracy. But for very low concentrations, a direct electrochemical measurement may well be the most accurate and is frequently the only method available.

Titration Methods

Potentiometric methods can be adapted to the detection of the endpoint of a titration. If either the titrant or the substance titrated is detected at an indicator electrode, a plot of the cell potential vs. the volume of titrant shows a sharp increase or decrease in potential at the endpoint. pH titration curves of acids and bases are familiar examples of this method.

Other ion–selective electrodes can be used to determine titration curves. For example, organic thiols may be determined by titration

---

[‡]   In multiplication or division, $z = xy$ or $z = x/y$, errors propagate as $(s_z/z)^2 = (s_x/x)^2 + (s_y/y)^2$.

with silver nitrate standard solution where the silver ion concentration is monitored with a Ag^+-selective electrode. The titration reaction is

$$Ag^+(aq) + RSH(aq) \rightarrow H^+(aq) + AgSR(s)$$

Typical solubility–product constants of silver thiolates are less than 10^{-16}; thus the silver ion concentration prior to the endpoint is very low and rises rapidly when the endpoint is passed, so that a sharp break in potential is obtained.

In many cases, nonselective indicator electrodes can be used to monitor a titration and to detect the endpoint.

Example 1.7 Compute the titration curve for the titration of 25 mL of 0.01 M $FeSO_4$ in 1 M H_2SO_4 with 0.01 M $Ce(SO_4)_2$ if the electrodes are Pt and a saturated calomel electrode.

Both the titrant and the sample form reversible couples at the platinum electrode with formal potentials:

$$Fe^{3+} + e^- \rightarrow Fe^{2+} \qquad\qquad E°' = 0.68 \text{ V}$$

$$Ce^{4+} + e^- \rightarrow Ce^{3+} \qquad\qquad E°' = 1.44 \text{ V}$$

The titration reaction is

$$Fe^{2+} + Ce^{4+} \rightarrow Fe^{3+} + Ce^{3+}$$

We first compute the equilibrium constant for the reaction. Consider the hypothetical cell (not the one being measured in the experiment) in which the two half–cells are the Fe^{3+}/Fe^{2+} couple and the Ce^{4+}/Ce^{3+} couple; the formal cell potential is $1.44 - 0.68 = 0.76$ V. Thus we have

$$\Delta G° = -FE° = 73 \text{ kJ mol}^{-1}$$

which corresponds to an equilibrium constant

$$K = \exp(-\Delta G°/RT) = 7.0 \times 10^{12}$$

The reaction will go to completion if the rate is fast enough.

Up to the endpoint, the potential of the Pt electrode will be determined by the Fe^{3+}/Fe^{2+} couple. The potential of the cell is

$$E = E°_{Fe^{3+}/Fe^{2+}} - \frac{RT}{F} \ln \frac{[Fe^{2+}]}{[Fe^{3+}]} - E_{sce}$$

$$E = 0.68 - 0.0592 \log \frac{[Fe^{2+}]}{[Fe^{3+}]} - 0.244$$

Consider the situation when 10 mL of titrant has been added. Since we have added $(0.01\ L)(0.01\ mol\ L^{-1}) = 10^{-4}$ mol of Ce^{4+}, there then must be 1×10^{-4} mol Fe^{3+} and 1.5×10^{-4} mol Fe^{2+} must remain. The concentration ratio then is

$$[Fe^{2+}]/[Fe^{3+}] = 1/1.5$$

and the cell potential is $E = 0.45$ V.

At the equivalence point, $[Fe^{3+}] = [Ce^{3+}] \simeq 0.005$ M and the equilibrium constant expression gives $[Fe^{2+}] = [Ce^{4+}] = 1.9 \times 10^{-9}$ M. The Fe^{2+}/Fe^{3+} concentration ratio is 3.8×10^{-7} and the cell potential is $E = 0.82$ V.

Beyond the endpoint, virtually all the Fe^{2+} has been oxidized; 2.5×10^{-4} mol of Ce^{3+} has been produced by the titration reaction and excess Ce^{4+} is being added. The cell potential is given by

$$E = E°'_{Ce^{4+}/Ce^{3+}} - \frac{RT}{F} \ln \frac{[Ce^{3+}]}{[Ce^{4+}]} - E_{sce}$$

$$E = 1.44 - 0.0592 \log \frac{[Ce^{3+}]}{[Ce^{4+}]} - 0.244$$

When 30 mL of titrant has been added, there will be 0.5×10^{-4} mol of Ce^{4+} present in excess. The Ce^{3+}/Ce^{4+} concentration ratio is 2.5/0.5 and the cell potential is $E = 1.15$ V.

Carrying out several such calculations, we can sketch the titration curve shown in Figure 1.16.

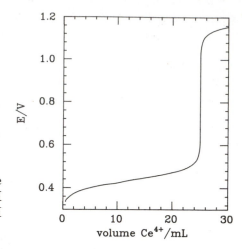

Figure 1.16 Titration curve for titration of 0.01 M $FeSO_4$ with 0.01 M $Ce(SO_4)_2$ in 1 M H_2SO_4.

1.8 BATTERIES AND FUEL CELLS

Throughout most of the nineteenth century, electrochemical cells provided the only practical source of electrical power and a great deal of effort was devoted to the development of inexpensive, efficient power cells and storage batteries.[‡] When steam–powered electrical generators came into use in the 1880's and electrical power began to be widely distributed, electrochemical power cells started into a long decline. In the recent past, galvanic cells have been used mostly as small portable power sources (*e.g.*, cells used in flashlights, children's toys, etc.) or as portable energy storage systems (*e.g.*, automobile batteries).

Interest in electrochemical cells as power sources has revived recently *(10,11)* as society has become more concerned for the environment and as fossil fuels have become scarcer and more expensive. Electric–powered automobiles offer the hope of substantially reduced environmental pollution and direct conversion of fossil fuel energy into an electrical energy via an electrochemical cell could in principle be achieved with much greater efficiency than is possible with a heat engine. In this section, we will review briefly the operation of some electrochemical cells which convert chemical energy into

---

[‡]   A battery is a collection of two or more cells connected in series so that the battery potential is the sum of the individual cell potentials.

electrical energy. For further information, see books by Mantell *(G2)*, Angrist *(G4)*, Bagotzky and Skundin *(G5)*, Pletcher *(G6)*, or Ventatasetty *(G9)*.

Electrochemical cells can be divided into two classes: *primary cells*, which are intended to convert chemical energy into electrical energy, and *secondary cells*, which are intended to store electrical energy as chemical energy and then resupply electricity on demand. A flashlight cell is used until all the chemical energy has been converted to electricity and is then discarded. A fuel cell in a spacecraft converts hydrogen and oxygen to water, extracting the energy as electricity. We usually do not attempt to recharge a flashlight cell and the fuel cell would not normally be run backwards to regenerate H_2 and O_2 gases. A storage battery, on the other hand, is intended to store electrical energy. Discharging and recharging are equally important parts of the operational cycle. The difference in function leads to differences in design.

We first consider three common primary cells which have a family resemblance, all having a zinc anode *(12)*.

The "Dry Cell"

The so-called "dry cell" used to power flashlights is descended from a cell invented by Georges Leclanché in 1868 and is sometimes called a Leclanché cell. The cell, which can be represented schematically by

$$Zn(s)\,|\,Zn^{2+}(aq),NH_4Cl(aq)\,|\,MnO_2(s),Mn_2O_3(s)\,|\,C(s)$$

is shown in Figure 1.17. The carbon rod cathode is surrounded by a

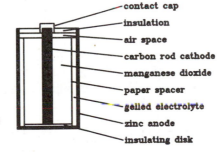

Figure 1.17
Construction of the Zn/MnO_2 dry cell.

- contact cap
- insulation
- air space
- carbon rod cathode
- manganese dioxide
- paper spacer
- gelled electrolyte
- zinc anode
- insulating disk

thick layer of MnO_2 (mixed with a little graphite to improve the conductivity). A paper barrier separates the MnO_2 from the aqueous electrolyte (which is gelled with starch or agar so that the cell is "dry") and the zinc anode which forms the cell container. A newly prepared cell has an open circuit potential of about 1.55 V, but under load the cell potential decreases to 1.3–1.5 V. The cell reaction

$$Zn(s) + 2\,MnO_2(s) + H_2O \rightarrow Zn(II) + Mn_2O_3 + 2\,OH^-(aq)$$

leads to an increase in pH which is buffered by the NH_4Cl. The cell reaction is only partially reversible, so that dry cells are not readily recharged. The exact nature of $Zn(II)$ depends on the pH and NH_4Cl concentration. It may be precipitated as a hydroxide or oxychloride or remain in solution as an ammine complex. On diffusion into the MnO_2 phase, however, precipitation as $ZnO\cdot Mn_2O_3$ occurs. There is some evidence that formation of this phase is responsible for the irreversibility of the cell. Zn/MnO_2 dry cells have a relatively short shelf life because of diffusion processes which amount to an internal short circuit. Because the cells are inexpensive, however, they have been widely used for a long time.

Because of the large market for this kind of cell, a considerable effort has been expended on improvements of the Leclanché cell. When NH_4Cl is replaced by KOH as the electrolyte, the cell reaction is transformed to

$$Zn(s) + 2\,MnO_2(s) \rightarrow Mn_2O_3(s) + Zn(OH)_2 \;[\text{or } Zn(OH)_4{}^{2-}]$$

Because the electrolyte is not gelled and is more corrosive than NH_4Cl, there are some packaging problems, the solutions to which increase the cost of the cell. However, the shelf life an an alkaline cell is much longer than that of the Leclanché cell, and the alkaline Zn/MnO_2 cell turns out to be well adapted to applications where a steady low–level drain of power is required, a situation where polarization of the traditional dry cell would lead to degradation of performance.

The Mercury Cell

One of the few commercially important cells which does not have nineteenth century origins is the alkaline zinc/mercury cell, developed for the U. S. Army during World War II. The cell,

$$Zn(Hg)\,|\,Zn(OH)_2(s)\,|\,KOH(aq),Zn(OH)_4{}^{2-}(aq)\,|\,HgO(s)\,|\,Hg(l)$$

has an open–circuit potential of 1.35 V and is very widely used to power transistor radios, watches, hearing aids, etc. Unlike the Zn/MnO_2 dry cell, the electrolyte is not consumed by the Zn/HgO cell reaction

$$Zn + HgO + H_2O \rightarrow Hg + Zn(OH)_2$$

and as a result, the cell potential is much more constant during discharge. Accordingly, mercury cells have often been used to provide a voltage reference in electronic instrumentation.

The cell is constructed with a mercuric oxide cathode (mixed with a little graphite to improve conductivity) pressed into the bottom of a nickel–plated steel case. Upon discharge, this layer becomes largely liquid mercury. The alkaline electrolyte is contained in a layer of adsorbent material and the zinc amalgam anode is pressed into the cell top. In addition to the relatively constant potential, mercury cells have a long shelf life and a high energy–to–volume ratio. A major disadvantage is that power output drops precipitously at temperatures below about $10°C$.

The Silver Cell

A close relative of the mercury cell is the Zn/Ag_2O_2 cell:

$$Zn(s)\,|\,Zn(OH)_2(s)\,|\,KOH(aq),Zn(OH)_4^{2-}(aq)\,|\,Ag_2O_2(s)\,|\,Ag(s)$$

This cell has many of the advantages of the mercury cell although the potential is less constant during discharge. The silver cell has a higher voltage than the mercury cell, about 1.5 V, and thus has a somewhat higher energy–to–weight ratio. It also operates successfully at significantly lower temperatures. Applications have been largely in situations where a very small, highly reliable cell is required, for example, in hearing aids and watches and in the guidance systems of rockets.

Fuel Cells

When hydrogen or a hydrocarbon fuel is burned in a heat engine, the heat of the combustion reaction is partially converted to work. The efficiency of the engine can be defined as the ratio of the work done on the surroundings ($-w$) to the heat evolved in the chemical reaction ($-q$):

$$\text{Efficiency} = w/q$$

The maximum theoretical efficiency of the process is limited. For an engine operating with a Carnot cycle (isothermal compression at T_1, adiabatic compression to T_2, isothermal expansion at T_2, and adiabatic expansion back to T_1), it can be shown that the maximum efficiency is determined by the temperature limits between which the engine operates:

$$\text{Efficiency} = (T_2 - T_1)/T_2 \qquad (1.29)$$

In practice, efficiencies are considerably less than theoretical because of friction and heat loss.

In contrast to a heat engine, the work obtained from an electrochemical cell is equal to $-\Delta G$ of the cell reaction, so that the efficiency, using the definition above, is

$$\text{Efficiency} = \frac{\Delta G}{\Delta H} = 1 - \frac{T\Delta S}{\Delta H} \qquad (1.30)$$

When ΔH and ΔS are of opposite sign, the "efficiency" of an electrochemical power cell, based on the w/q criterion, is greater than 1. When ΔH and ΔS are both negative (the more common situation), the efficiency decreases with increasing temperature but is usually much greater than that of a heat engine. This is hardly a new idea. Ostwald[‡] pointed out this difference in maximum efficiencies in 1894 *(13)* and suggested that electrochemical fuel cells should be developed to replace heat engines.

Example 1.8 Compute the maximum efficiency and the maximum work when one mole of $H_2(g)$ reacts with $O_2(g)$ to produce $H_2O(g)$ (a) in a Carnot engine operating between 600 K and 300 K, and (b) in an electrochemical cell operating under standard conditions at 298 K.

[‡] Friederich Wilhelm Ostwald (1853–1932) was Professor of Physical Chemistry at the University of Leipzig. Together with Arrhenius and van't Hoff, Ostwald is regarded as one of the founders of physical chemistry. His laboratory at Leipzig spawned a generation of physical chemists.

The standard enthalpy of formation of $H_2O(g)$ is -242 kJ mol^{-1}. According to eq (1.26), the maximum efficiency is 50%, so that the maximum useful work, assuming that ΔH is temperature independent, is 121 kJ mol^{-1}.

The standard free energy of formation of $H_2O(g)$ is -229 kJ mol^{-1}, and all this energy is theoretically convertible to electrical work. The efficiency then is (229/242) x 100% = 94.6%.

The Hydrogen–Oxygen Fuel Cell

The hydrogen–oxygen fuel cell could hardly be simpler in concept. The half–cell reactions

$$H_2(g) \;\rightarrow\; 2\,H^+ + 2\,e^-$$

$$\tfrac{1}{2}\,O_2(g) + 2\,H^+ + 2\,e^- \;\rightarrow\; H_2O$$

or

$$H_2(g) + 2\,OH^- \;\rightarrow\; 2\,H_2O + 2\,e^-$$

$$\tfrac{1}{2}\,O_2(g) + H_2O + 2\,e^- \;\rightarrow\; 2\,OH^-$$

sum to the formation of water

$$H_2(g) + \tfrac{1}{2}\,O_2(g) \;\rightarrow\; H_2O$$

and give a standard cell potential of 1.229 V. The H_2/O_2 system was studied very early in the history of electrochemistry (see Section 1.1) and the hydrogen–oxygen fuel cell was first described by Grove[‡] in 1839 *(14)*.

[‡] Sir William R. Grove (1811–1896) was a barrister by profession, but he maintained an active scientific career on the side. He is remembered for his work on galvanic cells and as a founder of The Chemical Society of London.

While it is relatively easy to construct a hydrogen–oxygen fuel cell in the laboratory, it is very difficult to extract much power. The major problem is that both half–cell reactions are very slow on most electrode surfaces. The H_2/H^+ reaction is relatively rapid on finely divided platinum or palladium surfaces (and fairly fast on other precious metals), but platinum is not a practical option for large–scale commercial use. The reduction of oxygen is an even more serious problem, as this rate is quite slow even on precious metal electrodes.

The H_2-O_2 fuel cell remained a gleam in the eye of electrochemists (and an occasional laboratory curiosity) until the U.S. space program developed a need for a lightweight, efficient, reliable, nonpolluting energy source for use on spacecraft. With a need defined and a customer willing to pay development costs, several companies began work on the problem.

The operation of a H_2-O_2 fuel cell is limited by the slow rate of the electrode reactions and by ohmic heating of the electrolyte solution. The rate problem can be attacked by increasing the electrode surface area and by increasing the temperature. While the reactions do go faster at higher temperatures, eq (1.30) predicts a decrease in theoretical efficiency with increasing temperature (ΔH and ΔS are both negative), so that a compromise is involved. Increasing the surface area is an effective strategy provided that the increased surface can be contacted by both the gas phase and the electrolyte solution.

The solution to the problem of the H_2-O_2 fuel cell in general has involved the use of electrode materials such as porous graphite with a small amount of precious metal catalyst imbedded in the pores. The interface between the H_2 and O_2 gas phases and the aqueous electrolyte phase occurs within the body of the porous electrode. In order to reduce ohmic heating of the electrolyte solution, most successful fuel cells use a thin film of solution or in some cases a thin film of an ionexchanger between the porous anode and cathode.

Cells Using Other Fuels

While the greatest development efforts thus far have been on the H_2-O_2 fuel cell, large–scale technological applications of fuel cells would depend on the use of other, more readily available fuels, such as carbon (coal), hydrocarbons (petroleum or natural gas), or carbohydrates (plant material). Research directed to the development of such fuel cells began in Germany during World War II and has continued, mostly in Europe, since then. While there has been significant progress and pilot–plant–scale fuel cells have been built, the processes are not yet competitive with traditional steam–powered turbines. For more details on fuel cells, see Bockris and Srinivasan *(G1)*, Pletcher *(G6)*, or reviews by Eisenberg *(15)* or Cairns *(16)*.

Storage Batteries

Energy storage cells (secondary cells) have somewhat different requirements than primary energy conversion cells. A storage cell must be capable of many charge/discharge cycles with high energy efficiency; thus there must be no irreversible side reactions or processes which convert components into unusable forms. Because storage cells are mostly used in situations requiring a portable power source, the energy available per unit weight should be as large as possible. A common figure of merit applied to storage cells is the energy density, expressible in units of $J\ g^{-1}$ or $kWh\ kg^{-1}$ (1 kWh = 3600 kJ).

By far the largest use of storage cells at present is in automobile batteries. Here the major requirement is that the battery be capable of delivering high power for a relatively short time (to start the engine). Storage cells are also used to power many small appliances. A potentially important application of storage cells is in powering electric cars and trucks. Electrically powered vehicles were popular in the period around World War I but did not successfully meet the competition of the internal combustion engine. With society's recently acquired awareness of the environmental damage of automobiles, there has been a revival of interest in electric cars and in efficient light–weight batteries to run them. Another potentially important application of storage cells is as load–leveling devices at power generating stations.

The Lead–Acid Storage Battery

The lead–acid cell *(17)*, familiar because of its use in automobile storage batteries, was invented by Planté in 1859.[‡] The cell can be represented by

$$Pb(s)\,|\,PbSO_4(s)\,|\,H_2SO_4(aq)\,|\,PbSO_4(s),PbO_2(s)\,|\,Pb(s)$$

[‡] Gaston Planté (1834–1889) was Professor of Physics in Paris; he is best known for his work on storage batteries.

The half–cell reactions are

$$Pb(s) + SO_4^{2-}(aq) \rightarrow PbSO_4(s) + 2\ e^-$$

$$PbO_2(s) + 4\ H^+(aq) + SO_4^{2-}(aq) + 2\ e^- \rightarrow PbSO_4(s) + 2\ H_2O$$

so that the overall cell reaction is

$$Pb(s) + PbO_2(s) + 2\ H_2SO_4(aq) \rightarrow 2\ PbSO_4(s) + 2\ H_2O$$

When the acid concentration is greater than about 2 M, the formal cell potential is about 2 V; thus a 12–V battery requires six cells. Both the lead sponge used as the anode material and the lead dioxide cathode (mixed with an "expander" such as $BaSO_4$ to increase the surface area) are packed into lead grids which provide the electrical contacts. When the battery is discharged, nonconducting $PbSO_4$ is formed at both the anode and cathode and the internal resistance of the cell increases. Some recovery of cell efficiency occurs as the H_2SO_4 electrolyte diffuses into the $PbSO_4$ cake and reachs unreacted Pb or PbO_2. The rate of recovery is thus tied to the diffusion rate and is much slower at low temperatures. Since the density of $PbSO_4$ is less than that of either lead or PbO_2, excessive discharge would rupture the lead grids. Thus a lead–acid battery is normally operated so that only 20–30% of the theoretically available energy is withdrawn before recharging. Since discharge of the cell reduces the concentration of sulfuric acid, the density of the H_2SO_4 electrolyte solution provides a convenient measure of the state of charge of the battery.

Given the half–cell potentials, recharging a lead–acid battery should result in generation of H_2 (instead of reduction of $PbSO_4$ to Pb) and O_2 (instead of oxidation of $PbSO_4$ to PbO_2). Fortunately, these processes are very slow at lead electrodes and thus are operationally irreversible. However, when the battery is fully charged, $H_2(g)$ and $O_2(g)$ production may occur if the charging potential is high enough. Since gas evolution tends to dislodge $PbSO_4$ from the grids, this is undesirable and thus regulation of the voltage of a battery charger is required.

Lead–acid storage batteries have undergone many generations of engineering improvement and are now reasonably efficient and reliable. Because of its low internal resistance, the lead–acid cell is capable of impressive bursts of power (up to about 10 kW for a few seconds), and can be charged and discharged for years. The principal disadvantage is their weight. Electric vehicles powered with lead–acid batteries thus must devote a large fraction of their load–carrying capacity to batteries, a restriction which has severely limited development.

Example 1.9 Compute the theoretical maximum energy density for a lead–acid cell, taking into account the weight of reactants in the cell reaction and neglecting the cell housing and electrolyte solution.

The electrical energy delivered at 2 V is

$$w_{elec} = nFE = 386 \text{ kJ mol}^{-1}$$

The weight of one mole of the reactants (Pb, PbO_2, and 2 H_2SO_4) is 642.5 g, so that

$$\text{Energy density} = (386 \text{ kJ mol}^{-1})/(642.5 \text{ g mol}^{-1}) = 601 \text{ J g}^{-1}$$

or 0.17 kWh kg^{-1}. Because the cell is usually not more than 30% discharged and the neglected weight is substantial, the practical energy density is in the range 80–200 J g^{-1}.

The Edison Cell

The alkaline iron–nickel cell can be represented as

$$Fe(s) \,|\, NiO_2(s), Ni_3O_4(s) \,|\, KOH(aq) \,|\, Fe_3O_4(s) \,|\, Fe(s)$$

The half–cell reactions

$$3 \text{ NiO}_2(s) + 2 \text{ H}_2O + 4 \text{ e}^- \rightarrow \text{Ni}_3O_4(s) + 4 \text{ OH}^-(aq)$$

$$3 \text{ Fe}(s) + 8 \text{ OH}^-(aq) \rightarrow \text{Fe}_3O_4(s) + 4 \text{ H}_2O + 8 \text{ e}^-$$

combine to give the overall cell reaction

$$6 \text{ NiO}_2(s) + \text{Fe}_3O_4(s) \rightarrow 3 \text{ Fe}(s) + 2 \text{ Ni}_3O_4(s)$$

This cell was perfected by Edison[‡] in 1910 as a power source for electric vehicles. The nominal cell potential is about 1.37 V but varies considerably with the state of charge of the cell. Since the electrolyte is not consumed in the cell reaction, a relatively small volume is required and the cell can be more compact and lighter in weight than a lead–acid cell. The theoretical maximum energy density is about 1300 $J\ g^{-1}$ (0.36 kWh kg^{-1}), which compares favorably with most other storage cells. The Edison cell has several other advantages: the basic solution is less corrosive than the concentrated acid used in a lead battery; the electrode assemblies are more rugged and less susceptible to damage by complete discharge or overcharging. However, the smaller cell potential requires more cells per battery than a lead–acid cell and, because nickel is more expensive than lead, the initial cost is greater. On the other hand, an Edison cell has a much longer service life than a lead–acid cell and so may be cheaper in the long run. Edison cells were used extensively in the heyday of the electric car and are quite well suited to that purpose. Because of its greater internal resistance, however, the Edison cell cannot match the lead–acid cell for peak power production and thus is not as satisfactory for turning starting motors on internal combustion engines.

The Nickel–Cadmium Cell

This cell *(18)* is a first cousin to the Edison cell, with iron and Fe_3O_4 replaced by cadmium and $Cd(OH)_2$. The cell, which has a nominal potential of 1.25 V, has many of the advantages and disadvantages of the Edison cell. It is intrinsically more expensive and has a lower energy density, but it has found an important niche in small rechargeable batteries used in transistor radios, tape recorders, pocket calculators, etc.

The Sodium–Sulfur Cell

The traditional energy storage cells discussed above employ moderately active metals and aqueous solutions. There have been

---

[‡]   Thomas Alva Edison (1847–1931) was a prolific inventor whose persistence made up for his lack of formal scientific training. Although Edison acknowledged a debt to Michael Faraday for his electrical and electrochemical inventions, his approach was less one of pure reason than an exhaustive trial of every possible solution to the problem at hand.

many attempts to develop cells based on the much more exoergic reactions of the alkali metals. A successful cell based on the reaction of lithium or sodium with fluorine or oxygen would combine high cell potential with light weight, but there are some obvious technical problems in building such a cell. A promising system with somewhat less severe problems is based on the reaction of sodium and sulfur

$$2\,Na + S \rightarrow 2\,Na^+ + S^{2-}$$

The cell is operated at 300°C with both sodium and sulfur in the liquid state. These two phases are separated by a β-alumina ($Na_2O \cdot 11Al_2O_3$) membrane which is very permeable to Na^+ at high temperature. Liquid sulfur supports the ionization of Na_2S. An inert metal such as molybdenum is used to contact the conducting phases and the cell can be represented as

$$Mo(s)\,|\,Na(l)\,|\,Na^+(\beta\text{-alumina})\,|\,Na_2S(s),S(l)\,|\,Mo(s)$$

The cell voltage is about 2 V, depending on the state of charge, and the theoretical maximum energy density is about 5000 J g^{-1} (1.4 kWh kg^{-1}). The major problem with the cell is in the membrane. Na atom diffusion into grain boundaries in the β-alumina membrane causes short circuits and structural failure of the membrane. The sodium–sulfur cell is probably not practical for electric vehicles (heating the cell to 300°C before operation is one problem), but as a power plant load-leveling device it has some promise.

REFERENCES

(Reference numbers preceded by a letter, *e.g.* *(A1)*, refer to a book listed in the Bibliography.)

1. A. A. Frost, *J. Am. Chem. Soc.* **1951**, *73*, 2680.
2. E. A. V. Ebsworth, *Educ. Chem.* **1964**, *1*, 123.
3. P. Delahay, M. Pourbaix, and P. van Rysselberghe, *J. Chem. Educ.* **1950**, *27*, 683.
4. F. Haber and Z. Klemenciewicz, *Z. phys. Chem.* **1919**, *67*, 385.
5. R. W. Murray, *Electroanalytical Chemistry* **1984**, *13*, 191.
6. M. A. Arnold and R. L. Solsky, *Anal. Chem.* **1986**, *58*, 84R.
7. N. H. Furman in *Treatise on Analytical Chemistry*, I. M. Kolthoff and P. J. Elving, eds, New York: Wiley, 1963. Part I, Vol. 4, p 2269.
8. S. Wawzonek in *Physical Methods of Chemistry*, Part IIA

(*Techniques of Chemistry*, Vol. I), A. Weissberger and B. W. Rossiter, eds, New York: Wiley, 1971, p 1.

9. R. P. Buck in *Physical Methods of Chemistry*, Part IIA (*Techniques of Chemistry*, Vol. I), A. Weissberger and B. W. Rossiter, eds, New York: Wiley, 1971, p 61.

10. D. P. Gregory, *Modern Aspects of Electrochemistry* **1975**, *10*, 239.

11. K. V. Dordesch, *Modern Aspects of Electrochemistry* **1975**, *10*, 339.

12. J. McBreen and E. J. Cairns, *Adv. Electrochem. Electrochem. Engin.* **1978**, *11*, 273.

13. W. Ostwald, *Z. Elektrochem.* **1894**, *1*, 129.

14. W. R. Grove, *Phil. Mag.* **1839**, *14*, 127.

15. M. Eisenberg, *Adv. Electrochem. Electrochem. Engin.* **1962**, *2*, 235.

16. E. J. Cairns, *Adv. Electrochem. Electrochem. Engin.* **1971**, *8*, 337.

17. J. Burbank, A. C. Simon, and E. Willihnganz, *Adv. Electrochem. Electrochem. Engin.* **1970**, *8*, 157.

18. P. C. Milner and U. B. Thomas, *Adv. Electrochem. Electrochem. Engin.* **1967**, *5*, 1.

PROBLEMS

1.1 An electrochemical cell is formed by placing platinum electrodes in separate beakers, connected through a salt bridge. One beaker contains a solution of $FeSO_4$, $Fe_2(SO_4)_3$, and H_2SO_4 (all 0.01 M); the other beaker contains $KMnO_4$, $MnSO_4$, and H_2SO_4 (all 0.01 M).

(a) Using standard notation, write a representation of the cell.

(b) Write the half–cell reactions.

(c) Write the cell reaction.

(d) Compute the standard potential of the cell.

(e) Compute the actual potential of the cell (neglect activity coefficients, but take account of the second ionization of H_2SO_4, $pK_{a2} = 1.99$).

(f) Compute the standard free energy change for the cell reaction and the corresponding equilibrium constant.

1.2 Calculate the potential of each of the cells (ignore activity coefficient corrections, but note that one of the half–cells is a standard reference electrode):

(a) $Zn(s) \,|\, ZnSO_4(0.01\ M) \,\|\, KCl(satd) \,|\, Hg_2Cl_2(s) \,|\, Hg$

(b) $Pt \,|\, FeCl_3(0.01\ M),\ FeCl_2(0.002\ M) \,\|\, KCl(3.5\ M) \,|\, AgCl(s) \,|\, Ag$

(c) $Pt \,|\, CrCl_3(0.1\ M),\ K_2Cr_2O_7(0.001\ M),\ HCl(0.001\ M) \,\|$
$\qquad\qquad\qquad\qquad KCl(3.5\ M) \,|\, AgCl(s) \,|\, Ag$

1.3 Use the data of Table A.4 to construct a free energy–oxidation state diagram for manganese in acidic solution. Given that the solubility product constants for $Mn(OH)_2$ and $Mn(OH)_3$ are, respectively, 2×10^{-13} and 4×10^{-43}, construct a free energy–oxidation state diagram for manganese in basic solution.

1.4 Given that the standard potential of the half–cell

$$SO_4{}^{2-}(aq)\,|\,PbSO_4(s)\,|\,Pb(s)$$

is -0.356 V, estimate the solubility–product constant of $PbSO_4$. You will need more data; see Table A.4.

1.5 If the standard potential for the reduction of $Sb_2O_5(s)$ to SbO^+ in acid solution is $+0.60$ V and the standard potential for the reduction of Sb_2O_5 to $Sb_2O_3(s)$ in basic solution is -0.13 V, what is the solubility product constant, $K_{sp} = [SbO^+][OH^-]$? What is the pH of a saturated solution of Sb_2O_3 at 298 K?

1.6 Devise general derivations of eqs (1.24), (1.25), and (1.26)

1.7 A silver electrode is immersed in 100 mL of 0.1 M KCl, and the solution titrated with 0.2 M $AgNO_3$ solution. The potential is determined *vs.* a standard Ag/AgCl electrode (with 3.5 M KCl electrolyte). Calculate the cell potential when 1, 10, 30, 45, 50, 55, and 70 mL of the silver solution has been added. The solubility product constant of AgCl is 2.3×10^{-10}. Neglect activity coefficients. Plot the calculated potential *vs.* volume of titrant.

1.8 Overman (*Anal. Chem.* **1971**, *43*, 616) found that an iodide–selective electrode can be used in the determination of Hg^{2+} by titration with a standard NaI solution. When he titrated 50 mL of unknown Hg^{2+} solution with 0.0100 M NaI, he obtained the titration curve shown in Figure 1.18. What was the concentration of mercuric ions in the original solution assuming that the titration reaction produced insoluble HgI_2?

This titration procedure worked for Hg^{2+} concentrations down to about 5×10^{-6} M, but at lower concentrations, endpoints were found which corresponded to ratios of iodide to mercury of more than 2:1. Why?

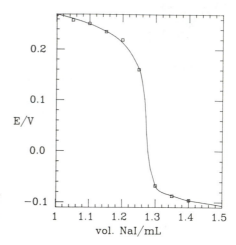

Figure 1.18 Titration of Hg^{2+} with iodide selective electrode.

1.9 The cell

$$Cd\,|\,Cd^{2+}(0.00972\ M)\,\|\,Cd^{2+}(0.00972\ M),CN^-(0.094\ M)\,|\,Cd$$

has a potential of -0.4127 V at 25°C. If the only significant reaction between Cd^{2+} and CN^- is the formation of $Cd(CN)_4^{2-}$, what is the equilibrium constant for the formation of the complex ion?

1.10 The quinhydrone electrode is sometimes used to measure pH. Quinhydrone is an easily prepared slightly soluble equimolar mixture of benzoquinone and hydroquinone. The reduction of the quinone is reversible on platinum:

$$C_6H_4O_2 + 2\ H^+ + 2\ e^- \rightleftarrows C_6H_4(OH)_2 \qquad E° = +0.700\ V$$

In practice, an unknown solution is saturated in quinhydrone and the potential measured with respect to a saturated calomel electrode.
(a) Show that the potential of the quinhydrone electrode gives the pH directly.
(b) Suppose that a cell potential of $+0.160$ V is measured *vs.* s.c.e. What is the pH of the solution?
(c) The quinhydrone electrode is not usable above pH 9. Why?

1.11 The adult human brain operates at a power of about 25 W. Most of the power is used to operate "sodium pumps" in the nerve cell membranes, which maintain the internal ionic concentration at about 0.015 M Na^+, while the external concentration is about 0.15 M Na^+. Assuming that *all* the power is used for such pumps and that they have an overall efficiency of 50%, calculate the total flux of Na^+ out of the brain cells per second. Assuming that the brain has 10^{10} cells and that each nerve impulse results in an uptake of 10^{-11} mole of Na^+, estimate the firing rate of the brain cells.

1.12 In the Krebs citric acid cycle, nicotinamide adenine dinucleotide (NAD^+) acts as an electron acceptor in three separate steps: (1) in the conversion of isocitric acid to a-ketoglutaric acid; (2) in the conversion of a-ketoglutaric acid to succinic acid; and (3) in the conversion of malic acid to oxaloacetic acid. From the data given in Table A.6, compute (a) the cell potential of each of these processes; (b) the standard free energy changes; and (c) the equilibrium constants (all at 25°C, pH 7).

1.13 From data given in Table A.4, compute
(a) The standard free energy change for the oxidation of Br^- by H_2O_2.
(b) The standard free energy change for the oxidation of H_2O_2 by Br_2.
(c) The standard free energy change for the disproportionation of H_2O_2 to H_2O and O_2.
(d) Knowing that solutions of hydrogen peroxide are stable for long times, what do you conclude?
(e) Assuming that the reactions described in parts (a) and (b) above are relatively fast, predict the effect of a trace of KBr on a hydrogen peroxide solution.

1.14 From data given in Table A.4, compute the standard free energy of disproportionation for the following reactions:

$$2\,Cu^+ \rightarrow Cu^{2+} + Cu$$

$$3\,Fe^{2+} \rightarrow 2\,Fe^{3+} + Fe$$

$$5\,MnO_4^{2-} + 8H^+ \rightarrow 4\,MnO_4^- + Mn^{2+} + 4\,H_2O$$

Can you generalize from these results to obtain a criterion for the stability of a species toward disproportionation based on half-cell potentials?

1.15 The potential of a glass electrode was 0.0595 V *vs.* s.c.e. when the electrodes were immersed in a pH 7.00 buffer solution and 0.2598 V when immersed in a solution of unknown pH.
(a) Calculate the pH of the unknown solution.
(b) If a possible junction potential introduces an uncertainty of ± 0.0005 V in comparing the potentials of the known and unknown solutions, what is the uncertainty in the pH of the unknown?

1.16 A sodium ion–selective electrode developed a potential of 0.2631 V *vs.* s.c.e. when immersed in 25.00 mL of an unknown solution. After addition of 5.00 mL of a 0.0100 M Na^+ standard solution, the potential decreased to 0.1921 V.
(a) Calculate the sodium ion concentration in the unknown solution.
(b) If the uncertainty in the potential measurements was ± 0.0005 V, what was the uncertainty in the sodium concentration of the unknown?

1.17 The potential of a commercial glass electrode in contact with a solution with $[Na^+] = 0.10$ M is -0.450 V at pH 8.00, -0.568 V at pH 10.00, and -0.690 V at pH 13.00.
(a) Compute the potentiometric selectivity coefficient, $k_{H,Na}$.
(b) At what solution pH would the apparent pH (computed assuming pH linear in potential without correction) be too low by 0.05 pH unit?

1.18 The silver–zinc cell,

$$Zn(s)\,|\,ZnO(s)\,|\,KOH(aq,40\%),K_2ZnO_2(aq)\,|\,Ag_2O(s)\,|\,Ag(s)$$

has had limited application in storage batteries.
(a) Write the half–cell reactions and the overall cell reaction.
(b) If the cell potential is 1.70 V, what is the maximum possible energy density?

2

THE ELECTRIFIED INTERFACE

In the last chapter, we focused on the equilibrium potential developed by an electrochemical cell. We relied on thermodynamics and paid no attention to the molecular details. We shall now look more closely at the interface between an electrode and an electrolyte solution. Virtually any surface in contact with an electrolyte solution acquires a charge and therefore an electric potential different from that of the bulk solution. There are four ways in which a surface may acquire a charge: (1) imposition of a potential difference from an external potential source; (2) adsorption of ions on a solid surface or on the surface of a colloidal particle; (3) electron transfer between a metallic conductor and the solution; and (4) for micelles and biological macromolecules and membranes, ionization of functional groups such as carboxylate, phosphate, or amino groups. Surface charge effects are particularly important in biological systems. The surface-to-volume ratio of a biological cell is large and most biochemical reactions occur at or near the surface of an immobilized enzyme.

We begin in Section 2.1 with the development of a mathematical model for a charged interface. We digress slightly in Sections 2.2–2.4 to a discussion of the effects of charged surfaces on the equilibrium and dynamic behavior of interfacial systems. Returning to electrodes in Section 2.5, we consider the phenomena of electrocapillarity and

69

double–layer capacitance. The ideas developed in our discussion of charged surfaces can be extended to the interaction of small ions in solution, and we conclude this chapter with a discussion of the Debye–Hückel theory of ionic activity coefficients.

2.1 THE ELECTRIC DOUBLE LAYER

A charged surface in contact with an electrolyte solution is expected to attract ions of opposite charge and to repel ions of like charge, thus establishing an *ion atmosphere* in the immediate vicinity of the surface. Two parallel layers of charge are formed—the charge on the surface itself and the layer of oppositely charged ions near the surface. This structure is called the *electric double layer*. For reviews of double layer theory, see Grahame *(1)*, Parsons *(2)*, Mohilner *(3)*, or Reeves *(4)*.

Helmholtz considered the problem of a charged surface in contact with an electrolyte solution in 1879[‡] *(5)*. He assumed that a layer of counter ions would be immobilized on the surface by electrostatic attraction such that the surface charge is exactly neutralized. In the Helmholtz model, the electric potential falls from its surface value, Φ_0, to zero in the bulk solution over the thickness of the layer of counter ions. This behavior is shown schematically in Figure 2.1a.

Later, Gouy[#] *(6)* and Chapman[§] *(7)* pointed out that ions are

----

[‡] Hermann Ludwig Ferdinand von Helmholtz (1821–1894) was trained as a physician and held chairs in physiology and anatomy at the Universities of Königsberg and Heidelberg. His interest in physics led to an appointment as Professor of Physics at the University of Berlin. Helmholtz made pioneering contributions in thermodynamics, physiological acoustics and optics, hydrodynamics, electrodynamics, and epistemology. It should be recalled that Arrhenius' theory of electrolyte solutions was put forward in 1887; in 1879 ions were thought to occur only in rather special circumstances; the Helmholtz layer was such a special case.

[#] Louis–Georges Gouy (1854–1926) was Professor of Physics at the University of Lyons. His contribution to the theory of the electric double layer was an extension of his interest in Brownian motion.

[§] David L. Chapman (1869–1958) was a Fellow of Jesus College,

subject to random thermal motion and thus would not be immobilized on the surface. They suggested that the ions which neutralize the surface charge are spread out into solution, forming what is called a *diffuse double layer*. According to the Gouy–Chapman model, the potential falls more slowly to the bulk solution value, as shown in Figure 2.1b.

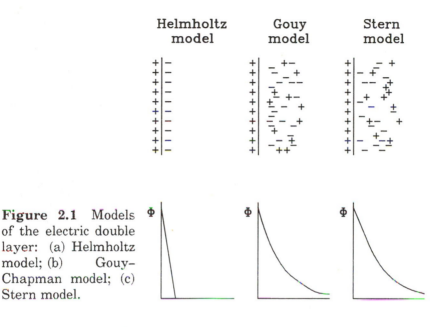

Figure 2.1 Models of the electric double layer: (a) Helmholtz model; (b) Gouy–Chapman model; (c) Stern model.

In 1924, Stern[‡] *(8)* observed that neither the Helmholtz model nor the Gouy–Chapman model adequately accounts for the properties of the double layer and suggested that the truth lies in a combination of the two models. Thus some ions are indeed immobilized on the surface (the *Helmholtz layer*), but usually not enough to exactly neutralize the charge; the remainder of the charge is neutralized by a diffuse layer (or *Gouy layer*) extending out into the solution. This

---

Oxford. Most of his scientific work was on gas–phase reaction kinetics.

[‡] Otto Stern (1888–1969) was Professor of Physics at the Universities of Frankfurt, Rostock, and Hamburg and moved to the Carnegie Institute of Technology in 1933. Stern is remembered primarily as the father of molecular beams research.

model is pictured schematically in Figure 2.1c.

In the following sections, we shall work through some of the details of the Gouy–Chapman theory, as modified by Stern, and discuss some of the implications.

The Poisson–Boltzmann Equation

The immediate goal of a theoretical description of the electrified interface is the derivation of an equation giving the electric potential Φ as a function of the distance from the surface.

The electric potential at any point in the solution is related to the net charge density at that point by a fundamental relationship derived from electrostatics and known as Poisson's equation:

$$\nabla^2 \Phi = -\rho / \epsilon \, \epsilon_0 \qquad (2.1)$$

In eq (2.1) ϵ is the dielectric constant of the medium and ϵ_0 is a constant called the permittivity of free space ($\epsilon_0 = 8.854 \times 10^{-12}$ $C^2 J^{-1} m^{-1}$). The operator ∇^2 is, in Cartesian coordinates,

$$\nabla^2 = \frac{\partial^2}{\partial x^2} + \frac{\partial^2}{\partial y^2} + \frac{\partial^2}{\partial z^2}$$

and in spherical polar coordinates,

$$\nabla^2 = \frac{1}{r^2} \frac{\partial}{\partial r} \left(r^2 \frac{\partial}{\partial r} \right) + \frac{1}{r^2 \sin\theta} \frac{\partial}{\partial \theta} \left(\sin\theta \, \frac{\partial}{\partial \theta} \right) + \frac{1}{r^2 \sin^2\theta} \frac{\partial^2}{\partial \phi^2}$$

The space charge density ρ is related to the concentrations and charges of the ions in solution and can be expressed as

$$\rho = \sum_i z_i F C_i \qquad (2.2)$$

where ρ has units of $C \, m^{-3}$, z_i is the charge on ion i, C_i is the concentration[‡] of i in mol m^{-3}, and F is the Faraday constant.

[‡] In this and subsequent chapters we will mostly use concentration units of mol m^{-3}. The major exceptions to this rule are those cases

Equations (2.1) and (2.2) could be combined to obtain a relation between ionic concentration and the electric potential. However, since neither of these quantities is known, we need yet another relation to achieve our initial goal. To obtain the necessary link, we recall the Boltzmann distribution law

$$N_1/N_0 = \exp[-(E_1 - E_0)/kT]$$

which gives the equilibrium ratio of the number N_1 of particles having energy E_1 to the number N_0 with energy E_0. In the equation, k is Boltzmann's constant (the gas constant divided by Avogadro's number, $k = R/N_A$), and T is the absolute temperature. The Boltzmann equation is applied here to calculate the relative concentrations of ions at positions of different electrostatic potential energy. The electrostatic potential energy of an ion with charge[‡] $z_i e$, located at a point where the electric potential is Φ, is given by

$$E_i = z_i e\, \Phi$$

Since the zero of electric potential is arbitrary, we will choose the electric potential of the bulk solution (far from the surface) as zero; thus the electrostatic potential energy of ions in the bulk solution is also zero. The Boltzmann equation may then be written as

$$N_i/N_i{}^\circ = \exp[-z_i e\, \Phi/kT]$$

where N_i is the number of ions of type i (per unit volume) at the point in the solution where the potential is Φ, and and $N_i{}^\circ$ is the number of these ions per unit volume in the bulk solution. Converting the N's to molar concentration units, and noting that $e/k = F/R$, we can rewrite the Boltzmann equation as

$$C_i/C_i{}^\circ = \exp[-z_i F\Phi/RT] \tag{2.3}$$

Notice that eq (2.3) predicts that a positive potential (due to a positive surface charge) would result in lower concentrations of positive ions

---

where the concentration is an approximation to an activity referred to the 1 M standard state.

[‡] The electronic charge e is taken as a positive constant, $e = 1.602 \times 10^{-19}$ C; z_i then includes both the magnitude (in units of e) and the sign of the charge on ion i.

($z_i > 0$) near the surface than in the bulk solution and higher concentrations of negative ions near the surface.

Combining eqs (2.2) and (2.3), we have

$$\rho = \sum_i z_i F C_i^{\,\circ} \exp[-z_i F\Phi/RT] \tag{2.4}$$

and finally, substitution of eq (2.4) into eq (2.1) gives the Poisson–Boltzmann equation:

$$\nabla^2\Phi = -\frac{F}{\epsilon\epsilon_0} \sum_i z_i C_i^{\,\circ} \exp[z_i F\Phi/RT] \tag{2.5}$$

The solution to this differential equation is the potential as a function of the position in the solution. In order to proceed further, we must specify the geometry of the problem and decide upon the method of solution. Integration of eq (2.5) is not a trivial exercise for any choice of geometry because of the sum of exponential terms. There are two methods which may be used in dealing with the summation: (1) a general, but approximate, solution obtained by expanding the exponential terms in a power series, and (2) an exact, but limited solution obtained by making a specific assumption about the nature of the electrolyte so that the summation can be eliminated.

Solutions of the Poisson–Boltzmann Equation

An exponential can be approximated by the power series

$$\exp(u) = 1 + u + \tfrac{1}{2} u^2 + \ldots \qquad u \ll 1$$

Thus, for $z_i\Phi \ll RT/F$, eq (2.5) can be written

$$\nabla^2\Phi = -\frac{F}{\epsilon\epsilon_0} \sum_i z_i C_i^{\,\circ} \left[1 + \frac{z_i F\Phi}{RT} + \frac{z_i^2 F^2\Phi^2}{2R^2T^2} + \ldots \right]$$

The first term in the sum vanishes since the bulk solution is electrically neutral. Retaining only the second term, we have

$$\nabla^2\Phi = +\frac{F^2\Phi}{\epsilon\epsilon_0 RT} \sum_i z_i^2 C_i^{\,\circ} \tag{2.6}$$

A convenient measure of the influence of electrolytes in problems like this is the ionic strength, defined by

$$I = \tfrac{1}{2} \sum_i z_i^2 C_i^\circ \tag{2.7}$$

Thus eq (2.6) can be written as

$$\nabla^2 \Phi = \frac{F^2 I \Phi}{\epsilon \epsilon_0 RT}$$

or, lumping all the constants together to define a new parameter

$$x_A = \left[\frac{\epsilon \epsilon_0 RT}{F^2 I} \right]^{\frac{1}{2}} \tag{2.8}$$

we have the linearized Poisson–Boltzmann equation

$$\nabla^2 \Phi = \Phi/x_A^2 \tag{2.9}$$

Let us assume that we are dealing with a surface in the y–z plane; it is reasonable to assume that the potential Φ is a function only of the distance from the surface x, so that eq (2.9) becomes

$$\frac{d^2 \Phi}{dx^2} = \frac{\Phi(x)}{x_A^2}$$

Differential equations of this form have the general solution

$$\Phi(x) = A\, e^{-x/x_A} + B\, e^{x/x_A}$$

where A and B are constants of integration to be determined from the boundary conditions. Since in our problem, $\Phi \to 0$ as $x \to \infty$, we see that B must be zero.

The other constant A is the potential at the surface $(x = 0)$ provided that the theory is adequate right up to the surface. However, the distance of closest approach to the surface must be at least as large as the ionic radius. If there are ions at this closest approach distance, then their position in space is more or less fixed and the assumption of random thermal motion will fail at very short distances. If this first layer of ions is essentially an immobilized Helmholtz layer, then we expect that the potential will drop more or less linearly from the surface potential Φ_s to the value Φ_a at the outer surface of the Helmholtz layer. We assume then that eq (2.9) applies to the outer or diffuse region (the Gouy layer). If a is the thickness of the immobilized Helmholtz layer, then the potential will be given by

$$\Phi(x) = \Phi_s (1 - x/a) + \Phi_a (x/a) \qquad x < a \qquad (2.10a)$$

$$\Phi(x) = \Phi_a \exp[(a - x)/x_A] \qquad x > a \qquad (2.10b)$$

The validity of eq (2.10b) depends on the correctness of eq (2.9) and the assumption that $z\Phi \ll RT/F$. Since $RT/F = 25.7$ mV at $25°C$, Φ_a apparently must be less than about $0.026/z$ volts.

If, instead of a planar surface, we consider the surface of a sphere (e.g., a macromolecule) and assume that the potential depends on the radial distance r from the center of the sphere but not on the angles θ or ϕ, the solution to eq (2.9) is

$$\Phi(r) = \Phi_a(a/r) \exp[(a - r)/x_A] \qquad (2.11)$$

where Φ_a is the potential at the outer surface of the Helmholtz layer $(r = a)$. Notice that this result differs from eq (2.10b) by the factor (a/r).

If we restrict our attention to a symmetrical electrolyte such as NaCl or $MgSO_4$, eq (2.5) can be integrated without approximations to obtain (see Problems)

$$\frac{[\exp(zF\Phi(x)/2RT) - 1][\exp(zF\Phi_a/2RT) + 1]}{[\exp(zF\Phi(x)/2RT) + 1][\exp(zF\Phi_a/2RT) - 1]} = \exp[(a - x)/x_A] \quad (2.12)$$

Equation (2.12) reduces to eq (2.10b) in the limit where $z\Phi_a \ll RT/F$.

Now that we have solutions to the Poisson–Boltzmann equation, let us examine some of the consequences. First, we look at the shape of the solutions represented by eq (2.12). In Figure 2.2 are plotted $\Phi(x)/\Phi_a$ vs. $(x - a)/x_A$ for $\Phi_a = 0.01$ and 0.1 V, assuming that $z = 1$ and $T = 298$ K. The curve for $\Phi_a = 0.01$ V is indistinguishable from the simple exponential dependence predicted by the approximate solution, eq (2.10b). Indeed, even for $\Phi_a = 0.1$ V, deviations from the approximate solution are not as great as we might have expected.

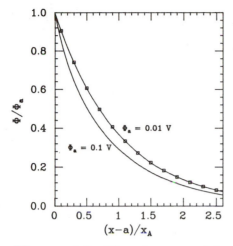

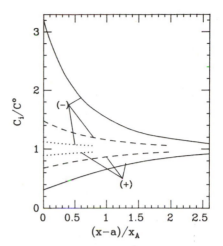

Figure 2.2 Electric potential calculated from eq (2.12) as a function of distance from the surface for Φ_a = 0.1 and 0.01 V. The points on the Φ_a = 0.01 V curve are calculated from eq (2.10b).

Figure 2.3 Concentrations (relative to bulk solution concentration) as a function of distance from the surface for a 1:1 electrolyte and Φ_a = 0.003 V (dots), 0.01 V (dashes), and 0.03 V (solid line).

The concentration ratios $C_+(x)/C°$ and $C_-(x)/C°$ are plotted as functions of $(x - a)/x_A$ for several values of Φ_a in Figure 2.3, assuming a 1:1 electrolyte (such as NaCl) and 298 K. Notice that the concentrations are quite different from the bulk values for large values of Φ_a at small distances.

Thickness of the Ion Atmosphere

Consider now the parameter x_A, defined by eq (2.8). Values of x_A are given in Table 2.1 for several concentrations of electrolytes of different types, calculated assuming a temperature of 25°C and the dielectric constant of water (ϵ = 78.54). Judging from the plots of Figures 2.2 and 2.3, x_A can be regarded as a measure of the thickness of the ion atmosphere at the charged surface.[‡] As we see in Table 2.1, this thickness depends strongly on the concentration and charge type of

---

[‡]   In the context of Debye–Hückel theory (Section 2.6), x_A is called the Debye length.

the electrolyte, varying from hundreds of water molecule diameters in dilute solutions to a distance on the order of one water molecule diameter in concentrated solutions of highly charged ions.

TABLE 2.1 Ion Atmosphere Thickness

C/mM	1:1	1:2	x_A/nm 1:3	2:2	2:3
0.1	30.4	24.8	21.5	15.2	7.8
1.0	9.6	7.8	6.8	4.8	2.5
10.0	3.0	2.5	2.15	1.52	0.78
100.0	0.96	0.78	0.68	0.48	0.25

Relation of Surface Potential to Surface Charge Density

We turn now to the relationship between the potential Φ_a and the net charge on the surface. If the system as a whole is electrically neutral, then the net surface charge (*i.e.*, charge on the surface adjusted for the charge of the Helmholtz layer) must be exactly balanced by the space charge in the Gouy layer. If σ is the net surface charge density (in C m^{-2}), then, for a planar surface,

$$\sigma = -\int_a^\infty \rho\, dx$$

Substituting for ρ from eq (2.1), we can carry out the integration:

$$\sigma = +\epsilon\epsilon_0 \int_a^\infty \frac{d^2\Phi}{dx^2}\, dx = +\epsilon\epsilon_0 \frac{d\Phi}{dx}\Bigg|_a^\infty$$

Since the derivative vanishes at the upper limit, we have

$$\sigma = -\epsilon \epsilon_0 \left(\frac{d\Phi}{dx}\right)_{x=a}$$

Differentiation of eq (2.10b) and evaluation of $d\Phi/dx$ at $x = a$ gives

$$\sigma = \epsilon \epsilon_0 \Phi_a / x_A \qquad (2.13)$$

A more complex expression results if we use eq (2.12) to evaluate $d\Phi/dx$ (see Problems), but it reduces to eq (2.13) in the limit that $z\Phi_a \ll RT/F$.

Example 2.1   Compute the potential at a charged surface in contact with 0.01 M (10 mol m^{-3}) KCl solution if the charges, $+e$, are spaced about 5 nm apart so that the average area per charge is 25 nm^2.

We can rearrange eq (2.13), putting σ in units of electronic charges per square nanometer and x_A in units of nanometers. Taking $\epsilon = 78.5$, we have

$$\Phi_a = 0.230 \, x_A \sigma$$

If $\sigma = 0.04 \, e \text{ nm}^{-2}$ and $x_A = 3.0$ nm, then

$$\Phi_a = (0.230)(0.04)(3.0) = 0.028 \text{ V}$$

Weaknesses of the Gouy–Chapman Theory

The theory as we have presented it can lead to absurd predictions. Consider the case of a 1 M NaCl solution in contact with a surface with $\Phi_a = 0.1$ V. According to eq (2.3), the concentration of chloride ions at the surface (i.e., at the outer surface of the Helmholtz layer) is 49 M! This corresponds to approximately 30 chloride ions per cubic nanometer—very nearly a closest packed layer of chloride ions. The ion–ion repulsion in such a structure would be far greater than the attraction to the positive surface and thus the prediction is ridiculous.

This absurdity points up the major weakness in the theory: we take account of electrostatic interaction between ions in solution and the charged surface, but neglect interactions between ions. This is a reasonable approximation in dilute solutions, but leads to gross errors in more concentrated solutions.

There are several less serious problems. We have tacitly assumed that the charge density ρ is a smoothly varying continuous function of the distance from the surface. While this is not true on an instantaneous microscopic level, this apparent weakness is not so bad if we regard the results as representing a time average where, through thermal motion of the ions, the charge is blurred more or less smoothly over the solution. A more serious defect results from the use of the bulk dielectric constant of the solvent in calculating electrical forces over short distances; since the water dipoles are oriented around ions or at the charged surface, the local properties are expected to be quite different than those of the bulk liquid.

In summary, the Gouy–Chapman treatment of the diffuse double layer should not be relied upon to give quantitatively accurate predictions for concentrated electrolyte solutions or for high surface charge densities or surface potentials. The theory is very useful, however, as a means of gaining a qualitative understanding of phenomena which are affected by the electrified interface. While the Gouy–Chapman approach continues to be used as a starting point for theoretical treatments of charged surfaces, some theorists have turned to molecular dynamics calculations where the motion of ions near a charged surface is followed in a computer simulation; potential and concentration distributions are then obtained by averaging over time. As it turns out, the qualitative results of the computer modeling studies are in good agreement with the insights obtained from Gouy–Chapman theory.

In the next few sections we will consider some important applications of the ideas obtained from Gouy–Chapman theory. In light of the weaknesses just discussed, we should not hope for high accuracy, but will be content with qualitative or, at best, semi-quantitative explanations of the phenomena discussed.

2.2 *SOME PROPERTIES OF COLLOIDS*

Particles in colloidal suspensions (*e.g.*, gold or ferric oxide sols) generally have electrically charged surfaces due to adsorption of ions. Macromolecular colloids such as proteins also have charged surfaces resulting from ionized functional groups. In general, therefore, colloidal particles have ion atmospheres. In this section we shall examine some of the characteristics of such particles which may be understood in terms of their electrified interfaces with the liquid solution.

The Stability of Colloids

We first consider the question of why large particles such as proteins should be stable in solution in the first place. Atoms or small molecules are subject to so–called London forces, which give rise to a van der Waals potential energy of interaction. London forces are short–range; the van der Waals energy varies with with internuclear separation as r^{-6} for interactions between atoms, but the situation is more complicated for macromolecules. One must consider the van der Waals attraction of each atom in one molecule for every atom in the other molecule. Some complicated integrals arise and we shall not go into the details of the calculation.[‡] The result for spherical particles of radius a is *(10)*

$$U_a(R) = -\frac{A}{12}\left[\frac{8a^2(R+a)}{R(R+2a)^2} + 2\ln\frac{R(R+4a)}{(R+2a)^2}\right] \quad (2.14)$$

where R is the distance between the particle surfaces. The parameter A is essentially empirical and usually is on the order of 10^{-19} to 10^{-20} J. The multiple attraction of the many atoms in a macromolecule results in the van der Waals energy falling off with distance much more slowly than might have been expected. With relatively long-range attractive forces, uncharged particles are expected to associate and eventually precipitate from solution.

So what keeps them apart? The free energy change, $\Delta G = \Delta H - T\Delta S$, governs the equilibrium behavior. We would expect that ΔS for

---

[‡] See Shaw *(9)* for further discussion of this problem.

coagulation would be negative, but for a relatively small number of large particles, the entropy term should be negligible and we must look for an explanation in the enthalpy term. Short–range repulsion sets in when the electron clouds overlap, but by the time this interaction is important, van der Waals attraction has already coagulated the particles.

If the two particles are electrically charged, then electrostatic repulsion will give a positive contribution to the potential energy of interaction, which rises exponentially as the particles approach one another. Notice, however, that according to eq (2.11) the distance parameter x_A determines the range of the potential and thus of the electrostatic repulsion. If x_A is very small, the diffuse layer is very compact and the particle's charge is shielded up to very short distances. In other words, when the ionic strength is high, the particles can approach quite closely before the electrostatic repulsion becomes significant, and by that time the van der Waals attraction may be quite large.

We expect therefore that when the ionic strength is increased, the double–layer thickness will decrease and the colloid will precipitate. This "salting–out" effect is commonly observed experimentally.

The calculation of the electrostatic repulsion between two particles, each of which has an ion atmosphere, is very difficult. All the difficulties attending Gouy–Chapman theory remain, of course, and the question of what happens when the two ion atmospheres begin to interpenetrate adds an important additional complication (6). Several attempts have been made and we will quote only one result. If it is assumed that $a \gg x_A$, that the surface charge density on the particles is constant, and that the surface potential adjusts with the ion atmosphere as the particles approach, the repulsive contribution to the potential energy is (11)

$$U_r(R) = - \frac{2\pi a^3}{(1 + a/x_A)^2} \frac{\sigma^2}{\epsilon \epsilon_0} \ln(1 - e^{-R/x_A}) \qquad (2.15)$$

The total potential energy is the sum of the attractive and repulsive contributions given by eqs (2.14) and (2.15), respectively.

To get a feel for the nature of these equations, the total potential energy is plotted as a function of distance between the particles in Figures 2.4 and 2.5. In both figures, it is assumed that the particles have radii of 50 nm and that the dielectric constant of pure water (78.5) is appropriate. The empirical parameter A was taken to be 5 × 10^{-20} J and the energy is given in units of kT. Since kT is a measure of the amount of thermal energy available, such a scale permits a rough judgment on the significance of a barrier to coagulation.

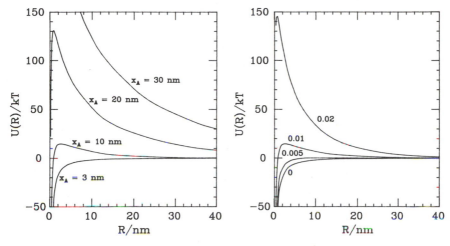

Figure 2.4 Potential energy of interaction (in units of kT) for two spherical particles of radius 50 nm and surface charge density 0.01 e nm^{-2} as a function of separation for x_A = 3, 10, 20, and 30 nm.

Figure 2.5 Potential energy of interaction (in units of kT) for two spherical particles of radius 50 nm in a solution with x_A = 10 nm as a function of separation for surface charge densities of 0, 0.005, 0.01, and 0.02 e nm^{-2}.

Figure 2.4 suggests that particles with a surface charge density of 0.01 e nm^{-2} should coagulate quickly when the electrolyte concentration is sufficiently high that the ion atmosphere is only 3 nm thick. When the thickness is 10 nm, there is a small barrier to coagulation (about 15 kT), but coagulation would be expected to occur, if slowly (e^{-15} = 3 × 10^{-7}). When the ion atmosphere is 20 nm thick, the barrier is about 130 kT and coagulation should be very slow (e^{-130} = 3 × 10^{-57}) and when the ion atmosphere is 30 nm thick, the barrier is even greater. Figure 2.5 shows that coagulation is very sensitive to surface charge density with no barrier at all for surface charges of 0 or 0.005 e nm^{-2}, a small barrier for 0.01 e nm^{-2}, and a barrier of about 145 kT for 0.02 e nm^{-2}.

According to eqs (2.8), (2.14), and (2.15), the potential energy also depends on the temperature and the dielectric constant of the medium as well as the size of the particle. We expect that salting–out should occur at lower ionic strength when either the dielectric constant or the temperature is reduced. Since the surface charge of a protein, for example, depends on the number of free carboxyl or amino groups present and thus on the pH of the medium, protein solubility is strongly pH dependent, going through a minimum at the isoelectric pH. Since protein solubilities are such intricate functions of ionic strength, temperature, dielectric constant, and pH, it is not surprising that protein mixtures can frequently be separated by selective precipitation techniques.

When salting–out occurs at electrolyte concentrations of 0.1 M or less, the effect is largely independent of the nature of the electrolyte and depends primarily on the ionic strength. Many proteins remain in solution under these conditions but are precipitated if much higher salt concentrations are used. Proteins are protected from coagulation in part by a solvation sheath of water molecules. When high concentrations of salt are used, the activity of water is reduced significantly by the solvation of the added salt. Thus in effect the surface of the protein is dehydrated and coagulation is facilitated. Under these conditions, the salting–out effect depends strongly on the nature of the added salt since some salts tie up more water than others. Protein and polymer chemists sometimes refer to the *lyotropic series* of cations and anions, which show decreasing salting–out effects:

$$Mg^{2+} > Ca^{2+} > Sr^{2+} > Ba^{2+} > Li^+ > Na^+ > K^+ > NH_4^+$$

$$SO_4^{2-} > Cl^- > NO_3^- > I^- > SCN^-$$

Salts with low solubility such as $CaSO_4$ are of little use in salting–out, of course; the most commonly used precipitants are very soluble salts with a component high on the lyotropic series, such as ammonium sulfate or magnesium chloride.

Surface pH

We have seen that ions are either concentrated in or excluded from the double layer, depending on the relative charges. This conclusion applies to hydrogen and hydroxide ions as well, so that the pH at the surface of a charged particle will be different from that of the bulk solution. According to eq (2.3), the concentration of H^+ ions at the surface is given by

$$[H^+]_{surface} = [H^+]_{bulk} \exp{\frac{-F\Phi_a}{RT}} \qquad (2.16)$$

or, converting to pH and setting T = 298 K,

$$pH_{surface} = pH_{bulk} + 16.9\,\Phi_a \qquad (2.17)$$

Thus a surface potential Φ_a = 59 mV would result in an increase in surface pH by one pH unit.

When the pH is less than the isoelectric point of a protein, the surface is positively charged (due to $-NH_3^+$ groups), and the surface pH should be higher than that of the bulk solution; above the isoelectric pH, the surface is negatively charged (because of $-CO_2^-$ groups), and the surface pH should be less than the bulk pH. Thus, to a degree, the surface pH on a protein tends to be self–buffered toward the isoelectric pH. While singly charged ions such as H^+ or OH^- can be significantly more (or less) concentrated near a charged surface than in bulk solution, such effects are even more dramatic for multiply charged ions. Thus eq (2.3) suggests that a surface potential Φ_a = -59 mV would result in a concentration magnification of 100 for Mg^{2+} and 1000 for Al^{3+}. Although we must be cautious about quantitative predictions from Gouy–Chapman theory, the qualitative effect is real and has important consequences, for example, in the action of metal ions as co–factors in enzymatic reactions.

Example 2.2   A dodecylsulfate micelle contains about 100 dodecylsulfate ions and has a radius of about 2 nm. Compute the surface charge density, the surface potential, Φ_a, and the difference between the surface pH and the pH of the bulk solution. Assume an ionic strength of 100 mM.

The surface charge density is

$$\sigma = \frac{(-100)(1.6 \times 10^{-19}\ C)}{(4\pi)(2 \times 10^{-9}\ m)^2} = -0.32\ C\ m^{-2}$$

The surface potential derived using the linearized Poisson–Boltzmann equation (see Problems) is

$$\Phi_a = \frac{\sigma}{\epsilon \epsilon_0} \frac{a}{1 + a/x_A}$$

so that with $x_A = 0.96$ nm and $a = 2$ nm,

$$\frac{a}{1 + a/x_A} = 0.65 \text{ nm}$$

and

$$\Phi_a = \frac{(-0.32 \text{ C m}^{-2})(0.65 \times 10^{-9} \text{ m})}{(78.5)(8.85 \times 10^{-12} \text{ C}^2 \text{J}^{-1} \text{m}^{-1})} = -0.30 \text{ V}$$

which is rather too large to justify use of the linearized Poisson–Boltzmann equation. Ignoring this problem, eq (2.17) predicts

$$pH_{surface} - pH_{bulk} = -5.1$$

While this result is doubtless an overestimate, it is in reasonable qualitative agreement with the following experimental result. When lauric acid is added to a solution of dodecylsulfate micelles, the long–chain carboxylic acid molecules are included in the micelles. A pH titration of the carboxylic acid yields an apparent pK_a (the pH of the half–neutralization point) of about 7 under these conditions. Since the pK_a of lauric acid is about 4.8, the surface pH is apparently about 2.2 pH units lower than that of the bulk solution.

2.3 ELECTROKINETIC PHENOMENA

Four rather peculiar effects, known as the electrokinetic phenomena, turn out to be relatively easily understood in terms of our theory of the diffuse double layer *(12,B8)*. These effects arise from the motion of an electrolyte solution past a charged surface.

The Phenomena

Consider an electrolyte solution flowing through a capillary as shown in Figure 2.6. Since in general there is a surface charge on the capillary walls due to adsorbed ions, there will be a diffuse space

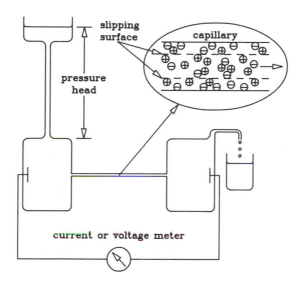

Figure 2.6 Apparatus for measuring the streaming current or streaming potential. With adsorbed anions, positive space charge is swept along by solution flow.

charge in the solution adjacent to the walls. As the solution flows through the capillary, some of the space charge is swept along with it and, if electrodes are provided at either end of the capillary, a current, called the *streaming current*, can be measured between them.

If, on the other hand, we were to apply a voltmeter across the electrodes rather than an ammeter, so that essentially no current flows, ions swept through the capillary will accumulate at one end, producing an electric potential difference across the capillary, called the *streaming potential*. If we were to start the experiment with a homogeneous solution, the streaming potential would grow with time until it was big enough that electric migration of ions upstream exactly canceled the flow of ions downstream.

The two effects, streaming current and streaming potential, may be jointly expressed by one phenomenological equation:

$$i = a_1 \Delta P + a_2 \Delta \Phi \tag{2.18}$$

Thus, when current i is allowed to flow without impedance in the external circuit, $\Delta \Phi = 0$, and the current is proportional to the pressure difference ΔP which generates the liquid flow. However, when the current is zero (*e.g.*, because of high external impedance or slow electrode kinetics), then a potential difference $\Delta \Phi$ is observed which is proportional but opposed to the pressure difference.

A related effect may be observed if, instead of applying a pressure difference, we impose a potential difference between the two electrodes. Now in the bulk solution, positive ions move toward the negative electrode and negative ions toward positive electrode. There is little drag on the bulk solution, which has zero net charge. Near the surface, however, the space charge is acted on by the field, tending to drag the solution along with it. If the cylindrical shell of solution near the capillary walls moves, then viscous drag will pull the solution in the center of the capillary along as well. The effect of liquid flow under the action of an electric potential difference is called *electroosmosis*.

If we carry out an electroosmosis experiment and allow flow to continue such that a pressure difference is built up, the flow will eventually slow and come to a halt as the pressure opposes and finally cancels the effect of the field. The relation of this steady–state phenomenon, called the *electroosmotic pressure*, to electroosmosis is analogous to the relation of streaming potential to streaming current. Electroosmosis and electroosmotic pressure may be expressed by the phenomenological equation

$$j = a_3 \Delta P + a_4 \Delta \Phi \tag{2.19}$$

where j is the flux of liquid through the capillary (in $m^3 s^{-1}$). Thus, in the absence of a pressure difference, the flux is proportional to the electric potential difference $\Delta \Phi$. When the flux goes to zero, the pressure difference is proportional but opposed to the electric potential difference.

The Zeta Potential

Electrokinetic effects arise because of motion of the diffuse layer of ions in solution relative to the solid surface. We might guess that the Helmholtz layer is stationary and that the Gouy layer moves with the bulk solution, but this is an oversimplification. The behavior is as if there were a slipping surface located in the diffuse part of the ion atmosphere. The potential at the slipping surface is given the symbol ζ (zeta) and is called the *electrokinetic potential* (or simply the *zeta potential*).

In a quantitative treatment of the electrokinetic effects, the appropriate form of the ∇^2 operator in the Poisson equation depends on the geometry of the problem. Thus for the symmetry of a capillary tube, cylindrical coordinates (r, θ, and z) should be used. If we can assume that Φ depends only on the distance from the capillary wall, then we can neglect terms in $d\Phi/d\theta$ and $d\Phi/dz$ and eq (2.1) can be written as

$$\rho(r) = - \frac{\epsilon \epsilon_0}{r} \frac{d}{dr}\left(r \frac{d\Phi}{dr}\right) \tag{2.20}$$

If we assume that a, the radius of the capillary, is much larger than the diffuse layer thickness x_A, then we need not rederive an equation for the potential but can use the result obtained for a planar surface,

$$\Phi(r) = \zeta \exp[(r - a)/x_A] \tag{2.21}$$

where the coordinate r is measured from the center of the capillary. We have assumed that a/x_A is large enough that the potential in the center of the capillary is nearly zero.

Calculation of the Coefficients

The coefficients in eqs (2.18) and (2.19) can be computed rather easily using the Gouy–Chapman model. The coefficient a_1 is the current per unit pressure difference when the potential difference is zero. Since the current results from sweeping space charge of charge density $\rho(r)$ along with the flowing solution, which moves with velocity $v(r)$, it must be given by the integral

$$i = \int_0^a v(r)\, \rho(r)\, 2\pi\, r\, dr$$

Substituting $\Phi(r)$ from eq (2.21) into eq (2.20) and differentiating, we obtain the charge density

$$\rho(r) = -(\epsilon \epsilon_0 \zeta / r x_A)(1 + r/x_A) \exp[(r - a)/x_A]$$

When a fluid of viscosity η flows through a cylindrical tube of radius a, the velocity is given as a function of r by Poiseuille's equation,

$$\mathbf{v}(r) = \Delta P (a^2 - r^2)/4\eta L \tag{2.22}$$

where ΔP is the pressure difference over the length L of the tube. Inserting these expressions into the integral and integrating, we get

$$i = -\frac{\pi \epsilon \epsilon_0 \zeta \Delta P}{2\eta L}\left[2a^2 - 4x_A{}^2(a/x_A - 1) + (a^2 - 4x_A{}^2)e^{-a/x_A}\right]$$

But since $a \gg x_A$, the last two terms are negligible and

$$a_1 = i/\Delta P = -\pi \epsilon \epsilon_0 \zeta a^2/\eta L \tag{2.23}$$

The coefficient a_2 is the current per unit potential difference when there is no pressure difference across the capillary. Under these conditions the current and potential difference are simply related by Ohm's law, $i = \Delta\Phi/R$. The resistance R of a cylinder of radius a, length L, and resistivity ρ is $R = \rho L/\pi a^2$, so that the current is

$$i = \pi a^2 \Delta\Phi/\rho L$$

and the coefficient is

$$a_2 = \pi a^2/\rho L \tag{2.24}$$

Similar calculations (see Problems) give

$$a_3 = \pi a^4/8\eta L \tag{2.25}$$

and

$$a_4 = a_1 \tag{2.26}$$

This last result turns out to be more general than we might have expected from our rather simple calculation. Indeed, the electrokinetic effects can be treated by the methods of nonequilibrium thermodynamics with the completely general result that $a_1 = a_4$

independent of the size or shape of the holes through which the solution flows.[‡]

Let us restate the four electrokinetic effects in terms of the phenomenological equations

$$i = a_1 \Delta P + a_2 \Delta \Phi \tag{2.18}$$

$$j = a_3 \Delta P + a_4 \Delta \Phi \tag{2.19}$$

The streaming potential is the potential difference per unit pressure difference at zero current, and thus may be expressed quantitatively as

$$\text{Streaming potential} = (\Delta \Phi / \Delta P)_{i=0} = -(a_1/a_2)$$

$$\text{Streaming potential} = \epsilon \epsilon_0 \zeta \rho / \eta \tag{2.27}$$

where ζ is the zeta potential, ρ is the resistivity, and η is the viscosity of the solution. The streaming current is the electric current per unit flux at zero potential difference:

$$\text{Streaming current} = (i/j)_{\Delta \Phi = 0} = (a_1/a_3)$$

$$\text{Streaming current} = -8\epsilon \epsilon_0 \zeta / a^2 \tag{2.28}$$

The electroosmotic pressure is the pressure difference per unit electric potential difference at zero flux and may be expressed as

$$\text{Electroosmotic pressure} = (\Delta P / \Delta \Phi)_{j=0} = -(a_4/a_3)$$

$$\text{Electroosmotic pressure} = +8\epsilon \epsilon_0 \zeta / a^2 \tag{2.29}$$

The electroosmotic flow is the flux per unit electric current at zero pressure difference:

$$\text{Electroosmotic flow} = (j/i)_{\Delta P = 0} = (a_4/a_2)$$

$$\text{Electroosmotic flow} = -\epsilon \epsilon_0 \zeta \rho / \eta \tag{2.30}$$

[‡]   This result is an example of Onsager's reciprocal relations *(13)*. For further discussion, see Bockris and Reddy *(B8)*.

Comparing these results we see that more remarkable conclusions have been obtained: the streaming potential, eq (2.27), is just the negative of the electroosmotic flow, eq (2.30); the streaming current, eq (2.28), is the negative of the electroosmotic pressure, eq (2.29).[‡] These equalities are also obtained from nonequilibrium thermodynamics and are thus quite general, independent of the nature of the holes through which the liquid flows.

Notice that the streaming potential and electroosmotic flow are predicted to be completely independent of the dimensions of the capillary tube. Indeed, it can be shown that the results we have obtained for a cylindrical capillary are independent not only of the dimensions, but also of the shape of the holes through which the liquid flows, provided only that (1) the flow is laminar (*i.e.*, nonturbulent), and (2) that the radii of curvature of the pores through which the liquid flows are much larger than the thickness of the double layer.

The electroosmotic pressure and the streaming current, on the other hand, depend on the radius of the capillary, and in general, on the average pore size when flow is through a porous plug or membrane.

Example 2.3 Compute the streaming potential expected when a solution having $\rho = 1\ \Omega\text{-m}$, $\epsilon = 78.5$, and $\eta = 10^{-3}$ Pa-s flows through a capillary tube with zeta potential $\zeta = -100$ mV under the influence of a pressure difference of 1 bar (10^5 Pa).

Substitution into eq (2.27) gives

$$\text{S.P.} = (78.5)(8.85 \times 10^{-12}\ \text{C}^2\text{J}^{-1}\text{m}^2)(-0.1\ \text{V})(1\ \Omega\text{-m})/(10^{-3}\ \text{Pa-s})$$

$$\text{S.P.} = -7.0 \times 10^{-8}\ \text{V Pa}^{-1}$$

[‡] These results are the more surprising since it would appear that the four quantities should have different units. However, the apparent units of the electroosmotic flow, $(\text{m}^3\text{s}^{-1})/(\text{C s}^{-1})$, reduce to m^3C^{-1} and the apparent units of the streaming potential, V Pa^{-1}, also reduce to m^3C^{-1}. Similarly, the apparent units of the streaming current, $(\text{C s}^{-1})/(\text{m}^3\text{s}^{-1})$, and the electroosmotic pressure, Pa V^{-1}, both reduce to C m^{-3}.

Thus

$$\Delta\Phi = (S.P.)\Delta P = -7.0 \times 10^{-3} \text{ V} \quad (-7.0 \text{ mV})$$

Example 2.4 Compute the streaming current expected when a solution with $\epsilon = 78.5$ flows through a bundle of capillaries of radius 0.1 mm and zeta potential -100 mV with a total flow of 0.1 cm^{3}s^{-1}.

Substitution into eq (2.28) gives

$$S.C. = -8(78.5)(8.85 \times 10^{-12} \text{ C}^2\text{J}^{-1}\text{m}^2)(-0.1 \text{ V})/(10^{-4} \text{ m})^2$$

$$S.C. = 5.6 \times 10^{-2} \text{ C m}^{-3}$$

or, with $j = 10^{-7} \text{ m}^3\text{s}^{-1}$,

$$i = (S.C.)j = 5.6 \times 10^{-9} \text{ A} \quad (5.6 \text{ nA})$$

Zeta Potentials at Glass–Water Interfaces

Glass–water interfaces appear to acquire a double layer through selective adsorption, usually of hydroxide ions. The surface potential is therefore usually negative and ζ is typically on the order of -150 mV in dilute solutions. The zeta potential is strongly dependent on the electrolyte concentration, however, with typical results shown in Figure 2.7.

The dependence of ζ on the concentration of electrolyte is a complicated combination of at least three effects. Equation (2.13) tells us that the surface potential is proportional to the surface charge density σ. It is also proportional to x_A and thus proportional to $C^{-1/2}$. In addition, if the slipping plane is located at a distance x into the diffuse layer, then ζ is less than the surface potential by the concentration–dependent factor $\exp(-x/x_A)$:

$$\zeta = \Phi_a \, e^{-x/x_A} = \frac{\sigma x_A}{\epsilon \epsilon_0} \, e^{-x/x_A}$$

Figure 2.7 The zeta potential at glass–aqueous solution interface as a function of electrolyte concentration for (a) KOH, (b) KNO_3, and (c) HNO_3 determined by measurement of the electroosmotic flow through a capillary tube *(14)*. The lines represent smooth curves drawn through data points rather than fits of data to a theoretical equation.

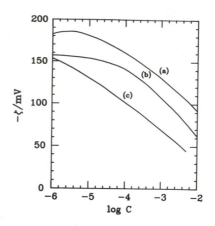

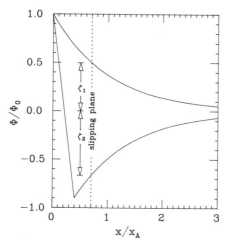

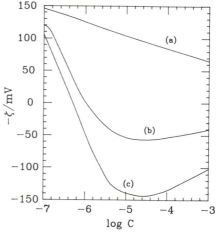

Figure 2.8 Reversal of zeta potential by adsorption of ions. The upper curve shows the normal behavior with Φ falling exponentially through the ion atmosphere. The lower curve shows the behavior expected when oppositely charged ions are adsorbed on the surface forming a "triple layer" and reversing the sign of the zeta potential.

Figure 2.9 Zeta potential at a glass–aqueous solution interface *vs.* concentration for (a) KNO_3, (b) $Ca(NO_3)_2$, and (c) $Al(NO_3)_3$, determined by electroosmotic flow through a glass capillary *(14)*.

Thus we would expect ζ to depend on the concentration of electrolyte according to

$$\zeta = a\sigma \ C^{-\frac{1}{2}} \exp(-\beta C^{\frac{1}{2}}) \qquad (2.31)$$

where a and β are constants. If σ is independent of electrolyte concentration, then ζ should be a monotonic decreasing function of C. The situation is somewhat more complicated if the surface charge density is affected by changes in electrolyte concentration. In Figure 2.7, we see that the zeta potential is higher for KOH solutions than for the other electrolytes, presumably because σ is higher for these solutions (hydroxide ions are preferentially adsorbed). Indeed, there is a maximum in the ζ vs. log(C) curves for KOH solutions which reflects a balance between σ increasing with increasing concentration and x_A decreasing. This interpretation is supported by the behavior of HNO_3, for which the zeta potential falls more rapidly with increasing concentration, presumably because of surface charge neutralization.

Sometime more positive ions are adsorbed on the negative surface than are required to neutralize the surface charge, in effect reversing the sign of the zeta potential. The origin of the behavior is shown in Figure 2.8, and some typical data are shown in Figure 2.9. As might be expected, reversal of the sign of the zeta potential is observed with the highly charged cations Al^{3+} and Th^{4+}.

2.4 ELECTROPHORESIS AND RELATED PHENOMENA

The electrokinetic phenomena involve stationary solid phases, the effects arising because of motion of the electrolyte solution past the electrified interface. Several related phenomena arise when the interface is also free to move.

The Electroviscous Effect

The viscosity of a colloidal solution is very sensitive to the ionic strength of the medium. For example, the relative viscosity of a solution of soluble starch, plotted in Figure 2.10, drops dramatically with increasing electrolyte concentration (15). The effect has two causes: (1) x_A is large at low ionic strength; the thick ion atmosphere increases the effective hydrodynamic size of the colloid and thus increases the viscous drag when the molecule moves in solution; and (2) electrostatic repulsion of charged functional groups results in an extended conformation of the starch molecule. When the ionic strength is higher, intergroup repulsion is shielded by the ion atmosphere and a more compact conformation is adopted. Again the effective hydrodynamic volume is reduced and the solution viscosity is lower. The first effect can be treated theoretically but it is usually considerably smaller than the second.

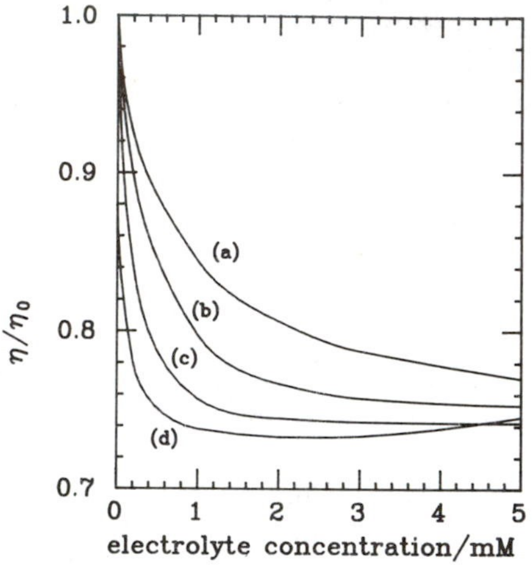

Figure 2.10 Relative viscosity of starch solutions as a function of electrolyte concentration *(15)* for (a) KCl, (b) $BaCl_2$, (c) $Co(NH_3)_6Cl_3$, and (d) $Pt(en)_3(NO_3)_4$.

Sedimentation Potential

When colloidal particles move through an electrolyte solution under the influence of gravity (as, for example, in an ultracentrifuge), the particles may leave some of their ion atmosphere behind, especially if the diffuse layer is very thick (low ionic strength). Thus charge separation may accompany sedimentation, leading to the development of an electric potential gradient along the length of the sedimentation tube. The phenomenon is thus analogous to the streaming potential discussed above. A sedimentation potential is generally an unwanted complication in an ultracentrifugation experiment. To reduce the magnitude of the effect, ultracentrifugation is usually carried out in high ionic strength media so that the diffuse layer is relatively compact and charge separation is minimized.

Electrophoresis

If an electric field (or potential difference) is applied across a solution containing charged particles, the particles are accelerated, rapidly attaining a terminal velocity where the electrical force is exactly balanced by the frictional retarding force as the particles move through solution. The phenomenon of electrophoresis is characterized by the electrophoretic mobility, u, defined as the velocity per unit electric field strength:

$$u = |\mathbf{v}| / |\mathbf{E}| \tag{2.32}$$

If $\mathbf{v}$ is expressed in m s^{-1} and $\mathbf{E}$ is in V m^{-1}, u apparently must have units of m^{2}V^{-1}s^{-1}. Typical mobilities of ordinary small ions in aqueous solution are on the order of 5×10^{-8} m^{2}V^{-1}s^{-1}; proteins generally have mobilities in the range $(0.1-1) \times 10^{-8}$ m^{2}V^{-1}s^{-1}. Although the phenomenon of electrophoresis can be understood qualitatively as the motion of charged particles in an electric field, the details of the motion depend upon the nature of the ion atmosphere surrounding the particles.

Consider first the motion of a small spherical particle of radius a through a medium of viscosity η. According to Stokes' law, the viscous force opposing the motion is

$$\mathbf{F}_{visc} = -6\pi\eta a \mathbf{v} \tag{2.33}$$

where $\mathbf{v}$ is the velocity of the particle. On the other hand, the electrical force propelling the particle through the solution is

$$\mathbf{F}_{elec} = Q\mathbf{E}$$

where Q is the charge on the particle and $\mathbf{E}$ is the electric field strength. If the surface charge density is σ, then the total charge on the particle is $Q = 4\pi a^2\sigma$. For a spherical particle moving through solution, the effective charge density is related to the zeta potential,

$$\sigma = \epsilon\epsilon_0\zeta(1 + a/x_A)/a$$

Combining these expressions, we have the electrical force on the particle:

$$\mathbf{F}_{elec} = 4\pi\epsilon\epsilon_0\zeta a(1 + a/x_A)\mathbf{E}$$

When the particle reaches terminal velocity, the acceleration is zero, and according to Newton's second law, the resultant of forces must be zero. Thus we have

$$4\pi\epsilon\epsilon_0\zeta a(1 + a/x_A)\mathbf{E} - 6\pi a\eta \mathbf{v} = 0$$

The electrophoretic mobility then is

$$u = |\mathbf{v}|/|\mathbf{E}| = \tfrac{2}{3}\,\epsilon\,\epsilon_0\zeta/\eta \tag{2.34}$$

where we have assumed that a $\ll x_A$. Equation (2.34) thus represents a limiting result for small spherical particles. When a/x_A approaches unity, the mobility should increase; however, the variation is more complicated than we might expect from this approach.

We can obtain an estimate of the electrophoretic mobility for large values of a/x_A by the following analysis. Suppose that the particle is very large compared with the size of its ion atmosphere. We then regard the surface as approximately planar and consider the relative motion of the solution past the surface. The calculation, which is very similar to that for the electrokinetic coefficient a_4, gives

$$u = \epsilon\,\epsilon_0\zeta/\eta \tag{2.35}$$

which should be the limiting value for a very large particle.

Charged particles of intermediate size distort the electrical lines of force and a more sophisticated approach is required. More exact calculations have shown that the electrophoretic mobility is (as we might expect) a function of the size, shape, and orientation of the particles. The mobility is, however, generally found to be proportional to the dielectric constant and to the zeta potential, and inversely proportional to the viscosity. The electrophoretic mobility is sometimes written as

$$u = f\epsilon\,\epsilon_0\zeta/\eta \tag{2.36}$$

where f is a unitless quantity which depends upon the geometrical factors. For spherical particles, f varies smoothly from $\tfrac{2}{3}$ for $a < x_A$ to 1 for $a > 100\,x_A$.

Example 2.5 Estimate the electrophoretic mobility of a particle of radius 3 nm and zeta potential 20 mV in a solution with $\epsilon = 78.5$, $\eta = 10^{-3}$ Pa–s, and ionic strength 0.001 M.

For 0.01 M ionic strength, $x_A = 9.6$ nm, so that $a/x_A = 0.31$ and $f \cong \tfrac{2}{3}$; eq (2.36) gives $u = 9.3 \times 10^{-9}$ m^{2}s^{-1}V^{-1}. This mobility is typical for charged macromolecules (see Section 3.3).

Practical Electrophoresis

Although eq (2.36) does not explicitly include the thickness of the diffuse layer, x_A, the zeta potential depends on x_A (or C) as shown by eq (2.31) and the geometric factor f depends on a/x_A. Thus the electrophoretic mobility is expected to decrease with increasing ionic strength of the electrolyte solution. The mobility is also proportional, of course, to the surface charge density on the moving colloidal particle and, for a protein, is expected to be pH dependent, even at constant ionic strength; indeed, the mobility is expected to go to zero at the isoelectric pH. Because of the dependence of the electrophoretic mobility on ionic strength and pH, a mechanism for control of protein mobilities is available which is used in practical separation techniques.

In our discussion thus far, we have tacitly assumed that electrophoresis experiments are done on a suspension of colloidal particles in liquid solution, and, indeed, the early experiments were. Thus, in classic experiments done in the 1930's, Tiselius[‡] found that electrophoretic mobilities of protein molecules in solution could be measured by a moving boundary method *(16)*. Tiselius' apparatus was essentially a U–tube, shown in very simplified form in Figure 2.11. The U–tube was constructed so that the solution of the colloid could be placed in the bottom of the tube and solvent (containing electrolyte) layered on top on both sides. Electrodes were then mounted in the two ends of the U–tube and a high voltage applied. When the protein molecules migrated in the solution, the boundary between the two regions of the solution remained intact and was seen to move with a velocity which was easily related to the electrophoretic mobility of the protein. To minimize convective mixing of the solutions and blurring of the boundary, it was necessary to very carefully thermostat the U–tube, and, since the rate of change of the density of water with temperature is smallest near the density maximum (4°C), best results were obtained when the system was thermostated at about 2–3°C. In practice, the boundary was observed by measurement of the refractive index. Fairly complex mixtures can be separated by means of Tiselius' electrophoresis techniques.

[‡]   Arne W. K. Tiselius (1902–1971) was a Professor at the University of Uppsala with research interests in biophysical chemistry. He received the Nobel Prize in Chemistry in 1948 for his work on electrophoresis.

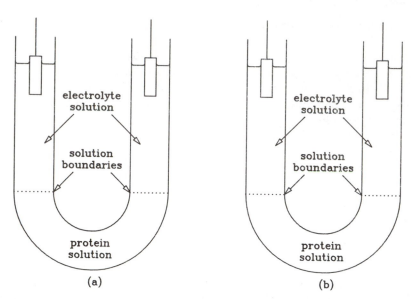

Figure 2.11 Moving boundary method for measurement of electrophoretic mobilities. (a) Initial condition with protein solution layered at the bottom of the U–tube. (b) After passage of current, boundary between the colloid and electrolyte solutions has moved.

A great advance in the utility of electrophoresis came with the marriage of electrophoretic migration with paper chromatography. By using a solid support of filter paper, most of the problems associated with convective mixing in solution can be eliminated and diffusive blurring of the zone boundaries can be reduced as well. Even further improvement can be obtained if a more homogeneous solid support such as starch gel or cellulose is used in place of the paper.

The manifold of electrophoresis techniques now available, including a host of procedures for detecting the protein molecules, goes far beyond the space available here. Several rather long chapters would be required to do the subject justice, and the reader is referred to a more specialized text *(17–19)* for further details.

2.5 *ELECTRODE DOUBLE–LAYER EFFECTS*

Unlike a protein molecule where the surface potential and the ion atmosphere are a response to the solution pH and ionic strength, the potential of an electrode, and thus the nature of the double layer, can be controlled by external circuitry. This control permits quantitative experimental studies of double–layer effects. In this section, we will discuss two effects which have important consequences in electrochemical experiments to be described in subsequent chapters.

Electrocapillarity

The interfacial tension (surface tension) at an electrode–solution interface turns out to be a function of the electrode–solution potential difference. When the electrode is a rigid solid, the effects of interfacial tension are difficult to discern, but when the electrode is nonrigid, a mercury drop for example, the effects are more obvious. In a dropping mercury electrode, mercury is allowed to flow through a fine–bore capillary immersed in the electrolyte solution. A mercury drop forms at the end of the capillary and grows in size until it becomes too heavy to be supported by the interfacial tension.[‡] Thus changes in interfacial tension are reflected in changes in the maximum mercury drop size, or, if the flow through the capillary is constant, in the lifetime of a mercury drop.

Consider a mercury drop attached to a mercury–filled capillary and suspended in an electrolyte solution. If a second electrode is provided, we can change the mercury–solution potential difference and thus induce a positive or negative charge on the mercury drop by passing a small current through the external circuit. Similarly, we can change the drop size by allowing mercury to flow through the capillary. Expansion of the drop involves an increase in surface area and thus work is done against the interfacial tension, $dw = \gamma\,dA$. Similarly, when the drop is charged electrical work is done, $dw = \Phi\,dQ$. Thus at constant temperature, pressure, and composition, the change in Gibbs free energy is

$$dG = \gamma\,dA + \Phi\,dQ$$

[‡] The dropping mercury electrode is discussed in more detail in Chapter 4.

This expression can be integrated

$$G = \gamma A + \Phi Q$$

and redifferentiated

$$dG = \gamma dA + A d\gamma + \Phi dQ + Q d\Phi$$

Subtracting the two expressions for dG leaves a form of the Gibbs–Duhem equation,

$$0 = A d\gamma + Q d\Phi$$

Introducing the surface charge density, $\sigma = Q/A$, we have

$$\left(\frac{\partial \gamma}{\partial \Phi}\right)_{T,P} = -\sigma$$

Substituting for σ from eq (2.13) gives

$$\left(\frac{\partial \gamma}{\partial \Phi}\right)_{T,P} = -\epsilon \epsilon_0 \Phi/x_A \tag{2.37}$$

Integration of eq (2.37) gives

$$\gamma = \gamma_0 - \epsilon \epsilon_0 \Phi^2/2x_A \tag{2.38}$$

Thus a plot of interfacial tension *vs. cell potential should be* parabolic with a maximum when the electrode–solution potential difference is zero.

If mercury flows through the capillary with constant rate u (mass per unit time), then the mass of the drop after time t is m = ut. The gravitational force (weight), corrected for the buoyancy of the solution, is

$$\mathbf{F} = utg(1 - d/d_{Hg})$$

where d and d_{Hg} are the densities of the solution and mercury, respectively, and g is the gravitational acceleration. The surface tension force holding the drop to the capillary is

$$\mathbf{F} = 2\pi \, r_c \gamma$$

where r_c is the radius of the capillary bore. These two forces are in balance the instant before the drop falls from the capillary ($t = t_d$). Thus if t_d, u, d, and r_c are measured, the interfacial tension can be computed from

$$\gamma = \frac{ut_d g(1 - d/d_{Hg})}{2\pi \, r_c} \tag{2.39}$$

Some typical data from the work of Grahame[‡] *(1)* are shown in Figure 2.12 for several aqueous electrolyte solutions of equal ionic strength.

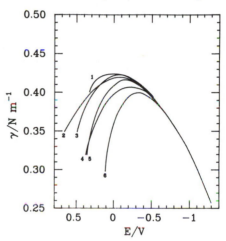

Figure 2.12 Interfacial tension of mercury in contact with various aqueous electrolyte solutions *(1)*: (1) KOH, (2) Ca(NO$_3$)$_2$, (3) NaCl, (4) KSCN, (5) NaBr, and (6) KI. The potential scale is adjusted so that zero potential occurs at the electrocapillary maximum for a capillary-inactive electrolyte.

The curves of Figure 2.12 show the qualitative behavior predicted by eq (2.39). However, the simple theory predicts identical curves for all electrolytes of equal ionic strength. The curves of Figure 2.12 coincide at negative potentials but show rather different behavior near zero or at positive potentials. When the mercury drop is positively charged, anions are expected to be concentrated near the electrode surface, and the different behavior suggests specific interactions of the various anions with the surface. The "soft" bases like I$^-$ and SCN$^-$

[‡] David C. Grahame (1912–1958) was Professor of Chemistry at Amherst College. He is remembered for his exceedingly careful and thorough studies of the electric double layer.

apparently interact more strongly than the "hard" bases like OH^-, Cl^-, and NO_3^-. Since mercury can be thought of as a soft acid, the interaction most likely involves some degree of chemical bond formation and we say that the anions are chemically adsorbed on the mercury surface. When the potential is made sufficiently negative, anions are desorbed and all curves merge together, suggesting that cations (at least Na^+, K^+, and Ca^{2+}) do not specifically interact with the mercury surface, even when it is highly negatively charged.

It is commonly observed that the drop time of a dropping mercury electrode depends on potential, usually varying by a factor of 2 or more over the experimentally accessible range. More complex behavior is sometimes observed when surfactants are absorbed on the electrode surface. Since adsorption is often potential dependent, very rapid changes in interfacial tension (and thus drop time) with potential are sometimes observed.

Double-Layer Capacitance

When the potential applied to an electrode immersed in an electrolyte solution is decreased from zero, the surface charge becomes negative and the net space charge of the double layer must increase to maintain overall electrical neutrality. Similarly, an increase in electrode potential must induce a net negative space charge. From the point of view of the external electric circuit, the double layer thus behaves as a capacitor, serving to store electric charge.

Capacitance is defined as the ratio of charge stored to voltage applied, or, more appropriately in this case, as the derivative (the differential capacitance)

$$C = \frac{dQ}{d\Phi} \qquad (2.40)$$

If we take C to be the capacitance per unit area (the capacity), then we can replace Q by the surface charge density σ. Consider first the capacity of the Gouy layer, C_G. Differentiating eq (2.13), we obtain

$$C_G = \epsilon \epsilon_0 / x_A \qquad (2.41)$$

A more exact calculation (see Problems) shows that C_G is a function of Φ_a and that eq (2.41) represents the minimum capacity. Minima are observed experimentally and provide a way to determine the potential of zero charge for an electrode *(20)*. The capacity *vs.* potential curves for some aqueous sodium fluoride solutions are shown in Figure 2.13.

Figure 2.13 Capacity of a mercury drop electrode in contact with aqueous sodium fluoride solutions as a function of potential measured relative to a normal (1 M KCl) calomel electrode *(1)*. The potential of minimum capacity corres ponds to the electrocapillary maximum (*i.e.*, to maximum interfacial tension, see Figure 2.12). The curves correspond to [NaF] = 0.001 M (dots), 0.01 M (dash–dot), 0.1 M (dashes), and 0.92 M (solid).

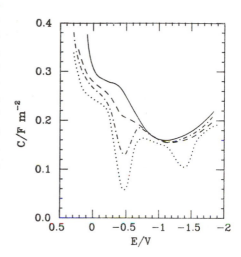

Example 2.6 Compute the Gouy layer capacity for a solution of a 1:1 electrolyte with concentrations of 0.001, 0.01, and 0.1 M.

The double–layer thicknesses for these concentrations are (Table 2.1) 9.6, 3.0, and 0.96 nm. Assuming the dielectric constant of pure water (ϵ = 78.5), eq (2.41) gives

$$C_G = 0.072, 0.23, \text{ and } 0.72 \text{ F m}^{-2}$$

for 0.001, 0.01, and 0.1 M solutions, respectively.

The curve for 0.001 M NaF solution in Figure 2.13 goes through a minimum of about 0.06 F m^{-2}, in excellent agreement with the value computed in Example 2.6 considering the simplicity of the theory. However, the minimum capacity of the 0.01 M NaF solution is about 0.1 F m^{-2} and for 0.1 M NaF the observed capacity at the electrocapillary maximum is only 0.2 F m^{-2}, rather less than the computed values, 0.23 and 0.72 F m^{-2}. Furthermore, the capacity minimum at Φ = 0 nearly disappears for concentrated solutions. The reason for these apparent discrepancies is not hard to find: we have neglected the effect of the immobilized ions and solvent dipoles on the surface which comprise the Helmholtz layer. These also contribute to the capacity. From the point of view of the external circuit, these two contributions—C_H for the Helmholtz layer and C_G for the Gouy layer—behave like capacities in series. Thus the total observed capacity is given by

$$1/C = 1/C_H + 1/C_G$$

or

$$C = \frac{C_H C_G}{C_H + C_G} \qquad (2.42)$$

For dilute solutions, the capacity of the diffuse layer C_G is small compared with the more or less constant Helmholtz layer contribution, so that $C \cong C_G$. When the diffuse layer becomes more compact and its capacity becomes large compared with that of the Helmholtz layer, then the latter contribution should dominate. Indeed, it is found that at zero potential the double–layer capacity initially increases with concentration and then levels off. In the case of NaF solutions in contact with mercury, the Helmholtz contribution to the double–layer capacity at $\Phi = 0$ appears to be approximately 0.29 F m^{-2}.

2.6 DEBYE–HÜCKEL THEORY

In 1923 Debye[‡] and Hückel[#] *(21)* found a way to calculate ionic

---

[‡] Peter J. W. Debye (1884–1966) was one of the outstanding physical chemists of this century. At one time or another, he held chairs at the Universities of Zurich, Utrecht, Göttingen, Leipzig, and Berlin, finally coming to rest at Cornell University during World War II. His contributions to electrolyte theory were largely made during his tenure at the Eidgenössische Technische Hochschule in Zurich in the 1920's.

[#] Erich Hückel (1896–1980) was a student of Debye's; their collaboration continued through most of the 1920's, whereupon Hückel went on to become a pioneer in quantum chemistry.

activity coefficients for dilute electrolyte solutions. The basic idea of the theory is that, just as in the case of the charged surfaces we have been considering, individual ions possess an atmosphere of ions of opposite charge. By calculating the electrostatic free energy of interaction of an ion with its atmosphere, an estimate of the activity coefficient can be obtained. Debye–Hückel theory starts with the Gouy–Chapman description of the ion atmosphere derived from the linearized Poisson–Boltzmann equation.

Calculation of an Ionic Activity Coefficient

We begin by outlining the thermodynamic part of the argument. The chemical potential of a species i is given by

$$\mu_i = \mu_i{}^\circ + RT \ln(a_i) \qquad (2.43)$$

where $\mu_i{}^\circ$ is the chemical potential of i in its standard state, and a_i is its activity. If the standard state is an ideal 1 M solution,[‡] then in an ideal solution, the activity is identical with the molar concentration. Assuming that the standard state chemical potential is unaffected by considerations of nonideality, the chemical potential in the hypothetical ideal solution is

$$\mu_i(\text{ideal}) = \mu_i{}^\circ + RT \ln(C_i)$$

If we now assume that the departure from ideality is due entirely to electrostatic interactions of the ion with its surroundings, then the electrical contribution to the chemical potential is

$$\mu_i(\text{elec}) = \mu_i - \mu_i(\text{ideal})$$

$$\mu_i(\text{elec}) = RT \ln(a_i) - RT \ln(C_i)$$

and, since the ratio of the activity to the concentration is the activity coefficient,

---

[‡]   The standard state in most treatments of activity coefficients is the 1 molal ideal solution. Since most practical work employs the molar concentration scale, we shall use the 1 M standard state here. The penalty for this choice is that most data tabulations must be converted to molar concentration units.

$$\gamma_i = a_i/C_i \tag{2.44}$$

we have

$$\mu_i(\text{elec}) = RT \ln(\gamma_i) \tag{2.45}$$

Thus if we can calculate the electrostatic contribution to the chemical potential of an ion, we will have the activity coefficient.

Consider an ion of charge $z_i e$ surrounded by its ion atmosphere and suppose that the electric potential at the surface of the ion is Φ. We now consider the hypothetical process of reversibly charging an initially uncharged ion up to its actual charge $z_i e$. The work done in this charging process is the change in the Gibbs free energy, which in this case is the electrical contribution to the chemical potential:

$$\mu_i(\text{elec}) = \int_0^{z_i e} \Phi dQ \tag{2.46}$$

where Φ is the contribution to the potential from the ion atmosphere. Notice that this chemical potential has units of energy per ion; we will convert to a molar basis later.

For a spherical particle of radius a, the surface potential Φ_a is related to the surface charge density σ by (see Problems)

$$\Phi_a = \frac{\sigma a}{\epsilon \epsilon_0 (1 + a/x_A)}$$

where ϵ is the dielectric constant of the medium and x_A is the thickness of the ion atmosphere (called the Debye length in this context). Substituting this expression into eq (2.11), we have the potential in the ion atmosphere:

$$\Phi(r) = \frac{\sigma}{\epsilon \epsilon_0} \frac{a^2}{r(1 + a/x_A)} \exp[(a - r)/x_A]$$

If Q is the total charge on the ion, the surface charge density is

$$\sigma = Q/4\pi a^2$$

We then have

$$\Phi(r) = \frac{Q}{4\pi \epsilon \epsilon_0} \frac{\exp[(a - r)/x_A]}{r(1 + a/x_A)} \qquad (2.47)$$

For the thermodynamic calculation, we need the potential at the ion site due to the ion atmosphere. Equation (2.47) contains contributions from both the ion atmosphere and the central or test ion. However, the latter contribution is easily computed:

$$\Phi_{central}(r) = \frac{Q}{4\pi \epsilon \epsilon_0 r}$$

Subtracting this from eq (2.47), we have the ion atmosphere contribution to the potential at the central ion:

$$\Phi_{atm}(r) = \frac{Q}{4\pi \epsilon \epsilon_0 r} \left[\frac{\exp[(a - r)/x_A]}{1 + a/x_A} - 1 \right]$$

and, at the surface of the ion, the potential is

$$\Phi_{atm}(a) = - \frac{Q}{4\pi \epsilon \epsilon_0} \frac{1}{a + x_A}$$

Substituting this expression in eq (2.46) and integrating, we have

$$\mu_i(elec) = - \frac{1}{4\pi \epsilon \epsilon_0} \frac{1}{a + x_A} \int_0^{z_i e} Q dQ$$

$$\mu_i(elec) = - \frac{z_i^2 e^2}{8\pi \epsilon \epsilon_0} \frac{1}{a + x_A}$$

Finally, we substitute this expression into eq (2.45), solving for $\ln(\gamma_i)$.

In doing so, we also multiply by Avogadro's number[‡] to convert to a molar basis. The resulting equation is

$$\ln(\gamma_i) = - \frac{z_i^2 e^2}{8\pi \epsilon \epsilon_0 kT} \frac{1}{a + x_A} \tag{2.48}$$

Mean Ionic Activity Coefficients

Unfortunately, there is no way of obtaining single ion activities or activity coefficients from experimental measurements. Thus we must pause for a moment to deduce the relationship between the single ion activity coefficients calculated with the Debye–Hückel theory and the activity coefficients which are actually measurable.

Consider an electrolyte, which, on dissolution of one mole, gives ν_+ moles of cations and ν_- moles of anions. We define the mean chemical potential of the electrolyte as

$$\mu_\pm = (\nu_+ \mu_+ + \nu_- \mu_-)/\nu \tag{2.49}$$

where $\nu = \nu_+ + \nu_-$ and $\mu_\pm$ is related to the mean activity $a_\pm$ by

$$\mu_\pm = \mu_\pm^\circ + RT \ln(a_\pm) \tag{2.50}$$

Substituting the analogous equations for the single ion chemical potentials, eq (2.43), into eq (2.49) gives

$$\mu_\pm = (\nu_+ \mu_+^\circ + \nu_- \mu_-^\circ)/\nu + RT \ln(a_+^{\nu_+/\nu} a_-^{\nu_-/\nu}) \tag{2.51}$$

Comparing eqs (2.50) and (2.51), we see that the mean ionic activity is related to the single ion activities by

$$a_\pm = a_+^{\nu_+/\nu} a_-^{\nu_-/\nu} \tag{2.52}$$

Since $a_\pm = C_\pm \gamma_\pm$, $a_+ = C_+ \gamma_+$, and $a_- = C_- \gamma_-$, eq (2.52) implies the following relationships:

$$a_\pm^\nu = a_+^\nu + a_-^\nu - \tag{2.53a}$$

[‡] Multiplication by Avogadro's number is equivalent to conversion of the gas constant in the denominator to Boltzmann's constant.

$$C_\pm{}^\nu = C_+{}^{\nu}{}_+ + C_-{}^\nu{}_- \qquad (2.53b)$$

$$\gamma_\pm{}^\nu = \gamma_+{}^{\nu}{}_+ + \gamma_-{}^\nu{}_- \qquad (2.53c)$$

But, since $C_+ = \nu_+ C$ and $C_- = \nu_- C$, $C_\pm$ is not the ordinary molar electrolyte concentration C, but is

$$C_\pm = C(\nu_+{}^{\nu}{}_+ + \nu_-{}^{\nu}{}_-)^{1/\nu} \qquad (2.54)$$

Example 2.7 Determine the relationship between the mean ionic concentration $C_\pm$ and the ordinary molar concentration C for binary electrolytes with ionic charges from 1 to 4.

For a symmetrical electrolyte (1:1, 2:2, etc.), $\nu_+ = \nu_- = 1$, $\nu = 2$. Thus

$$(\nu_+{}^{\nu}{}_+ + \nu_-{}^\nu{}_-)^{1/\nu} = (1 \cdot 1)^{\frac{1}{2}} = 1$$

and $C_\pm = C$. For a 1:2 electrolyte, $\nu_+ = 2$, $\nu_- = 1$, $\nu = 3$ and

$$(\nu_+{}^{\nu}{}_+ + \nu_-{}^\nu{}_-) = (2^2 \cdot 1)^{\frac{1}{3}} = 1.587$$

so that $C_\pm = 1.587\ C$. Similar calculations generate the following table:

Electrolyte	$C_\pm/C$	Electrolyte	$C_\pm/C$
1:1	1.000	2:3	2.551
1:2	1.587	2:4	3.175
1:3	2.280	3:3	1.000
1:4	3.031	3:4	3.536
2:2	1.000	4:4	1.000

We can now relate the mean ionic activity coefficient $\gamma_\pm$ to the computed ionic activity coefficients γ_+ and γ_-. Taking logs of eq (2.53c) we have

$$\nu \ln(\gamma_\pm) = \nu_+ \ln(\gamma_+) + \nu_- \ln(\gamma_-)$$

and, substituting eq (2.48) into this expression, we have

$$\ln(\gamma_\pm) = - \frac{e^2}{8\pi\epsilon\epsilon_0 kT} \frac{1}{a + x_A} \frac{\nu_+ z_+^2 + \nu_- z_-^2}{\nu_+ + \nu_-}$$

But, since $|\nu_+ z_+| = |\nu_- z_-|$ the third factor on the right reduces to $|z_+ z_-|$, and we are left with

$$\ln(\gamma_\pm) = - \frac{e^2 |z_+ z_-|}{8\pi\epsilon\epsilon_0 kT} \frac{1}{a + x_A}$$

Recalling that x_A is inversely proportional to the square root of the ionic strength, eq (2.8), we can rewrite this expression in terms of the ionic strength:

$$\log(\gamma_\pm) = \frac{A |z_+ z_-| I^{\frac{1}{2}}}{1 + BaI^{\frac{1}{2}}} \tag{2.55}$$

In eq (2.55), A and B are constants, I is the ionic strength in concentration units of mol L^{-1}, and we have converted to common logs. Inserting the values of the constants and assuming a temperature of 25°C and the dielectric constant of water (78.54), we can write

$$\log(\gamma_\pm) = - \frac{0.5092 |z_+ z_-| I^{\frac{1}{2}}}{1 + 3.29 a I^{\frac{1}{2}}} \tag{2.56}$$

In eq (2.56), the ionic strength and $|z_+ z_-|$ are known for any given electrolyte solution. The parameter a (in nm) entered the theory as the ionic radius. However, in going from single ion activity coefficients to mean ionic activity coefficients, it has lost its simple meaning, and should to be regarded as a parameter adjustable to fit the theoretical activity coefficients to experiment.

At low ionic strength, the second term in the denominator of eq (2.56) can be neglected compared with unity and we have the limiting form of Debye–Hückel theory:

$$\log(\gamma_\pm) = -0.5092 |z_+ z_-| I^{\frac{1}{2}} \tag{2.57}$$

Equation (2.57) contains no adjustable parameters and thus is an absolute prediction of the activity coefficients.

Experimental Determination of Activity Coefficients

Thermodynamic quantities such as enthalpies or free energies of formation, half–cell potentials, acid–base dissociation constants, solubility product constants, complex formation constants, etc., are generally determined as functions of concentration and extrapolated to infinite dilution. Since activity coefficients approach unity at infinite dilution, these tabulated quantities refer to ideal solutions. Once the free energy change for an ideal solution process is known, comparison with values of $\Delta G°$ determined for finite concentrations can yield the activity coefficients.

The activity of the solvent or of a volatile solute can be determined from vapor pressure measurements and the activity coefficients computed from the departure of the measured pressure from ideal behavior—Raoult's law or Henry's law. Solute activity coefficients can also be determined from freezing point or osmotic pressure data. A text on physical chemistry should be consulted for further details on these methods.

In a cell potential measurement, the potential is given by the Nernst equation, eq (1.18). With

$$a_i = C_i \gamma_i$$

the Nernst equation can be written in the form

$$E + \frac{RT}{nF} \ln[F(C_i)] = E° - \frac{RT}{nF} \ln[F(\gamma_i)]$$

If the concentrations C_i and the cell potential E are known, the left–hand side of this expression can be plotted, usually *vs.* the square root of ionic strength. Extrapolation to $I = 0$ gives the standard cell potential $E°$ and, knowing this quantity, the activity coefficients can be determined.

Example 2.8 Consider the cell

$$Pt \,|\, H_2(g) \,|\, HCl(aq) \,|\, AgCl(s) \,|\, Ag$$

with cell reaction

$$2 \, AgCl(s) + H_2(g) \rightarrow 2 \, Ag(s) + 2 \, H^+(aq) + 2 \, Cl^-(aq)$$

The potential is given by

$$E = E° - \frac{RT}{2F} \ln \frac{[H^+]^2[Cl^-]^2}{P(H_2)} - \frac{RT}{2F} \ln(\gamma_{H^+})^2(\gamma_{Cl^-})^2$$

where we have assumed ideal behavior for $H_2(g)$. Since

$$\gamma_\pm^2 = \gamma_{H^+} \, \gamma_{Cl^-}$$

the Nernst equation can be rearranged to

$$F(I) = E + \frac{RT}{2F} \ln \frac{[H^+]^2[Cl^-]^2}{P(H_2)} = E° - \frac{2RT}{F} \ln \gamma_\pm$$

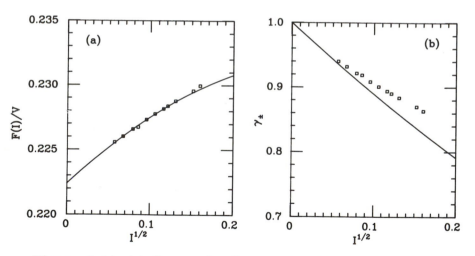

Figure 2.14 (a) Corrected cell potential F(I) for the cell of Example 2.7 as a function of the square root of ionic strength. The solid line corresponds to a least–squares fit of the data to a parabola. (b) Mean activity coefficients, $\gamma_\pm$, computed from the cell potentials. The solid line corresponds to activity coefficients computed using eq (2.57). Data from Harned and Ehlers *(22)*.

The left–hand side of this expression, $F(I)$, is plotted $vs. \sqrt{I}$ in Figure 2.14. Extrapolation to $I = 0$ yields $E° = 0.2224$ V. Subtraction of $E°$ from each value of $F(I)$ and rearrangement gives

$$\gamma_{\pm} = \exp\{F[E° - F(I)]/2RT\}$$

The values of $\gamma_{\pm}$ are plotted $vs. \sqrt{I}$ in Figure 2.14 along with values computed from eq (2.57).

Another way of determining activity coefficients is to measure the solubility of a slightly soluble salt in solutions with the ionic strength adjusted with an electrolyte which has no ion in common with the insoluble salt. For a salt $M_{\nu_+} A_{\nu_-}$ the solubility equilibrium expression is

$$K_{eq} = a_+{}^{\nu_+} a_-{}^{\nu_-} = a_{\pm}{}^{\nu}$$

or

$$K_{eq} = C_{\pm}{}^{\nu} \gamma_{\pm}{}^{\nu}$$

If the solubility is measured at several ionic strengths, K_{eq} can be determined by extrapolation to zero ionic strength and, knowing K_{eq}, the mean ionic activity coefficients $\gamma_{\pm}$ can be computed for each solution. Some data from such a study (23) are shown in Figure 2.15 for the 1:1 electrolyte,

$$Co(NH_3)_4(C_2O_4)^+ \quad Co(NH_3)_2(NO_2)_2(C_2O_4)^-$$

the 1:2 electrolyte,

$$[Co(NH_3)_4(C_2O_4)^+]_2 \, S_2O_6{}^{2-}$$

and the 3:1 electrolyte,

$$Co(NH_3)_6{}^{3+} \quad [Co(NH_3)_2(NO_2)_2(C_2O_4)^-]_3$$

For the data shown, extending up to ionic strengths of about 0.01 M, the agreement with theory is seen to be quite good, probably within experimental error. At higher ionic strengths, the plots begin to curve. In most cases, plots of activity coefficients $vs.$ ionic strength go through minima and the activity coefficients increase with concentration at high ionic strength. Examples of this behavior are shown in Figure 2.16 for HCl, KCl, $CaCl_2$, and $LaCl_2$. Even for the two 1:1 electrolytes, the behavior in concentrated solutions is rather different.

Inclusion of the parameter in the denominator of eq (2.56) helps the theory match experimental results to somewhat higher ionic strengths, but it cannot ever produce a minimum in the activity coefficient – ionic strength curve. Some experimental data for NaCl solutions are compared with theoretical predictions from eqs (2.56) and (2.57) in Figure 2.17. In fitting these data, the parameter a was taken to be 0.397 nm. With inclusion of the adjustable parameter, Debye-Hückel theory activity coefficients fit the experimental data up to an ionic strength of about 0.04 M. Beyond this the one parameter can no longer fit the data by itself and some further development is necessary.

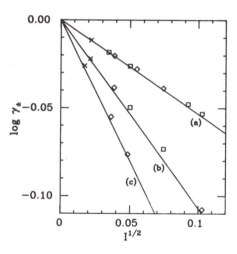
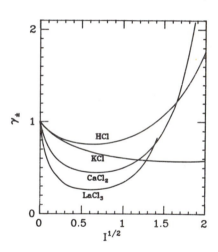

Figure 2.15 Activity coefficients computed from the solubilities of (a) 1:1, (b) 1:2, and (c) 3:1 electrolytes at 15°C in pure water (crosses) and in solutions with ionic strength adjusted with NaCl (diamonds) and KNO_3 (squares). The straight lines correspond to eq (2.57) (with the coefficient adjusted for the different temperature). Data from Brönsted and LaMer (23).

Figure 2.16 Mean ionic activity coefficients for several electrolytes at 25°C. Data from Robinson and Stokes (B9).

Figure 2.17 Mean ionic activity coefficients of NaCl solutions. Points are experimental data *(B9)*. The solid line is calculated from eq (2.57), the dashed line from eq (2.56), and the dotted line from eq (2.68). The parameter a in eqs (2.56) and (2.68) was taken to be 0.397 nm and n_S in eq (2.68) was 3.5.

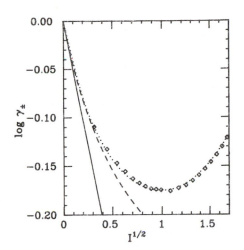

Conversion to Other Concentration Scales

Comparison of activity coefficients appropriate to the molar concentration scale with literature values given for molal concentrations or mole fractions can be confusing. As an aid to conversion, we give here the necessary equations. The details of the derivations are given by Robinson and Stokes *(B9)*.

Molal concentrations, m, are defined as the number of moles of solute per kilogram of solvent. Conversion to molar concentrations C is as follows:

$$C = \frac{md}{1 + mM_B/1000} \tag{2.58}$$

$$m = \frac{C}{d - CM_B/1000} \tag{2.59}$$

where M_B is the solute molecular weight and d is the solution density.

If an electrolyte of molality m ionizes to ν_1 cations and ν_2 anions, then the mole fractions X_i are

$$X_i = \frac{\nu_i m}{\nu m + 1000/M_A} \tag{2.60a}$$

where $\nu = \nu_1 + \nu_2$, and M_A is the molecular weight of the solvent. In terms of molar concentrations, the mole fractions are given by

$$X_i = \frac{\nu_i C}{\nu C + 1000d/M_A - CM_B/M_A} \tag{2.60b}$$

If the molal concentration scale is used, then the mean ionic activity is defined by eq (2.51) with

$$a_{\pm} = m_{\pm}\gamma_{\pm}{}^m$$

The mean ionic activity coefficient $\gamma_{\pm}{}^m$ is related to the activity coefficient for the molar concentration scale by

$$\gamma_{\pm}{}^m = (C/md_0)\gamma_{\pm} \tag{2.61}$$

where d_0 is the density of the solvent. If the mole fraction scale is used, the corresponding activity coefficients are

$$\gamma_{\pm}{}^x = \frac{d + 0.001 \, C(\nu M_A - M_B)}{d_0} \gamma_{\pm} \tag{2.62}$$

Solvent Activity

The standard state of the solvent in an electrolyte solution is usually taken to be the pure solvent at 1 bar total pressure and 25°C. In simple treatments of solution equilibrium, it is usually assumed that the solvent activity is 1, but a more exact value is sometimes required. If the solution were ideal, the solvent activity would be equal to the mole fraction, but in solutions where the mole fraction of solvent is significantly less than unity, nonideality is usually sufficiently important that the mole fraction is a poor estimate of the activity.

The most straightforward way to determine solvent activity is to measure the partial pressure of solvent vapor. We define the activity of the solvent in solution in terms of the chemical potential

$$\mu_A = \mu_A{}^\circ + RT \ln(a_A) \tag{2.63}$$

where $\mu_A{}^\circ$ is the chemical potential of the pure solvent. When the pure solvent is in equilibrium with its vapor,

$$\mu_A{}^\circ = \mu_A(\text{vapor}) = \mu_A{}^\circ(\text{vapor}) + RT \ln(P_A{}^\circ)$$

where $P_A{}^\circ$ is the equilibrium partial pressure.[‡] When a solution is in equilibrium with vapor, we have

$$\mu_A = \mu_A{}^\circ(\text{vapor}) + RT \ln(P_A)$$

Subtracting these expressions gives

$$\mu_A - \mu_A{}^\circ = RT \ln \frac{P_A}{P_A{}^\circ}$$

so that

$$a_A = P_A/P_A{}^\circ \tag{2.64}$$

While the vapor pressure method is straightforward, it is not always convenient. If the activity coefficient of the solute is known as a function of concentration, the solvent activity can be computed. Consider the following fundamental equation for the Gibbs free energy:

$$dG = PdV - TdS + \sum_i \mu_i dn_i$$

Integrating and redifferentiating, we have

$$G = PV - TS + \sum_i \mu_i n_i$$

$$dG = PdV + VdP - TdS - SdT + \sum_i \mu_i dn_i + \sum_i n_i d\mu_i$$

Subtracting the two expressions for dG gives the Gibbs–Duhem equation

---

[‡]   In very careful work, account must be taken of nonideality of the vapor phase as well. We ignore that complication here.

$$0 = VdP - SdT + \sum_i n_i d\mu_i$$

which at constant temperature and pressure reduces to

$$\sum_i n_i d\mu_i = 0$$

If we have only two components, the solvent (A) and the solute (B), then

$$n_A d\mu_A = - n_B d\mu_B$$

Equation (2.63) gives μ_A and a similar expression can be written for μ_B. Differentiating and substituting into the Gibbs–Duhem equation gives

$$n_A RT \, d \ln(a_A) = - n_B RT \, d \ln(a_B)$$

But since

$$a_B = \gamma_{\pm} C$$

and

$$n_B = C V$$

we have

$$d \ln(a_A) = -(C V/n_A) \, d \ln(\gamma_{\pm} C)$$

which can be integrated to give

$$\ln a_A = - \frac{1}{n_A} \int_0^{\gamma_{\pm} C} C V \, d \ln(\gamma_{\pm} C) \tag{2.65}$$

where we have assumed a constant number of moles of solvent. With sufficient solute activity coefficient data, eq (2.65) can be integrated numerically to obtain the solvent activity.

Beyond Debye–Hückel Theory

Debye–Hückel theory involves the calculation of the free energy associated with the formation of an ion atmosphere about ions in an electrolyte solution. There are details of the theory which leave much to be desired, but perhaps the greatest limitation is the implicit assumption that the formation of the ion atmosphere accounts entirely for deviations of the activity coefficients from unity. Clearly, the assumption must be a good one at high dilution because the limiting law form of Debye–Hückel theory is essentially perfect. Yet we know that at higher concentrations, the activity coefficients begin to increase again.

We can obtain a qualitative understanding of the behavior of electrolyte activity coefficients at high concentrations (and indeed derive a semiquantitative theory) from the following considerations.[‡] When a salt dissolves in water, the ions are solvated with several water molecules held rather tightly to the ions. We should regard those water molecules held in the solvation sheath of the ions as part of the ionic species rather than as "free" solvent; thus the concentration of "free" solvent molecules is lowered. This effect is very small in dilute solutions but is quite significant at high electrolyte concentrations. For example, in NaCl solutions, it appears from other experimental evidence that about seven water molecules are involved in solvation of the two ions. Thus in 0.01 M NaCl, about 0.07 mole of water per liter is removed from the free solvent. However, since the concentration of water is about 55.5 M, this is quite negligible. In 1 M NaCl, on the other hand, 7 moles per liter of water is immobilized, leaving 48.5 moles of free water, a more significant change.

Consider an electrolyte solution containing n_A moles of solvent per mole of solute. The solute ionizes to give ν_1 moles of cations with chemical potential μ_1 and ν_2 moles of anions with chemical potential μ_2. Apparently, the total free energy of the solution is

$$G = n_A \mu_A + \nu_1 \mu_1 + \nu_2 \mu_2$$

If we take into account the n_S moles of solvent immobilized by the ions, we can also express the total free energy by

---

[‡] See Robinson and Stokes *(B9)* for a discussion of other approaches to this problem.

$$G = (n_A - n_S)\mu_A + \nu_1\mu_1' + \nu_2\mu_2'$$

where μ_1' and μ_2' are the chemical potentials of the solute ions corrected for bound solvent. Both expressions for G should be correct, so that equating them gives

$$n_S\mu_A + \nu_1(\mu_1 - \mu_1') + \nu_2(\mu_2 - \mu_1') = 0$$

It is convenient to change (temporarily) to the mole fractions as measures of the activities with activity coefficients $\gamma_1{}^X$ and $\gamma_2{}^X$. The chemical potentials then are

$$\mu_A = \mu_A{}^\circ + RT \ln a_A$$

$$\mu_1 = \mu_1{}^\circ + RT \ln X_1 + RT \ln \gamma_1{}^X$$

$$\mu_1' = \mu_1{}^{\circ'} + RT \ln X_1' + RT \ln \gamma_1{}^{X'}$$

with similar equations for μ_2 and μ_2'. Substitution of these expressions into the above equation yields

$$[n_S\mu_A{}^\circ + \nu_1(\mu_1{}^\circ - \mu_1{}^{\circ'}) + \nu_2(\mu_2{}^\circ - \mu_2{}^{\circ'})]/RT =$$

$$-n_S \ln a_A - \nu_1 \ln \frac{X_1}{X_1'} - \nu_2 \ln \frac{X_2}{X_2'} - \nu_1 \ln \frac{\gamma_1{}^X}{\gamma_1{}^{X'}} - \nu_2 \ln \frac{\gamma_2{}^X}{\gamma_2{}^{X'}} \qquad (2.66)$$

The mole fractions are

$$X_1 = \frac{\nu_1}{n_A + \nu}$$

$$X_1' = \frac{\nu_1}{n_A - n_S + \nu}$$

where $\nu = \nu_1 + \nu_2$. Thus we have

$$\frac{X_1}{X_1'} = \frac{n_A - n_S + \nu}{n_A + \nu} = \frac{X_2}{X_2'}$$

When the solute concentration is very small, n_A is very large (we have assumed one mole of solute). Thus as the solute concentration approaches zero, $X_1/X_1' \to 1$, $a_A \to 1$, all the activity coefficients approach 1, and all the log terms on the right–hand side of eq (2.66) go to zero; thus the sum of standard state terms on the left–hand side of the equation must be zero, independent of the solute concentration. Rewriting eq (2.66) with the new information included, and using eq (2.53c) to convert the activity coefficient terms to mean ionic activity coefficients, we have

$$n_S \ln a_A + \nu \ln \frac{n_A - n_S + \nu}{n_A + \nu} + \nu \ln \gamma_\pm^x = \nu \ln \gamma_\pm^{x'} \quad (2.67)$$

We now convert back to the molar concentration scale. First we replace n_A, the number of moles of solvent per mole of solute, by

$$n_A = \frac{1000\, d}{CM_A} - \frac{M_B}{M_A}$$

where M_A and M_B are the molecular weights of the solvent and solute, respectively, d is the solution density, and C is the molar concentration. Substituting eq (2.62) for $\ln \gamma_\pm^x$ and taking the Debye–Hückel result, eq (2.55), for $\ln \gamma_\pm^x$, eq (2.67) yields (after some algebra)

$$\log \gamma_\pm = - \frac{A|z_+ z_-| I^{\frac{1}{2}}}{1 + BaI^{\frac{1}{2}}}$$

$$- \frac{n_S}{\nu} \log a_A - \log \left[\frac{d - 0.001C[M_B - M_A(\nu - n_S)]}{d_0} \right] \quad (2.68)$$

The parameters A and B are given in eq (2.56) and the solvent activity is measurable, at least in principle. Thus a and n_S are the only adjustable parameters. Equation (2.68) does give a very good fit to activity coefficients for concentrated solutions. Some typical results are shown in Figure 2.17 for NaCl solutions. Best–fit values of the parameters a and n_S for several electrolytes are given in Table 2.2. The values of n_S, although not grossly out of line with solvation numbers obtained from other experimental approaches, are generally somewhat small and certainly should not be taken as reliable estimates of the degree of solvation.

TABLE 2.2 Empirical Values of n_S and a

Electrolyte	n_S	a/nm
HCl	8.0	0.447
HBr	8.6	0.518
NaCl	3.5	0.397
NaBr	4.2	0.424
KCl	1.9	0.363
$MgCl_2$	13.7	0.502
$MgBr_2$	17.0	0.546

From Robinson and Stokes *(B9)*.

REFERENCES

(Reference numbers preceded by a letter, *e.g. (B2)*, refer to a book listed in the Bibliography.)

1. D. C. Grahame, *Chem. Rev.* **1941,** *41,* 441.
2. R. Parsons, *Modern Aspects of Electrochemistry* **1954,** *1,* 103.
3. D. M. Mohilner, *Electroanalytical Chemistry* **1966,** *1,* 241.
4. R. M. Reeves, *Modern Aspects of Electrochemistry* **1974,** *9,* 239.
5. H. von Helmholtz, *Wied. Ann.* **1879,** *7,* 337.
6. L.-G. Gouy, *Compt. Rend.* **1909,** *149,* 654; *J. Phys.* **1910,** *9,* 457.
7. D. L. Chapman, *Phil. Mag.* **1913,** *25,* 475.
8. O. Stern, *Z. Elektrochem.* **1924,** *30,* 508.
9. D. J. Shaw, *Introduction to Colloid and Surface Chemistry*, 3rd ed, London: Butterworths, 1980.
10. H. C. Hamaker, *Physica,* **1937,** *4,* 1058.
11. R. Hogg, T. W. Healy, and D. W. Fürstenau, *Trans. Faraday Soc.* **1966,** *62,* 1638; G. R. Wiese and T. W. Healy, *ibid.,* **1970,** *66,* 490.
12. S. S. Dukhin and B. V. Deryagin, in *Surface and Colloid Science,*

Vol. 7, E. Matijevic, ed, New York: Wiley–Interscience, 1974.

13. L. Onsager, *Phys. Rev.* **1931,** *37,* 405, *38,* 2265.
14. A. J. Rutgers and M. de Smet, *Trans. Faraday Soc.* **1945,** *41,* 258.
15. H. G. Bungenberg de Jong, *Rec. trav. chim.* **1924,** *43,* 189.
16. A. Tiselius, *Kolloid–Z.* **1938,** *85,* 129.
17. M. Bier, ed, *Electrophoresis,* New York: Academic Press, 1959.
18. L. P. Cawley, *Electrophoresis and Immunoelectrophoresis,* Boston: Little, Brown, 1969.
19. D. J. Shaw, *Electrophoresis,* New York: Academic Press, 1969.
20. R. S. Perkins and T. N. Anderson, *Modern Aspects of Electrochemistry* **1969,** *5,* 203.
21. P. Debye and E. Hückel, *Phys. Z.* **1923,** *24,* 185.
22. H. S. Harned and R. W. Ehlers, *J. Am. Chem. Soc.* **1932,** *54,* 1350.
23. J. N. Brönsted and V. K. LaMer, *J. Am. Chem. Soc.* **1924,** *46,* 555.

PROBLEMS

2.1   Derive eq (2.11). Hint: Use ∇^2 in spherical polar coordinates and make the substitution $\Phi(r) = u(r)/r$.

2.2   Derive eq (2.12). Hint: With $u = zF\Phi/RT$, show that eq (2.5) can be written

$$\nabla^2 u = (1/x_A{}^2) \sinh(u)$$

This expression is then integrated to obtain

$$du/dx = -(2/x_A) \sinh(u/2)$$

A second integration, followed by rearrangement, yields eq (2.12).

2.3   Use the exact solution to the Poisson–Boltzmann equation for a symmetrical electrolyte to show that the surface charge density, eq (2.13), is the first term of a power series:

$$\sigma = (\epsilon \epsilon_0 \Phi_a/x_A)[1 + \tfrac{1}{6}(zF\Phi_a/2RT)^2 + \ldots\,]$$

2.4 Use eq (2.11) to derive an expression for the surface charge density σ at a spherical surface as a function of the surface potential Φ_a, the radius of the sphere a, and the ion atmosphere thickness x_A.

2.5 Derive eq (2.25) for the electrokinetic coefficient a_3 by integration of Poiseuille's equation.

2.6 Derive eq (2.26) for the electrokinetic coefficient a_4 by equating the electrical and viscous forces on a cylindrical volume of radius r and integrating to obtain $\mathbf{v}(r)$ and then integrating again to determine the flux j. Hint: The electrical force on a volume element dV is $d\mathbf{F} = \mathbf{E} \rho(r)dV$, where $\mathbf{E}$ is the field strength, $\mathbf{E} = -\Delta\Phi/L$, and $\rho(r)$ is the space charge density. Integrate $d\mathbf{F}$ to obtain the force on the cylindrical volume. Use Newton's law of viscous flow to compute the viscous force.

2.7 Use the result of Problem 2.3 to show that eq (2.41) is actually the minimum on a C vs. Φ_a curve.

2.8 Consider an electrochemical cell with a planar electrode of area 1 cm^2 and a double–layer capacity of $C_d = 0.20$ F m^{-2} in contact with an electrolyte solution. A reference electrode is placed in the solution at a point such that the solution resistance between the electrodes is $R_s = 1000$ Ω. Suppose that the electrode-solution interfacial potential is initially 0 V and that at time t = 0, the cell is polarized so that the electrode is at $+0.1$ V relative to the bulk solution. Neglecting charge transfer at the electrode and the variation of C_d with potential, compute:
(a) the charge stored by the double layer at t = ∞.
(b) the initial charging current i_0.
(c) the time required for the charging current to fall to 0.01 i_0. Hint: The potential across the cell can be written $E = iR_s + Q/C_d$, where $i = dQ/dt$ and C_d is the double–layer capacitance.

2.9 Butler and Huston (*Anal. Chem.*, **1970**, *42*, 1308) used electrochemical measurements to obtain activity coefficients in mixed electrolyte solutions. Using solutions containing known concentrations of NaCl and NaF, they measured potentials of the following cells:

Ag| AgCl(s)| Na$^+$(aq), Cl$^-$(aq), F$^-$(aq)| Na$^+$–selective glass electrode

Ag| AgCl(s)| Na$^+$(aq),Cl$^-$(aq),F$^-$(aq)| F$^-$–selective electrode

Na$^+$–electrode| Na$^+$(aq),Cl$^-$(aq),F$^-$(aq)| F$^-$–selective electrode

What combinations of activity coefficients were obtained from each of the three cells?

2.10 Using eq (2.65) and the Debye–Hückel limiting law, eq (2.57), derive an expression for the solvent activity at low electrolyte concentrations (assume constant volume).

2.11 Harned and Geary (*J. Am. Chem. Soc.* **1937**, *59*, 2032) measured the potential of the cell

$$Pt| H_2(g)| HCl(0.01 M), BaCl_2| AgCl(s)| Ag(s)$$

at 25°C as a function of the $BaCl_2$ concentration. The potential of the cell is given by the Nernst equation, corrected for activity coefficients:

$$E = E° - (RT/F) \ln([H^+][Cl^-]/P_{H_2}) - (RT/F) \ln(\gamma_\pm{}^2)$$

Potentials, corrected to the standard state pressure of hydrogen, are as follows:

[BaCl$_2$]	E/V
0.0	0.46411
0.00333	0.45273
0.00667	0.44529
0.01333	0.43539
0.02000	0.42864
0.03000	0.42135

By a clever extrapolation procedure, the authors found that $E° = 0.22234$ V. Use these data to compute the mean ionic activity coefficient for HCl in $BaCl_2$ solutions for the six concentrations given. Compute the ionic strengths of these solutions and use eqs (2.57) and (2.56) to estimate the activity coefficients [assume that a = 0.447 nm in eq (2.56)].

2.12 The mean ionic activity coefficient of 1.00 molal KOH is 0.735. The solution density is 1.0435 g cm^{-3} and the density of pure water is 0.9971 g cm^{-3}. What is the molar concentration of

KOH? What is the mean ionic activity coefficient on the molar
scale?

2.13 At the half–neutralization point in the titration of acetic acid with
sodium hydroxide, the concentrations of acetic acid and sodium
acetate, by definition, are equal, and it is customary to say that
the pH equals the pK_a of the acid. If, however, we take account
of the activity coefficient of sodium acetate, estimating $\gamma_\pm$ with
the Debye–Hückel limiting law, eq (2.57), by how much does the
pH actually differ from pK_a at the half–neutralization point?

2.14 Use the Debye–Hückel limiting law, eq (2.57), to estimate the
potential of the Fe^{3+}/Fe^{2+} half–cell when $[Fe(NO_3)_2] =
[Fe(NO_3)_3] = [HNO_3] = 0.002$ M.

3 ELECTROLYTIC CONDUCTANCE

With this chapter we begin consideration of experiments where current passes through an electrochemical cell. We will focus first on the measurement of the conductance of an electrolyte solution under conditions where the electrode–solution interface can be ignored. The interpretation of conductance data will lead to consideration of the contributions of individual ions to solution conductance and to conductance as a transport property related to diffusion. Finally we will consider membrane and liquid junction potentials which arise because of differences in ion transport rates.

3.1 CONDUCTIVITY

The resistance of any current–carrying device, whether it is a piece of wire or an electrolyte solution, is defined by Ohm's law as the ratio of the voltage across the device to the direct current flowing through it, $R = \Delta\Phi/i$. We are accustomed to think of resistance as a constant, dependent on temperature perhaps, but independent of the applied voltage and current flow; such a resistance is referred to as *ohmic*. Although some resistances encountered in electrochemistry are non-ohmic and depend on voltage, it is usually a good approximation to consider the resistance of an electrolyte solution as ohmic. Resistance is an extensive property of the system since it depends on the size (and shape) of the current–carrying device. For a device of uniform cross-sectional area A and length L, we can define an intensive quantity called the resistivity, ρ:

$$\rho = RA/L$$

In discussing electrolyte solutions, it is more convenient to speak of the conductance, the reciprocal of resistance, and of the conductivity, κ, the reciprocal of resistivity,

$$\kappa = L/RA \tag{3.1}$$

The unit of resistance is the ohm (Ω), so that resistivity has units of Ω-m. The unit of conductance is the siemens (S), $1 \text{ S} = 1 \, \Omega^{-1}$, so that the units of conductivity are $S \, m^{-1}$.

Example 3.1 Compute the conductivity of an electrolyte solution if the resistance is measured in a cell with parallel electrodes 1 cm^2 in area and spaced 1 cm apart and found to be 1000 Ω.

Substituting numbers into eq (3.1), we have

$$\kappa = \frac{0.01 \text{ m}}{(1000 \, \Omega)(0.0001 \text{ m}^2)} = 0.1 \, \Omega^{-1} \text{m}^{-1} = 0.1 \text{ S m}^{-1}$$

Before proceeding with an analysis of conductance data, we must give some attention to the measurement of the conductance (or resistance) of an electrolyte solution. This is a nontrivial problem and we must first ask how the resistance of a solution can be measured at all without running afoul of the resistance and capacitance of the electrode–solution interfaces which are present whenever we put electrodes into the solution. Thus, we first consider the electrical equivalent circuit of an electrochemical cell.

Electrochemical Cell Equivalent Circuit

There is a theorem in electrical circuit analysis which states in effect that any two–terminal device may be represented (at least approximately) by some combination of potential sources, resistances, capacitances, and inductances. We saw in Chapter 1 that an electrode–solution interface is in fact a potential source, and, in Chapter 2, that the electrode double layer behaves like a capacitance. The electrode–solution interface may also be thought of as having an impedance which represents the finite rate of mass transport and electron transfer at the electrode surface.[‡] If we add to these the series resistance due to the solution itself, we can draw the

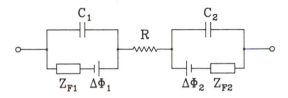

Figure 3.1 An electrochemical cell equivalent circuit.

electrochemical cell equivalent circuit shown in Figure 3.1; $\Delta\Phi_1$ and $\Delta\Phi_2$ are the two electrode–solution potential differences, the difference between which is the equilibrium cell potential, E. C_1 and C_2 are the double–layer capacitances at the two electrodes, Z_{F1} and Z_{F2} are the faradaic impedances, and R is the solution resistance. Our goal is the measurement of R without interference from the other equivalent circuit parameters.

If we were simply to attach an ammeter to the cell and measure the direct current, i, we might use Ohm's law to find the cell resistance

$$R_{cell} = (\Delta\Phi_1 - \Delta\Phi_2)/i$$

but we see from Figure 3.1 that the resistance thus obtained includes a contribution from the faradaic impedances. Even if we were able to separate the cell resistance into its components, there are other problems. The resistive component of a faradaic impedance turns out to be nonohmic, that is, it is not constant but depends on the potential; furthermore, when a finite current is drawn, the cell potential does not have its equilibrium value. In any case, it is most convenient to use identical electrodes and a homogeneous solution so that $\Delta\Phi_1 = \Delta\Phi_2$

[‡] This so–called *faradaic impedance* has both a resistive and a capacitive component; see Section 5.4.

and the net cell potential is zero. We could pass current using an external potential source, but then electrolysis would change the solution composition and there would still be problems with the faradaic impedances.

The solution to these problems is to use alternating current in the resistance measurement. If we apply an a.c. voltage across the cell and measure only the a.c. component of the current, considerable simplification is obtained. Since we are only concerned with the alternating current which flows as a result of an applied sinusoidal voltage, the d.c. potential sources, $\Delta\Phi_1$ and $\Delta\Phi_2$, can be ignored. Furthermore, the impedance of a capacitor toward flow of a sinusoidal current is $1/\omega C$, where ω is the angular frequency ($\omega = 2\pi$ times the frequency in Hz). If the capacitance of the electrode double layer is on the order of magnitude of 10 μF and the frequency of the a.c. signal is 1000 Hz, then the impedance of the capacitor is $1/(2\pi)(1000 \text{ s}^{-1})(10^{-5} \text{ F}) = 16 \ \Omega$. Faradaic resistances depend on the d.c. potential across the cell and are often much larger than the impedance of the double-layer capacitance. Referring to the equivalent circuit of Figure 3.1, we see that the path for a.c. current is primarily through the double–layer capacitance. Thus to a good approximation, the equivalent circuit of

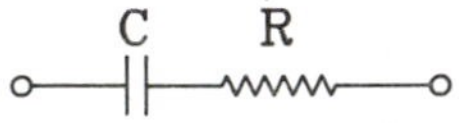

Figure 3.2 Equivalent circuit for large electrodes and alternating current.

the cell may be reduced to a capacitor and resistor in series, as shown in Figure 3.2. The equivalent circuit capacitance C is the series equivalent of the two double–layer capacitances, $C = C_1 C_2/(C_1 + C_2)$, and the equivalent circuit resistance R is nearly exactly the solution resistance.

Measurement of Solution Resistance

The classical method for measurement of an unknown resistance R_x involves the Wheatstone bridge (shown in Figure 3.3a). The bridge is balanced by adjusting the variable resistance R_s to zero the galvanometer G. Analysis of the circuit then shows that the unknown resistance R_x is given by

$$R_x = R_s (R_1/R_2)$$

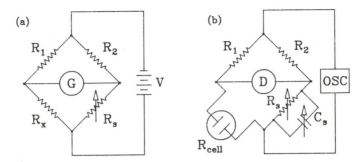

Figure 3.3 Wheatstone bridge circuits. (a) Direct–current bridge for measurement of resistance. (b) Alternating–current bridge for measurement of the resistance and capacitance of an electrochemical cell.

The d.c. Wheatstone bridge must be somewhat modified for measurement of solution resistances as shown in Figure 3.3b. The battery is replaced by an audio–frequency oscillator (usually operating at about 1000 Hz) and the galvanometer is replaced by an a.c. detector, perhaps an oscilloscope or simply a set of earphones. The bridge is provided with an adjustable capacitor C_s so that both the capacitive and resistive parts of the impedance can be balanced.

From our discussion so far, it should be clear that the design of the conductance cell is of considerable importance. It is necessary to have relatively large electrodes in order that the double–layer capacitance be as large as possible. Since the chemistry of the electrode process should not interfere with the conductance measurement, an inert electrode material is used; the almost universal choice is platinum. A fine layer of amorphous platinum (platinum black) is usually deposited on the electrodes so that the effective surface area is much larger than the apparent geometrical area and the double–layer capacitance is correspondingly increased. The electrodes must be rigidly mounted so that the distance between them is constant. Referring to eq (3.1), we see that the conductivity depends not only on the measured resistance, but also on the distance between the electrodes and the electrode area. Since these geometrical factors are not easily measured, it is customary to calibrate a conductance cell by first measuring the resistance when filled with a solution of known conductivity. In this way the cell constant, L/A, is obtained experimentally and the geometry need not be accurately measured.

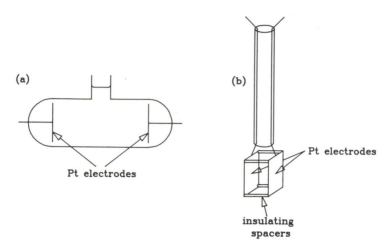

Figure 3.4 Two types of conductance cells. (a) Cell with electrodes rigidly mounted at a fixed separation; (b) an electrode assembly which can be dipped into a solution contained in a beaker.

Two common conductance cell designs are shown in Figure 3.4. See reviews by Shedlovsky and Shedlovsky *(1)* or by Holler and Enke *(2)* for further details.

Molar Conductivity

Development of the techniques for measuring solution conductivities occupied the last half of the nineteenth century (with further elaborations as electronic instrumentation evolved), but the first observations of the phenomenon were made by Erman[‡] in 1801 *(3)*. In experiments on the electrolysis of water (discovered by Nicholson and Carlisle in 1800), Erman found that the current increased with increasing salt concentration. This observation was confirmed by later workers and it became customary to define another quantity, called the *molar conductivity* to approximately factor out the concentration dependence. Molar conductivity is defined by

---

[‡] Paul Erman (1764–1851) was Professor of Physics at the University of Berlin. His studies on electrical phenomena, although limited by the crude instruments available at the time, foreshadowed much of the nineteenth–century development of the physics of electricity and magnetism.

$$\Lambda = \kappa/C \tag{3.2}$$

where C is the concentration in mol m^{-3}, (1 mol m^{-3} = 10^{-3} mol L^{-1} = 1 mM). The units of Λ are S m^{2}mol^{-1}.[‡]

Example 3.2 A 0.100 M solution of KCl, known to have a conductivity of 1.2896 S m^{-1}, was found to have a resistance of 89.3 Ω when placed in a conductance cell. What is the value of L/A? A 0.200 M solution of NaCl, measured in the same cell, gave R_x = 56.6 Ω. What is the conductivity of the solution? What is the molar conductivity?

According to eq (3.1),

$$L/A = \kappa R = (1.2896 \text{ S m}^{-1})(89.3 \ \Omega) = 115.2 \text{ m}^{-1}$$

so that for the NaCl solution,

$$\kappa = (115.2 \text{ m}^{-1})/(56.6 \ \Omega) = 2.035 \text{ S m}^{-1}$$

$$\Lambda = (2.035 \text{ S m}^{-1})/(200 \text{ mol m}^{-3}) = 0.0102 \text{ S m}^2\text{mol}^{-1}$$

The molar conductivities of several salts are plotted as functions of concentration in Figure 3.5. The molar conductivity depends on concentration; the magnitude of the dependence varies from one salt to another, but in every case the variation is greatest at low concentrations. Kohlrausch,[#] who collected most of the early

---

[‡] In the older literature κ is given the units of Ω^{-1}cm^{-1} (or the whimsical units, mho cm^{-1}), and Λ, with units of Ω^{-1}cm^{2}mol^{-1}, is called the *equivalent conductivity* and defined as $1000\kappa/zC$, where C is the molar concentration and $z = \nu_+z_+ = \nu_-|z_-|$ is the number of equivalents per mole. The new definition is rather more straightforward, but of course leads to different numerical values; thus beware when using older tabulations of data.

[#] Friedrich Wilhelm Kohlrausch (1840–1910) was Professor of Physics at Göttingen, Zurich, Darmstadt, Würzburg, Strasbourg,

conductance data, tried various empirical equations to fit this concentration dependence. Finally in 1900 *(4)*, he established that the limiting behavior for strong (*i.e.*, completely ionized) electrolytes is

$$\Lambda = \Lambda^\circ - s\sqrt{C} \tag{3.3}$$

where Λ° is the molar conductivity extrapolated to infinite dilution and the slope s depends on the electrolyte and solvent.

Figure 3.5 Molar conductivities of aqueous solutions of NaCl (crosses), KCl (diamonds), BaCl$_2$ (squares), and LaCl$_3$ (pluses) at 298 K *vs.* the square root of molar concentration. The dashed lines represent the linear extrapolation to infinite dilution using eq (3.3). Data from *(B9)*.

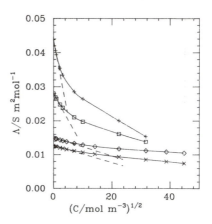

Ionic Conductivities

If, at infinite dilution, the behavior of electrolytes is indeed ideal (*i.e.*, we can neglect interionic interactions), and if electrolytes are completely ionized, then the molar conductivity of an electrolyte should be the sum of the molar conductivities of the individual ions.[‡] If one mole of the salt produces ν_+ moles of cations with molar conductivity Λ_+° and ν_- moles of anions with molar conductivity Λ_-°, then

$$\Lambda^\circ = \nu_+\Lambda_+^\circ + \nu_-\Lambda_-^\circ \tag{3.4}$$

---

and Berlin. He developed the a.c. methods for measurement of solution conductance and spent his career on experimental and theoretical studies of conductance.

[‡] This idea, known as Kohlrausch's law of independent ionic migration, was first proposed in 1876.

The molar conductivity of an individual ion cannot be measured directly, but if eq (3.4) is correct, then subtracting the molar conductivities of NaCl and KCl, for example, should give the difference between the molar conductivities of the sodium and potassium ions.

$$\Lambda°(KCl) - \Lambda°(NaCl) = (149.8 - 125.4) \times 10^{-4} \text{ S m}^2\text{mol}^{-1}$$

$$\Lambda°(K^+) - \Lambda°(Na^+) = 23.2 \times 10^{-4} \text{ S m}^2\text{mol}^{-1}$$

Comparing KI and NaI, the difference is 23.8×10^{-4}, and for $KBrO_3$ and $NaBrO_3$, the difference is 23.3×10^{-4}. Although there is a little scatter, the agreement is quite good.

Transference Numbers

When current passes through an electrolyte solution, it is carried in part by cations moving toward the cathode and in part by anions moving toward the anode. The fractions of the total current carried by positive and negative ions are called the transference numbers

$$t_+ = \nu_+\Lambda_+/\Lambda \qquad (3.5a)$$

$$t_- = \nu_-\Lambda_-/\Lambda \qquad (3.5b)$$

where, since

$$\nu_+\Lambda_+ + \nu_-\Lambda_- = \Lambda$$

the transference numbers must sum to 1:

$$t_+ + t_- = 1$$

There are several experimental methods for the determination of transference numbers (5,B9), the most straightforward being the classical method introduced by Hittorf[‡] in 1853 (6). In Hittorf's experiment, current is passed through a cell such as that shown in

---

[‡] Johann W. Hittorf (1824–1914) was a professor at the University of Münster. He is best known for his measurements of transference numbers; his work played an important part in the formulation of electrolyte theory by Arrhenius in the 1880's.

Figure 3.6. In the two end compartments, electrodes are placed which show uncomplicated behavior in the experimental solutions. For example, if we wish to measure transference numbers in a solution of

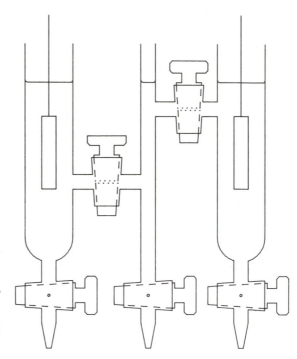

Figure 3.6 Hittorf cell for the determination of transference numbers.

silver nitrate, we would use silver electrodes. At the cathode, silver ions are plated out when current flows; at the anode, silver ions go into solution. Now suppose that one Faraday of charge is passed through the cell. The current in the bulk solution is carried by Ag^+ and NO_3^- ions with fractions t_+ and t_-, respectively. Thus t_+ moles of Ag^+ must have gone from the anode compartment to the central compartment, and the same number from the central compartment to the cathode compartment. Meanwhile t_- moles of NO_3^- must migrate from the cathode compartment to the central compartment and from the central compartment to the anode compartment. Thus the anode compartment loses t_+ moles of Ag^+ by transference but gains one mole of Ag^+ by dissolution of the anode, for a net increase of $1 - t_+$ moles. There is of course a corresponding increase of $t_- = 1 - t_+$ moles of NO_3^-. The cathode compartment has corresponding decreases in both Ag^+ and NO_3^-.

If the solutions in each of the three compartments are analyzed after passing a known charge through an initially homogeneous solution, the number of moles of Ag^+ gained in the anode compartment and the number of moles of Ag^+ lost in the cathode compartment can be calculated. Dividing by the number of Faradays of charge passed gives the transference numbers. The cell must be designed to minimize convective mixing and the experiment should be of relatively short duration to minimize diffusive mixing. Analysis of the central compartment solution should show no change in the concentrations and thus provides a check on the influence of convection and diffusion.

A moving boundary method, analogous to that used by Tiselius to measure the electrophoretic mobilities of proteins (Section 2.4), has also been used to measure transference numbers.

With transference numbers available, eqs (3.5) can be used to separate molar conductivities into the contributions of the positive and negative ions. If transference numbers are known over a range of concentration, they can be extrapolated to infinite dilution to obtain the infinite dilution ionic conductivities, Λ_+° and Λ_-°.

A selection of molar ionic conductivities at infinite dilution is given in Table A.7. These data can be used to compute the molar conductivities of a wide variety of electrolytes at infinite dilution.

Example 3.3 Compute the molar conductivities of potassium chloride, hydrochloric acid, and barium chloride solutions at infinite dilution.

$$\Lambda^\circ(KCl) = \Lambda^\circ(K^+) + \Lambda^\circ(Cl^-)$$

$$\Lambda^\circ = (73.5 + 76.3) \times 10^{-4} = 149.8 \times 10^{-4}\ S\ m^2 mol^{-1}$$

$$\Lambda^\circ(HCl) = \Lambda^\circ(H^+) + \Lambda^\circ(Cl^-)$$

$$\Lambda^\circ = (349.8 + 76.3) \times 10^{-4} = 426.1 \times 10^{-4}\ S\ m^2 mol^{-1}$$

$$\Lambda^\circ(BaCl_2) = \Lambda^\circ(Ba^{2+}) + 2\,\Lambda^\circ(Cl^-)$$

$$\Lambda^\circ = (127 + 2 \times 76) \times 10^{-4} = 279 \times 10^{-4}\ S\ m^2 mol^{-1}$$

With a table of molar ionic conductivities *(B2,B9,H2,H3)* such as Table A.7, we can also compute transference numbers.

Example 3.4 Compute the infinite dilution transference numbers in solutions of KCl, HCl, and $BaCl_2$.

For KCl,

$$t_+{}^\circ = \Lambda^\circ(K^+)/\Lambda^\circ(KCl) = 73.5/149.8 = 0.491$$

$$t_-{}^\circ = 1 - t_+{}^\circ = 0.509$$

for HCl,

$$t_+{}^\circ = \Lambda^\circ(H^+)/\Lambda^\circ(HCl) = 349.8/426.1 = 0.821$$

$$t_-{}^\circ = 1 - t_+{}^\circ = 0.179$$

and for $BaCl_2$,

$$t_+{}^\circ = \Lambda^\circ(Ba^{2+})/\Lambda(BaCl_2) = 127.7/280.3 = 0.456$$

$$t_-{}^\circ = 1 - t_+{}^\circ = 0.554$$

Ionic Mobilities

It is sometimes useful to think of the transport of an ion in terms of its mobility. (We encountered the electric mobility in our discussion of electrophoresis in Section 2.4.) The mobility of an ion is defined as the velocity per unit electric field strength,

$$u_i = |\mathbf{v}_i| / |\mathbf{E}| \tag{3.6}$$

To see how mobility is related to molar ionic conductivity, we start with Ohm's law, written in the form $i = \Delta\Phi/R$. If the potential difference across a conductance cell of length L is $\Delta\Phi$, the field strength is $\mathbf{E} = -\Delta\Phi/L$. With eq (3.1) rearranged to $R = L/A\kappa$, we have

$$i = -\kappa A \mathbf{E} \tag{3.7}$$

Now consider 1 cubic meter of solution containing C_i moles of ions carrying charge $z_i F$ coulombs mol^{-1} and moving with an average velocity v_i. The contribution to the current density across the end of the cubic meter is the product of these three factors:

$$(i/A)_i = C_i z_i F v_i \qquad (3.8)$$

Combining eqs (3.7) and (3.8), we have

$$v_i = - \frac{\kappa_i E}{C_i z_i F}$$

or, with eq (3.6),

$$u_i = \frac{\kappa_i}{C_i |z_i| F}$$

Finally, with the definition of molar conductivity, eq (3.2), we have

$$u_i = \frac{\Lambda_i}{|z_i| F} \qquad (3.9)$$

Thus ionic mobilities at infinite dilution can be obtained from molar ionic conductivities such as those given in Table A.7.

Example 3.5 Compute the mobilities of the sodium and barium ions at infinite dilution.

The ionic mobility of the sodium ion ($z = 1$) is

$$u°(Na^+) = \Lambda°(Na^+)/F$$

$$u°(Na^+) = 5.2 \times 10^{-8} m^2 V^{-1} s^{-1}$$

The ionic mobility of the barium ion ($z = 2$) is

$$u°(Ba^{2+}) = \Lambda°(Ba^{2+})/2F$$

$$u°(Ba^{2+}) = 6.6 \times 10^{-8} m^2 V^{-1} s^{-1}$$

Frictional Coefficients

When an ion moves through solution, it is subject to a viscous drag force proportional to the ion's velocity,

$$\mathbf{F}_i = f_i \mathbf{v}_i$$

where the proportionality constant f_i is called a frictional coefficient. Under steady-state conditions, the ion moves at constant velocity so that the viscous drag force exactly cancels the electrical driving force, $z_i e \mathbf{E}$. Equating these forces, we have

$$f_i \mathbf{v}_i = z_i e \mathbf{E}$$

or

$$u_i = \frac{|\mathbf{v}_i|}{|\mathbf{E}|} = \frac{|z_i| e}{f_i} \tag{3.10}$$

or, using eq (3.9),

$$\Lambda_i = \frac{z_i^2 F e}{f_i} \tag{3.11}$$

Thus the motion of an ion through solution can be described equivalently in terms of molar ionic conductivities, ionic mobilities, or frictional coefficients.

Stokes' Law Radii

Still another way of thinking about the rate at which ions move through solution is by imagining the ion to be a hard sphere of effective hydrodynamic radius r_i and frictional coefficient given by Stokes' law:

$$f_i = 6\pi \eta r_i \tag{3.12}$$

Substituting eq (3.12) into eq (3.10), we can solve for the Stokes' law radius r_i:

$$r_i = \frac{|z_i| F}{6\pi \eta N_A u_i} \tag{3.13}$$

where we have replaced e by F/N_A.

Example 3.6 Compute the Stokes' law radii of Na^+ and Ba^{2+}.

The viscosity of water at 25°C is 0.890×10^{-3} kg m^{-1}s^{-1}; inserting the mobilities from Example 3.5, eq (3.13) gives

$$r(Na^+) = 1.84 \times 10^{-10} \text{ m} \quad (184 \text{ pm})$$

$$r(Ba^{2+}) = 2.89 \times 10^{-10} \text{ m} \quad (289 \text{ pm})$$

Faster moving ions, of course, have smaller Stokes' law radii, and slower ones have larger radii. While Stokes' law is not really valid for small ions in ordinary solvents,[‡] Stokes' law radii do give a rough measure of the effective size of ionic species as they move through a solution. Some Stokes' law radii, calculated as in Example 3.6, are compared with ionic radii from crystal structure data in Table 3.1. Several insights can be obtained from examination of these data. In the first place, it is hard to imagine a hydrodynamic radius being smaller than the crystal radius of the same ion. Comparison of the radii of I^-, for example, suggests that the Stokes' law radii are probably systematically underestimated, perhaps by about a factor of 2. However, within a related series of ions, e.g., the alkali metal cations, the trends are interesting. The crystal radii increase, as expected, in the series $Li^+ < Na^+ < K^+$, but the Stokes' law radii go in the opposite direction. This must mean that the small, highly polarizing Li^+ ion tightly binds a lot of solvent molecules which must move with the ion as a unit. The same effect is seen in comparison of the alkaline earth cations Mg^{2+}, Ca^{2+}, and Ba^{2+}. The three halide ions, Cl^-, Br^-, and I^- all have about the same Stokes' law radii despite an increase in the crystal radii; this again is consistent with larger, less polarizing ions binding fewer solvent molecules or binding them

---

[‡] The law was derived in 1845 by Sir George Stokes assuming a rigid sphere moving through a continuous medium; using Stokes' law for a charged deformable ion moving through a solvent of discrete dipolar molecules of size comparable to the ion must introduce some serious errors.

TABLE 3.1　Stokes' Law and Crystal Radii of Some Ions

Ion	r_S/pm	r_C/pm	Ion	r_S/pm	r_C/pm
Li^+	238	90	OH^-	47	119
Na^+	184	116	F^-	168	119
K^+	125	152	Cl^-	121	167
Mg^{2+}	348	86	Br^-	118	182
Ca^{2+}	310	114	I^-	120	206
Ba^{2+}	289	149	ClO_4^-	136	226
Al^{3+}	439	68	SO_4^{2-}	231	244

Crystal radii from Shannon (7).

less tightly so that fewer move as a unit. Comparing the isoelectronic ions Na^+, Mg^{2+}, and Al^{3+} we see an even more dramatic effect which has the same qualitative explanation. Extending the comparison to F^-, which is isoelectronic with Na^+ and has about the same crystal radius, we see that the solvation sheath is apparently greater for Na^+ than for F^-, consistent with the idea that cations are rather more specifically solvated (formation of coordinate covalent bonds) than are anions (orientation of water dipoles and/or hydrogen bonding).

From crystal structure data, OH^- and F^- would be expected to be nearly the same size, but the Stokes' law radius of OH^- is 47 pm, much less than the radius of any other anion. A similar discrepancy is found for H^+, which has a Stokes' law radius of 26 pm. Structural evidence suggests that H^+(aq) is better represented by H_3O^+, which might be expected to have a Stokes' law radius comparable to Li^+ or Na^+, say about 200 pm. The anomalously high conductivities of H^+ and OH^- in water can be understood if we recall that these water-related ions are part of an extensive hydrogen–bonded network involving solvent water molecules. A small shift in hydrogen bonds thus can move the charge from one oxygen center to another and several such shifts can move the charge much faster than any of the nuclei travel. Thus the conductivity is much higher than might have been expected from the size of the H_3O^+ or OH^- ions.

Theoretical Treatment of Conductivity

Kohlrausch's discovery that molar conductivity varies as the square root of concentration, eq (3.3), attracted the attention of several theoreticians,[‡] most notably Onsager,[#] who derived a conductance limiting law starting from the Debye–Hückel treatment of the ion atmosphere *(8)*. Onsager's result can be written as

$$\Lambda = \Lambda^\circ - \frac{|z_1 z_2| F^2 \Lambda^\circ}{12\pi \epsilon \epsilon_0 x_A RT N_A} \frac{q}{1 + q^{\frac{1}{2}}} - \frac{(|z_1| + |z_2|)F^2}{6\pi \eta N_A x_A} \tag{3.14}$$

where x_A is the ion atmosphere thickness, η is the solvent viscosity, z_1 and z_2 are the positive and negative ion charges, and

$$q = \frac{z_1 z_2}{(z_1 - z_2)(z_2 t_1 - z_1 t_2)}$$

where t_1 and t_2 are the transference numbers. When $z_1 = -z_2$ (a 1:1 or 2:2 electrolyte), $q = \frac{1}{2}$ and is independent of the transference numbers. The other parameters in eq (3.14) have their usual significance. Both correction terms in eq (3.14) are proportional to $1/x_A$, which in turn is proportional to the square root of the ionic strength, eq (2.8), and thus matchs the $\sqrt{C}$ dependence found by Kohlrausch.

The first term in the Onsager limiting law arises from the so-called "ion atmosphere relaxation effect." When an ion is attracted by an electric field, it is also subject to an opposite force exerted by its ion atmosphere, which tends to restrain the ion and thus lowers its

---

[‡] See Robinson and Stokes *(B9)* for a review of the various approaches.

[#] Lars Onsager (1903–1976) taught at Brown University in the early 1930's, later moving to Yale. He is best known for his work in nonequilibrium thermodynamics for which he received the Nobel Prize in 1968, but he made important contributions in many other areas of theoretical chemistry.

contribution to the solution conductance. The effect increases with the density of ions in the atmosphere, *i.e.*, inversely proportion to x_A. The second correction term results from the "electrophoretic effect." When the ion in question moves through the solution, it tends to take its ion atmosphere with it, resulting in a viscous drag force opposing the motion, thus the dependence on viscosity.

Like Debye–Hückel theory from which it is derived, eq (3.14) is a limiting law valid only at very low concentrations, usually less than 0.001 M.

Example 3.7 Compute the limiting law slope for a plot of the molar conductivity of KCl ($\Lambda^\circ = 149.8 \times 10^{-4}$ S m^{2}mol^{-1}) *vs.* $\sqrt{C}$ assuming the dielectric constant and viscosity of pure water ($\epsilon = 78.54$, $\eta = 0.890 \times 10^{-3}$ kg m^{-1}s^{-1}), and compare with the slope of the KCl line in Figure 3.5.

Inserting the values of the physical constants and T = 298 K into eq (3.14), we have

$$\Lambda = \Lambda^\circ - (6.97 \times 10^{-11})\Lambda^\circ/x_A - (1.84 \times 10^{-12})/x_A$$

Substitution of constants in eq (2.8) gives

$$x_A^{-1} = (1.04 \times 10^8)C^{\frac{1}{2}}$$

where C is in mol m^{-3}. Substituting x_A and Λ°, we have

$$\Lambda = \Lambda^\circ - (3.0 \times 10^{-4})C^{\frac{1}{2}}$$

which is in satisfactory agreement with the experimental result:

$$\Lambda = \Lambda^\circ - (2.7 \times 10^{-4})C^{\frac{1}{2}}$$

Effects of Frequency and Electric Field Strength

According to Onsager's description of ionic motion in an electrolyte solution, the motion of an ion is restrained by the attraction of its atmosphere. An ion atmosphere is not formed instantaneously since it requires the arrangement of many ions in or near a minimum potential energy configuration. The time required to configure an ion atmosphere is on the order of $10^{-7}/C$ seconds. Thus in a 1 mM solution, ionic motion is subject to an ion atmosphere relaxation time on the order of 10^{-7} s. If the a.c. potential applied across the conductance cell is in the radio–frequency range, above 10 MHz say, ions move faster than their atmospheres can be rearranged. Under these conditions, ionic motion escapes the retarding effect of the ion atmosphere and the measured conductivity is increased. This effect was predicted by Debye and Falkenhagen *(9)* and first observed by Sack *(10)* and is known as the *Debye–Falkenhagen effect* or the *Sack effect.*

A related effect occurs when the a.c. potential across the conductance cell is very large and the electrode spacing is small so that the oscillating electric field is large. With a field on the order of 10^6 V m^{-1}, an ion is accelerated to a sufficiently high velocity that it effectively escapes its ion atmosphere. That is, the electrical force on the ion greatly exceeds the retarding force of the ion atmosphere. The ion may leave behind some of the coordinating solvent molecules as well, so that its effective Stokes' law radius is smaller. For both these reasons, the conductivity increases for large electric field strengths. This effect was first observed by Wien *(11)* in 1922 and is called the *Wien effect.*

Both the Debye–Falkenhagen and the Wien effects can be understood quantitatively by extensions of Onsager's theory of ionic motion *(B5)*.

Ultrasonic Vibration Potential

When a high–frequency sound wave travels through an electrolyte solution, solvent molecules are forced into an oscillatory motion. Ions in the solution tend to move with the solvent but may lag a little behind. If the lag in cation motion is different from the lag in anion motion (because of different frictional coefficients), a small potential difference will develop at the frequency of the ultrasonic wave. If a pair of electrodes is placed in the solution with a spacing of half the wavelength of the sound wave, this potential can be detected. This effect was predicted by Debye in 1933 *(12)* but was not observed experimentally until after World War II. Debye originally suggested this experiment as a way of estimating the effective masses of the solvated ions. Subsequent experimental work showed that it is really the ionic partial molar volumes which are determinable, and there has been a substantial volume of work directed to this end. Work on ultrasonic vibration potentials has been reviewed recently by Zana and Yeager *(13)*.

3.2 CONDUCTANCE APPLICATIONS

The measurement and interpretation of conductance has given us many important qualitative insights into the properties of electrolyte solutions and has provided a means by which quantitative theories of electrolytes can be tested. Many of the experimental methods are now primarily of historical interest, at least for aqueous solutions, but some applications remain useful in our arsenal of techniques for extracting information from nature.

Measurement of Dissociation Constants

In the development of our understanding of electrolyte solutions in the late nineteenth and early twentieth centuries, conductance measurements provided much of the proof of the existence of ionized salts in solution. Conductance measurements played a particularly important role in classifying acids, bases, and salts into strong electrolytes (*i.e.*, completely ionized salts such as NaCl or strong acids such as HCl) and weak electrolytes (*i.e.*, weak acids such as acetic acid). Thus strong electrolytes exhibit the linear dependence of Λ on $\sqrt{C}$ which we have seen in Section 3.1. Weak electrolytes, on the other hand, typically have very small molar conductivities at moderate concentrations which increase rapidly as the concentration is reduced.

In explaining the dependence of Λ on concentration, Arrhenius[‡] assumed that, for a partially dissociated electrolyte at low concentration, the fraction of the electrolyte which is dissociated is simply the ratio of the molar conductivity to the molar conductivity extrapolated to infinite dilution:

[‡] Svante A. Arrhenius (1859–1927) was a lecturer at the Technical High School of Stockholm and later director of the physical chemistry department of the Nobel Institute. Arrhenius is generally regarded as one of the founders of the discipline of physical chemistry and is best remembered for his work on electrolyte solutions.

$$a = \Lambda/\Lambda^\circ. \tag{3.15}$$

Consider then the dissociation of acetic acid,

$$HOAc + H_2O \rightleftarrows H_3O^+ + OAc^-$$

with the corresponding equilibrium constant,

$$K_a = \frac{[H_3O^+][OAc^-]}{[HOAc]}$$

The individual species concentrations can be written in terms of the degree of dissociation, a, and the total acetic acid concentration C:

$$[H_3O^+] = [OAc^-] = aC$$

$$[HOAc] = (1 - a)C$$

Substituting these into the equilibrium constant expression, we have

$$K_a = \frac{a^2C}{1 - a}$$

where C must have units of mol L^{-1} if K_a is to refer to the usual standard states. Using eq (3.15) to eliminate a, we have

$$K_a = \frac{(\Lambda/\Lambda^\circ)^2C}{1 - \Lambda/\Lambda^\circ} \tag{3.16}$$

This expression, which is known as Ostwald's dilution law, may be rearranged to

$$C\Lambda = K_a(\Lambda^\circ)^2(1/\Lambda) - K_a\Lambda^\circ$$

so that a plot of $C\Lambda$ *vs.* $1/\Lambda$ should give a straight line with the slope and intercept yielding both the infinite dilution molar conductivity and the equilibrium constant.

This method works well and many of the earlier acid–base equilibrium constants were obtained in this way. Easier ways of obtaining such data are now available, but it is sometimes useful to remember that the conductance method exists.

Example 3.8 In 1894, Kohlrausch and Heydweiller *(14)* measured the conductivity of extremely pure water, obtaining $\kappa = 6.2 \times 10^{-6}$ S m^{-1}. Estimate the self–dissociation constant of water from this measurement.

The sum of the molar ionic conductivities of $H^+(aq)$ and $OH^-(aq)$ at infinite dilution is

$$\Lambda° = \Lambda°(H^+) + \Lambda°(OH^-) = 0.0549 \text{ S m}^2\text{mol}^{-1}$$

The ionic concentrations should be sufficiently low that we can substitute this value, together with κ, in eq (3.2), and solve for C.

$$C = \kappa/\Lambda$$

$$C = (6.2 \times 10^{-6} \text{ S m}^{-1})/(0.0549 \text{ S m}^{-2}\text{mol}^{-1})$$

$$C = 1.13 \times 10^{-4} \text{ mol m}^{-3}$$

$$[H^+] = [OH^-] = 1.13 \times 10^{-7} \text{ mol L}^{-1}$$

$$K_{eq} = [H^+][OH^-] \cong 1.3 \times 10^{-14}$$

Determination of Electrolyte Charge Type

When new compounds are discovered and found to be electrolytes, determination of the charges carried by the positive and negative ions may be an important part of the characterization. If the molar conductivity is determined at various concentrations, and a plot of Λ *vs.* $\sqrt{C}$ constructed, the intercept, $\Lambda°$, gives a hint of the charge type since the conductivities of salts of a given type tend to fall within certain limits. The limiting slope gives a better indication, however, since it depends upon the charge type of the electrolyte; see eq (3.14).

Conductometric Titrations

One of the most generally applicable electroanalytical techniques consists of following a titration reaction conductometrically *(15)*.

Consider, for example, the titration of hydrochloric acid with sodium hydroxide. Initially, the conductance of the HCl solution is due to H_3O^+ and Cl^- ions. As the titrant is added, the H_3O^+ ions are replaced by Na^+, and, since the molar ionic conductivity of Na^+ is much less than that of H_3O^+ (see Table A.7), the conductance of the solution falls. Beyond the equivalence point, on the other hand, excess NaOH is being added, and since the OH^- molar ionic conductivity is very large, the conductance of the solution rises again. The conductance of the solution is due, of course, to all the ions present. This sum is demonstrated graphically in Figure 3.7. In practice, a conductometric titration is carried out by adding aliquots of titrant, measuring the conductance and plotting the measured conductance *vs.* the volume of titrant added. The points can usually be fitted to two straight–line segments which intersect at the equivalence point.

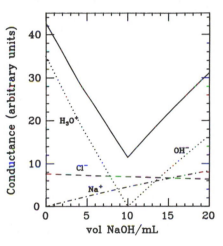

Figure 3.7 Conductometric titration of HCl with NaOH. Dashed lines indicate contributions of individual species to the solution conductance, which is indicated by the solid line.

This technique can be extended to any titration reaction which leads to a change in the slope of the conductance *vs.* concentration plot at the equivalence point. Any ionic reaction which produces a precipitate, a gas, or any other nonelectrolyte is well suited for use in a conductometric titration. Although the technique is very general, there are practical limitations: (1) the change in slope at the equivalence point is often small; (2) the analyte and titrant concentrations usually must be greater than 0.001 M; and (3) nonparticipating electrolytes should be absent.

3.3 DIFFUSION

The ionic conductivity and the electric mobility are measures of the rate of charge transport through solution under the influence of an electric potential gradient. Electrical conductivity is one of a class of physical properties called transport properties. Other transport properties include thermal conductivity (transport of heat energy under the influence of a temperature gradient), viscosity (transport of momentum under the influence of a velocity gradient), and diffusion (transport of mass under the influence of a chemical potential or concentration gradient). Because an understanding of diffusion will be particularly important in Chapters 4 and 5 and because electric mobility and diffusion can be related easily, we will digress for a few pages to a discussion of diffusion.

Fick's Laws of Diffusion

Consider the motion of molecules or ions in a liquid solution. In the absence of electric or gravitational fields, the motion is simply a random thermal agitation where the particles respond to bumps they receive from their neighbors to go shooting off in one direction or another. Although classical mechanics would have us believe that the Newtonian equations of motion at least in principle govern this motion, in practice it is just as well to regard the motion as entirely random and unpredictable. If we focus on an individual molecule, we will see it undergo a series of displacements in random directions and of random length. However, we are interested in the average behavior of a large number of molecules. Suppose that, in a time interval δt, a particle moves a distance δx in the x–direction. The average displacement, $<\delta x>$, is zero because a molecule is as likely to go in the positive x–direction as in the negative. On the other hand, if we square the displacements before averaging, the mean square displacement (in time δt) $<(\delta x)^2>$ is nonzero. Now consider a section of the solution where there is a linear concentration gradient in the x–direction, and focus on a small section of cross–sectional area A and length (along the x–axis) $2\delta x$. Divide this volume in half as shown in Figure 3.8, and suppose that the average concentration of particles in the left–hand volume is n_1 and in the right–hand volume n_2 (particles per unit volume). Thus, since the volume of each element is $A\delta x$, the number of particles in the two volumes are $n_1 A\delta x$ and $n_2 A\delta x$, respectively. Now on the average in time interval δt, half the particles will move to the right and half will move to the left. Thus the number crossing the central dividing plane from left to right will be $\frac{1}{2}n_1 A\delta x$ and the number crossing from

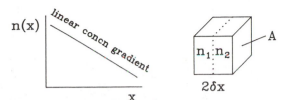

Figure 3.8 Model for the mathematical description of diffusion.

right to left will be $\frac{1}{2}n_2 A\delta x$. Thus the net number moving across the barrier (from left to right) is $\frac{1}{2}(n_1 - n_2)A\delta x$. Dividing by the time interval then gives the net rate of transfer:

$$\text{Rate} = \frac{(n_1 - n_2)A\delta x}{2\delta t} \tag{3.17}$$

Recalling our original assumption that the concentration gradient is linear, we can express the concentration difference, $n_1 - n_2$, in terms of the concentration gradient dn/dx:

$$n_1 - n_2 = -\frac{dn}{dx}\delta x$$

and, substituting this into eq (3.17), we have (noting that n is apparently a function of both x and t, so that we ought to write the derivative as a partial derivative):

$$\text{Rate} = -\frac{A(\delta x)^2}{2\delta t}\frac{\partial n}{\partial x} \tag{3.18}$$

The rate in eq (3.18) is the number of particles passing through a reference plane of cross–sectional area A per unit time. If we divide eq (3.18) by A and by Avogadro's number, we will have the flux J in moles per unit cross–sectional area per unit time

$$J = -D\frac{\partial C}{\partial x} \tag{3.19}$$

where the *diffusion coefficient* D is given by

$$D = \frac{(\delta x)^2}{2\delta t} \tag{3.20}$$

with SI units of $m^2 s^{-1}$. This relationship was first established empirically by Fick in 1855,[‡] and is known as Fick's first law of diffusion.

Notice that eq (3.19) tells us that the flux is positive (a net flow of particles from left to right) only if the concentration gradient is negative (there are fewer particles on the right to begin with). Thus this simple equation contains the elements of the phenomenon of diffusion: flow of particles into regions of lower concentration.

Example 3.9 An integral form of eq (3.20) is sometimes used to estimate the distance a particle might diffuse in a certain time:

$$<x^2> = 2Dt$$

where $<x^2>$ is the mean–square displacement in the x–direction in time t. Estimate the distance a particle with $D = 10^{-9}$ $m^2 s^{-1}$ will travel in 10^{-6} s.

$$<x^2> = 2(10^{-9}\ m^2 s^{-1})(10^{-6}\ s) = 2 \times 10^{-15}\ m^2$$

Thus the root–mean–square displacement in the x–direction in 1 μs is 4.5×10^{-8} m (45 nm). In an isotropic solution, we would expect $<x^2> = y^2> = <z^2>$ so that $<r^2> = 3<x^2>$ or

$$<r^2> = 6Dt$$

Thus the total r.m.s. distance travelled in 1 μs is 135 nm.

[‡] Adolf Eugen Fick (1829–1901) was Professor of Physiology at the University of Würzburg. Fick's work on diffusion was part of a wide–ranging effort to apply the principles of physics to physiology.

Fick also discovered a second empirical law, which was later found to be derivable from the first law. Consider an infinitesimal volume element of unit cross–sectional area and thickness dx. The flux into this volume element is $J(x)$, and the flux out of the volume element is $J(x + dx) = J(x) + (\partial J/\partial x)dx$. The difference between the flux in and the flux out is the net increase in number of moles of particles in the volume element per unit time. Dividing this by the volume of the element dx, we have the time rate of change of the concentration $\partial C/\partial t$.

$$\frac{\partial C}{\partial t} = \frac{J(x) - J(x + dx)}{dx} = -\frac{\partial J}{\partial x}$$

Differentiating eq (3.19), we get

$$\frac{\partial J}{\partial x} = -D\frac{\partial^2 C}{\partial x^2}$$

so we are left with

$$\frac{\partial C}{\partial t} = D\frac{\partial^2 C}{\partial x^2} \qquad\qquad (3.21)$$

which is Fick's second law. This result, of course, is a second–order partial differential equation, which might not seem to be a particularly simple way of expressing what we know about diffusion. However, it turns out that the equation can be solved relatively easily for a wide variety of boundary conditions.

Before we go on, it is instructive to write down the solution to eq (3.21) for a particularly interesting case. Suppose that we bring two solutions into contact at time $t = 0$. In one solution the concentration of the diffusing species is C_0 and in the other solution, the concentration is zero; the initial interface is at $x = 0$. We suppose that for $t > 0$, $C \to 0$ as $x \to \infty$ and $C \to C_0$ as $x \to -\infty$. With these initial and boundary conditions, the solution to eq (3.21) is (see Appendix 4)

$$C(x,t) = \tfrac{1}{2}C_0[1 - \text{erf}(x/2D^{\frac{1}{2}}t^{\frac{1}{2}})] \qquad\qquad (3.22)$$

The function erf (ψ) is called the error function and defined by the integral

$$\text{erf}\,(\psi) = (2/\sqrt{\pi}) \int_0^{\psi} \exp(-u^2)du \tag{3.23}$$

The error function is a transcendental function like logarithms and exponential functions and is found in many mathematical tables. As seen in Figure 3.9, the error function goes to zero as its argument goes to zero and approaches unity as its argument increases. The error function is an odd function of the argument, *i.e.*, $\text{erf}(-\psi) = -\text{erf}(\psi)$. The derivative of an error function is an exponential function:

$$\frac{d}{dx}\,\text{erf}[\psi(x)] = (2/\sqrt{\pi})\,\exp(-\psi^2)\,\frac{d\psi}{dx} \tag{3.24}$$

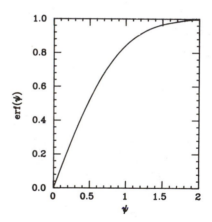

Figure 3.9 The error function.

The concentration profile, given by eq (3.22), and the concentration gradient, computed using eq (3.24), are shown for various times in Figure 3.10. The results are as we would expect intuitively: the concentration gradient, which initially is a sharp spike at x = 0, broadens and decreases in height as time passes, *i.e.*, as the solute diffuses into the right–hand solution. Since the flux of solute is proportional to the concentration gradient through eq (3.19), Figure 3.10c is equivalent to a plot of flux *vs.* distance at successive times.

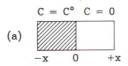

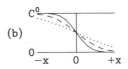

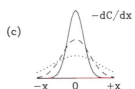

Figure 3.10 (a) Initial distribution of solute in a diffusion experiment. (b) Concentration profiles at successive times after beginning of diffusion. (c) Concentration gradient at successive times after beginning of diffusion.

Example 3.10 Compute the concentration gradient and flux at the position of the original boundary 1, 10, and 100 s after the boundary is formed if $D = 10^{-9}$ m^{2}s^{-1} and $C_0 = 0.1$ M (100 mol m^{-3}).

Differentiating eq (3.22), using eq (3.24), we obtain

$$\frac{\partial C(x,t)}{\partial x} = -\frac{C_0}{2(\pi Dt)^{\frac{1}{2}}} \exp \frac{-x^2}{4Dt}$$

At $x = 0$, the exponential term reduces to 1. Substituting numbers, we obtain

$$-\frac{\partial C(0,t)}{\partial x} = 8.92 \times 10^5, \ 2.82 \times 10^5, \ 8.92 \times 10^4 \ \text{mol m}^{-4}$$

at $t = 1$, 10, and 100 s. Substituting the concentration gradient into eq (3.19), we obtain the flux

$$J = 892, \ 282, \ 89.2 \ \mu\text{mol m}^{-2}\text{s}^{-1}$$

Relation of Diffusion and Mobility

We have seen that diffusion may be regarded as the net movement of particles in solution due to a concentration gradient. Another more general way of stating this is to say that particles will move in solution with a velocity proportional to the chemical potential gradient. This may be expressed mathematically as

$$\mathbf{v}_i = -\frac{1}{f_i N_A} \frac{\partial \mu_i}{\partial x} \tag{3.25}$$

where $\mathbf{v}_i$ is the average velocity of particles of type i, f_i is the frictional coefficient, and N_A is Avogadro's number. Suppose that the particles are ions so that the chemical potential depends not only on their activity, but also on the electric potential.[‡] Thus we have

$$\mu_i = \mu_i{}^\circ + RT \ln(a_i) + z_i F \Phi \tag{3.26}$$

Differentiating eq (3.26) and substituting into (3.25), we have the Nernst–Planck equation

$$\mathbf{v}_i = -\frac{1}{f_i} \left[kT \frac{d \ln(a_i)}{dx} + z_i e \frac{d\Phi}{dx} \right] \tag{3.27}$$

We now consider two limiting cases. First we assume that the activity is uniform, but the electric potential nonuniform. Equation (3.27) then becomes

$$\mathbf{v}_i = + z_i e \mathbf{E} / f_i$$

where we have recalled that the negative gradient of the electric potential is the electric field strength, $\mathbf{E} = -d\Phi/dx$. Dividing through by $\mathbf{E}$, we have the mobility:

$$u_i = |\mathbf{v}_i| / |\mathbf{E}| = |z_i| e / f_i \tag{3.28}$$

which is identical to eq (3.10).

[‡] This generalized chemical potential is sometimes called the *electrochemical potential*.

For the second limiting case, we assume that the electric potential is constant so that eq (3.27) contains only the first term on the right–hand side. If the concentration of species i is C_i and the average velocity is $\mathbf{v}_i$, then the flux J_i is $C_i\mathbf{v}_i$ or

$$J_i = -\frac{C_i kT}{f_i}\frac{d\ln(a_i)}{dx}$$

Noting that the activity can be written $a_i = C_i\gamma_i$, we have

$$J_i = -\frac{kT}{f_i}\left[C_i\frac{d\ln(C_i)}{dx} + C_i\frac{d\ln(\gamma_i)}{dx}\right] \tag{3.29}$$

and, if we assume that the activity coefficient is independent of x, either because the solution is dilute and nearly ideal or because of the presence of a homogeneous excess of some other electrolyte which mostly determines the activity coefficient, then the second term may be dropped and eq (3.29) becomes

$$J_i = -\frac{kT}{f_i}\frac{dC_i}{dx}$$

which is Fick's first law, with

$$D_i = kT/f_i \tag{3.30}$$

Since the frictional coefficient occurs in both eqs (3.28) and (3.30), we can eliminate it to obtain a relation between the diffusion coefficient and the electric mobility,

$$D_i = \frac{kTu_i}{|z_i|\,e} = \frac{RTu_i}{|z_i|\,F} \tag{3.31}$$

Equation (3.31) is sometimes called the Einstein relation. Combining eqs (3.9) and (3.31) gives the connection between the diffusion coefficient and the ionic conductivity, the Nernst–Einstein equation,

$$D_i = \frac{kT\Lambda_i}{z_i^2 e^2} \tag{3.32}$$

Example 3.11 Compute the diffusion coefficients of Na^+ and Ba^{2+} at 25°C in dilute solutions.

At 25°C, the factor RT/F has the value 0.0257 V. Thus, using the mobilities computed in Example 3.6, we have

$$D°(Na^+) = (0.0257 \text{ V})(5.2 \times 10^{-8} \text{ m}^2\text{V}^{-1}\text{s}^{-1})$$

$$D°(Na^+) = 1.34 \times 10^{-9} \text{ m}^2\text{s}^{-1}$$

$$D°(Ba^{2+}) = (0.0257 \text{ V})(6.6 \times 10^{-8} \text{ m}^2\text{V}^{-1}\text{s}^{-1})/2$$

$$D°(Ba^{2+}) = 0.85 \times 10^{-9} \text{ m}^2\text{s}^{-1}$$

We now have three essentially equivalent measures of the transport properties of an ion in solution: the molar ionic conductivity Λ, the ionic mobility u, and the diffusion coefficient D. The three quantities are compared for Na^+, Mg^{2+}, and Al^{3+} in Table 3.2. Notice that whereas $\Lambda°$ increases with ionic charge, $u°$ is nearly independent of charge and $D°$ decreases with charge. This is a reflection of two effects; as the charge increases, the size of the solvation shell increases; each of these ions has a primary coordination of six water molecules, but with increasing charge, outer–sphere water molecules are also oriented about the ion. Thus a large aggregate such as Al^{3+} with its solvation sheath moves through solution more slowly than Na^+ and its smaller solvation sheath. The dependence of $D°$ on ionic charge is thus qualitatively reasonable. When we look at the velocity of particles in a field, we expect the retarding effect of the solvation sheath to be opposed by a larger electrical force on ions with higher charge. Thus the two effects more or less cancel in the case of the mobilities. Conductivities, on the other hand, measure not only the velocity of ions in the field, but also the amount of charge carried and so increase with increasing charge.

TABLE 3.2 Comparison of Ionic Conductivities, Mobilities, and Diffusion Coefficients

Ion	Λ° $S\ m^2mol^{-1}$	u° $m^2V^{-1}s^{-1}$	D° m^2s^{-1}
Na^+	50×10^{-4}	5.2×10^{-8}	13.4×10^{-10}
Mg^{2+}	106×10^{-4}	5.5×10^{-8}	7.1×10^{-10}
Al^{3+}	189×10^{-4}	6.5×10^{-8}	5.6×10^{-10}

3.4 MEMBRANE AND LIQUID JUNCTION POTENTIALS

As we noted in Chapter 1, there is in general a potential difference across the boundary between any two dissimilar phases. In this section, we will consider three cases of potentials arising across phase boundaries.

Liquid Junction Potentials

Consider the junction of two electrolyte solutions. The two solutions may be carefully layered or they may be separated by a fritted glass disk or an asbestos fiber. In either case the system as a whole is not at equilibrium and mixing will eventually take place. The role of the glass frit or asbestos fiber is simply to slow the rate of diffusive mixing. If the rate of mixing is slow enough, we can regard the two solutions as essentially constant in composition and can deal with a steady–state model of the boundary between the two phases.

Now consider what happens when the liquid junction is formed between phases a and β and diffusion is allowed to proceed. Suppose for the moment that a single electrolyte is present with concentrations $C^a > C^\beta$. Since the diffusion coefficients of the anions and cations in general are not equal, one species is expected to diffuse more rapidly than the other. If, for example, the cations diffuse from the a–phase into the β–phase more rapidly than do the anions, then a charge separation results and with it an electric potential difference which ultimately will oppose the continued diffusion of the two species at unequal rates. The steady–state condition then is diffusion of the two species at the same rate, but with the positive ions slightly ahead of the negative ions; see Figure 3.11.

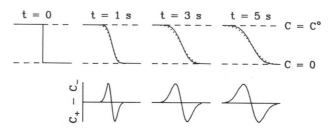

Figure 3.11 Diffusion of a 1:1 electrolyte solution in a tube. (a) Concentration profiles at several times; the solid lines represent the cations and the dashed lines represent the anions. (b) Concentration differences, $C_+ - C_-$, at several times.

Consider now the more general case of a liquid junction between two electrolyte solutions of different compositions and concentrations. Suppose that the liquid junction has reached a steady–state condition, and consider a charge δQ which moves reversibly through a distance dx in the phase boundary region. There will be contributions to δQ from cations moving in the $+dx$ direction and from anions moving in the $-dx$ direction:

$$\delta Q = F \sum_i z_i \delta n_i$$

where z_i is the charge on ion i and δn_i is the number of moles of i moving from x to x + dx. If i is a cation, z_i and δn_i will be positive, but if i is an anion, both quantities will be negative. The fraction of the charge carried by ion i (the transference number, t_i) then will be

$$t_i = \frac{F z_i \delta n_i}{\delta Q}$$

so that δn_i can be written

$$\delta n_i = (t_i/z_i) \, \delta Q \, /F$$

In moving from x to x + dx, ion i moves through a chemical potential difference, given by eq (3.26):

$$d\mu_i = RT \, d \ln(a_i) + Fz_i d\Phi$$

where $d\Phi$ is the electric potential difference across the distance dx. The total change in free energy then is

$$\delta G = \sum_i d\mu_i \delta n_i$$

or

$$\delta G = \left[\frac{RT}{F} \sum_i \frac{t_i}{z_i} \, d \ln(a_i) + \sum_i t_i d\Phi \right] \delta Q \qquad (3.33)$$

However, since we assumed that charge transport through the phase boundary is reversible, δG must be zero; the bracketed expression must then be zero, independent of δQ. Furthermore, by definition

$$\sum_i t_i = 1$$

so that we can rearrange the bracketed expression in eq (3.33) to obtain

$$d\Phi = -\frac{RT}{F} \sum_i \frac{t_i}{z_i} \, d \ln(a_i)$$

We then integrate over the phase boundary to obtain

$$\Delta\Phi = -\frac{RT}{F} \sum_i \int_a^\beta \frac{t_i}{z_i} \, d \ln(a_i) \qquad (3.34)$$

Equation (3.34) is a general expression for the liquid junction potential, but the integral is generally intractable. To proceed further, we must build some simplifying assumptions into the model.

When only one electrolyte is present and the two phases differ only in concentration, then it is reasonable to assume that the transference numbers are constant. When these are factored out, the integrals of eq (3.34) are easy and we obtain for a 1:1 electrolyte ($z_1 = 1$, $z_2 = -1$)

$$\Delta\Phi = -\frac{RT}{F}\left[t_1 \ln\frac{a_1^{\beta}}{a_1^{a}} - t_2 \ln\frac{a_2^{\beta}}{a_2^{a}}\right]$$

or if we further assume that $a_1^{a} = a_2^{a} = a^{a}$ and $a_1^{\beta} = a_2^{\beta} = a^{\beta}$,

$$\Delta\Phi = \frac{RT}{F}(t_1 - t_2)\ln\frac{a^{a}}{a^{\beta}} \qquad (3.35)$$

Thus we predict that when two solutions, differing only in the concentration of a 1:1 electrolyte, are in contact the liquid junction potential is proportional to the difference in transference numbers of the positive and negative ions and to the log of the activity ratio.

Example 3.12 Compute the liquid junctions potentials for the contact of 0.1 and 0.01 M KCl solutions and for the contact of 0.1 and 0.01 M HCl solutions neglecting activity coefficient differences and assuming that the transference numbers are independent of concentration.

Computing the transference numbers from ionic conductivities (Example 3.5), we have, for KCl, $t_1 = 0.491$ and $t_2 = 0.509$, so that substitution in eq (3.35) gives ($RT/F = 25.7$ mV)

$$\Delta\Phi = (25.7 \text{ mV})(-0.018)\ln(0.1) = 1.1 \text{ mV}$$

For HCl, $t_1 = 0.821$ and $t_2 = 0.179$, so that

$$\Delta\Phi = (25.7 \text{ mV})(0.642)\ln(0.1) = -38.0 \text{ mV}$$

To deal with the more generally interesting case where mixtures of several electrolytes are present, we must make an approximation to allow evaluation of the transference numbers. One approach to this problem, suggested by Henderson in 1907 (16), is to assume a linear concentration gradient for each ionic species. Thus if we define a dimensionless distance parameter x, scaled so that x ranges from 0 in phase a to 1 in phase β, the concentration of ion i can be written

$$C_i(x) = C_i^{\beta}x + C_i^{a}(1 - x)$$

We will also assume that the activity coefficients are constant in the phase boundary region, so that we can write

$$\frac{\partial \ln(a_i)}{\partial x} = \frac{\partial \ln(C_i)}{\partial x}$$

Thus we have

$$d \ln(a_i) = \frac{\partial \ln(a_i)}{\partial x} dx = \frac{1}{C_i} \frac{\partial C_i}{\partial x} dx$$

or since $\partial C_i / \partial x = C_i^\beta - C_i^a$,

$$d \ln(a_i) = \frac{C_i^\beta - C_i^a}{C_i(x)} dx$$

The transference number of ion i can be written in terms of the concentrations and mobilities:

$$t_i = \frac{|z_i| C_i(x) u_i}{\sum\limits_j |z_j| C_j(x) u_j}$$

where we assume that the mobilities are independent of concentration and thus of x. Substituting the expressions for t_i and $d \ln(a_i)$ into eq (3.34), we obtain

$$\Delta\Phi = - \frac{RT}{F} \sum\limits_i \frac{|z_i| u_i}{z_i} (C_i^\beta - C_i^a) \int\limits_0^1 \frac{dx}{a + bx}$$

where

$$a = \sum\limits_j |z_j| u_j C_j^a$$

and

$$b = \sum\limits_j |z_j| u_j (C_j^\beta - C_j^a)$$

The integral is readily evaluated:

$$\int_0^1 \frac{dx}{a + bx} = \frac{\ln(a + bx)}{b} \Bigg|_0^1 = \frac{1}{b} \ln \frac{a + b}{a}$$

Thus

$$\Delta\Phi = -\frac{RT}{F} \frac{\sum (|z_i|/z_i)u_i(C_i^\beta - C_i^a)}{\sum |z_i| u_i(C_i^\beta - C_i^a)} \ln \frac{\sum |z_i| u_i C_i^\beta}{\sum |z_i| u_i C_i^a} \tag{3.36}$$

Equation (3.36), sometimes called the Henderson equation, is a general (albeit somewhat approximate) result for liquid junction potentials. When a only a single 1:1 electrolyte is present, eq (3.36) reduces to eq (3.35). Another relatively simple situation can be set up experimentally. Suppose that only one cation is present and that it has the same concentration in both phases (*e.g.*, NaCl in one solution, NaBr in the other). The prelogarithmic term in eq (3.36) then reduces to RT/F. Alternatively, if a single common anion is present, the prelogarithmic term reduced to $-RT/F$. Thus we have

$$\Delta\Phi = \pm \frac{RT}{F} \ln \frac{\sum |z_i| u_i C_i^a}{\sum |z_i| u_i C_i^\beta} \tag{3.37}$$

Example 3.13 Compute the liquid junction potential for the contact of 0.1 M HCl and 0.1 M KCl solutions.

Equation (3.37) can be used in this case since we have a common anion; since the concentrations are all 0.1 M, we have

$$\Delta\Phi = -\frac{RT}{F} \ln \frac{u(H^+) + u(Cl^-)}{u(K^+) + u(Cl^-)}$$

Using mobilities calculated as in Example 3.6 (and assuming these quantities to be independent of concentration), we have

$$\Delta\Phi = -(25.7 \text{ mV}) \ln \frac{36.3 + 7.9}{7.7 + 7.9} = -26.8 \text{ mV}$$

Example 3.14 Compute the liquid junction potential for 0.1 M HCl
and 0.1 M KCl solutions contacted through a saturated KCl (4.2 M)
salt bridge.

In this case, we must deal separately with two liquid junctions, one
between the HCl solution and the salt bridge, the other between the
salt bridge and the KCl solution. In the first case, we must use eq
(3.36), but eq (3.35) will handle the second junction. Starting with
the simpler case, we can use the method of Example 3.12 to obtain
for the salt bridge/KCl junction:

$$\Delta\Phi = (25.7 \text{ mV})(-0.018) \ln(42) = -1.7 \text{ mV}$$

Inserting the mobilities (computed as above from the infinite dilution
conductivities) and concentrations into eq (3.36), we obtain for the
HCl/salt bridge junction:

$$\Delta\Phi = +4.2 \text{ mV}$$

for a total liquid junction potential of $+2.9$ mV.

Comparing the results of Examples 3.13 and 3.14, we see that the use
of a saturated KCl salt bridge is expected to reduce the junction
potential for 0.1 M KCl and HCl solutions by nearly a factor of 10.
The secret, of course, is that the K^+ and Cl^- ions, which have very
nearly equal mobilities, carry most of the current at the salt
bridge/solution junctions. The result is generally applicable, however,
and explains in part the popularity of KCl as an electrolyte,
particularly when liquid junctions are unavoidable.

Donnan Membrane Equilibria

In a system containing a semipermeable membrane, an interphase
potential superficially like a liquid junction potential may develop.
Suppose that two solutions, one containing an ordinary electrolyte with
small ions and the other containing a solution of charged
macromolecules and small counterions, are separated by a membrane
which allows the free passage of solvent molecules and small ions but
which is impermeable to the macromolecules. At equilibrium, the

chemical potentials of the small ions and solvent[‡] will be equal in the two phases. However, since the condition of electrical neutrality requires that the ionic concentrations be unequal in the two solutions, an electric potential difference must develop across the membrane. Unlike a liquid junction potential, however, the potential in this case does not depend on the relative mobilities of the ions and the system is in a true equilibrium state. The development of potentials across semipermeable membranes was first studied by Donnan[‡] in 1911 *(17)* and the phenomenon is usually referred to with his name.

Consider a 1:1 electrolyte in equilibrium across the membrane; equilibrium requires that the ionic chemical potentials be equal:

$$\mu_i{}^{\alpha} = \mu_i{}^{\beta}$$

or, since $\mu_i{}^{\circ}$ is the same for the α and β phases,

$$RT \ln(a_i{}^{\alpha}) = RT \ln(a_i{}^{\beta}) \pm F\Delta\Phi$$

where the upper sign applies to cations, the lower sign to anions and $\Delta\Phi = \Phi_{\beta} - \Phi_{\alpha}$. If the cation (i = 1) and anion (i = 2) equations are added, we get

$$RT \ln(a_1{}^{\alpha} a_2{}^{\alpha}) = RT \ln(a_1{}^{\beta} a_2{}^{\beta})$$

or

$$a_1{}^{\alpha} a_2{}^{\alpha} = a_1{}^{\beta} a_2{}^{\beta}$$

which may be regarded as the condition of Donnan membrane equilibrium. If the activity coefficients in the α- and β-phases are equal, we can also write

---

[‡]   The equalization of the solvent chemical potential leads to the phenomenon of osmotic pressure.

[‡]   Frederick G. Donnan (1870–1956) was Professor of Physical Chemistry at the University of Liverpool and later at University College, London. Donnan was a student of Ostwald and played an important role in introducing the then new ideas of physical chemistry in Britain. He is best remembered for his work on membrane equilibria.

$$C_1{}^a C_2{}^a = C_1{}^\beta C_2{}^\beta \tag{3.38}$$

Electroneutrality of the a phase requires:

$$C_1{}^a - C_2{}^a = 0$$

If the β–phase also contains a macromolecule M with concentration C_M and charge z_M, then the electroneutrality condition is

$$C_1{}^\beta - C_2{}^\beta + z_M C_M = 0$$

In general there will also be conservation relations governing the total number of moles of anions and cations. These relations, together with the Donnan equilibrium expression, eq (3.38), can be used to compute the individual concentrations. Since the activity of an ion is different in the two phases, a potential difference develops:

$$\Delta\Phi = \frac{RT}{F} \ln \frac{a_1{}^a}{a_1{}^\beta} = -\frac{RT}{F} \ln \frac{a_2{}^a}{C_2{}^\beta} \tag{3.39}$$

Example 3.15   Compute the Donnan membrane potential for a system initially containing 100 mL of 0.10 M KCl (phase a) and 100 mL of 10^{-3} M macromolecules with $z_M = 100$ with 0.10 M Cl⁻ counterions (phase β).

Electroneutrality requires:

$$C_1{}^a = C_2{}^a \qquad C_2{}^\beta = 0.1 + C_1{}^\beta$$

and conservation of sodium ions requires:

$$C_1{}^a + C_1{}^\beta = 0.1$$

Substitution in eq (3.38) gives the quadratic equation

$$C_1{}^\beta (0.1 + C_1{}^\beta) = (0.1 - C_1{}^\beta)^2$$

the solution to which is

$$C_1{}^\beta = 0.033 \text{ M}$$

Thus

$$C_1{}^a = C_2{}^a = 0.067 \text{ M}$$

$$C_2{}^\beta = 0.133 \text{ M}$$

and

$$\Delta\Phi = (25.7 \text{ mV})\ln(2) = 17.8 \text{ mV}$$

Ion–Selective Membranes

Now that we have semiquantitative theories of the liquid junction potential and the Donnan membrane potential, we can turn to the operation of an ion–selective membrane such as the glass membrane of a H^+–sensitive glass electrode which was discussed qualitatively in Section 1.5.

We assume that the interior of the membrane is permeable only to Na^+ ions, but the inner and outer surfaces of the membrane are hydrated and can contain other cations, in particular H^+. We can divide the system into five regions which we imagine as separate phases:

a – the outer electrolyte solution
β – the outer surface of the membrane
γ – the interior of the membrane
β' – the inner surface of the membrane
a' – the inner electrolyte solution

The total membrane potential then has four contributions:

$$\Delta\Phi = \Delta\Phi_{a\beta} + \Delta\Phi_{\beta\gamma} + \Delta\Phi_{\gamma\beta'} + \Delta\Phi_{\beta'a'} \tag{3.40}$$

We will assume that the membrane surfaces are in equilibrium with the adjacent electrolyte solutions so that the arguments used in discussing Donnan membrane equilibria can be used for the $a-\beta$ and $\beta'-a'$ interfaces. If H^+ is in equilibrium across the $a-\beta$ interface, then the H^+–chemical potentials must be equal for the two phases:

$$\mu_H{}^\circ(a) + RT \ln(a_H{}^a) + F\Phi_a$$

$$= \mu_H{}^\circ(\beta) + RT \ln(a_H{}^\beta) + F\Phi_\beta \qquad (3.41)$$

or

$$\Delta\Phi_{a\beta} = \Phi_\beta - \Phi_a = \frac{\mu_H{}^\circ(a) - \mu_H{}^\circ(\beta)}{F} + \frac{RT}{F} \ln \frac{a_H{}^a}{a_H{}^\beta}$$

Similarly for the $\beta'-a'$ interface

$$\Delta\Phi_{\beta'a'} = \frac{\mu_H{}^\circ(\beta') - \mu_H{}^\circ(a')}{F} + \frac{RT}{F} \ln \frac{a_H{}^{\beta'}}{a_H{}^{a'}}$$

Since the a and a' phases are both aqueous solutions and the β and β' phases are both hydrated glass,

$$\mu_H{}^\circ(a) = \mu_H{}^\circ(a') \text{ and } \mu_H{}^\circ(\beta) = \mu_H{}^\circ(\beta')$$

Thus the standard state terms cancel when $\Delta\Phi_{a\beta}$ and $\Delta\Phi_{\beta'a'}$ are added to obtain the solution/membrane surface contributions to the membrane potential:

$$\Delta\Phi_{a\beta} + \Delta\Phi_{\beta'a'} = \frac{RT}{F} \ln \frac{a_H{}^a a_H{}^{\beta'}}{a_H{}^\beta a_H{}^{a'}} \qquad (3.42)$$

On the other hand, the Na^+ ions are also in equilibrium between the outer membrane regions and the solutions. Thus for the $a-\beta$ interface,

$$\mu_{Na}{}^\circ(a) + RT \ln(a_{Na}{}^a) + F\Phi_a$$

$$= \mu_{Na}{}^\circ(\beta) + RT \ln(a_{Na}{}^\beta) + F\Phi_\beta \qquad (3.43)$$

Subtracting eq (3.43) from eq (3.41) and rearranging, we have

$$\mu_H{}^\circ(a) + \mu_{Na}{}^\circ(\beta) - \mu_H{}^\circ(\beta) - \mu_{Na}{}^\circ(a) = RT \ln \frac{a_H{}^\beta a_{Na}{}^a}{a_H{}^a a_{Na}{}^\beta}$$

but this is just the standard free energy change for the equilibrium

$$Na^+(a) + H^+(\beta) \rightleftharpoons Na^+(\beta) + H^+(a)$$

so that the argument of the logarithm term is just an equilibrium constant expression

$$K = \frac{a_H{}^a \, a_{Na}{}^\beta}{a_H{}^\beta \, a_{Na}{}^a} \tag{3.44}$$

By a similar argument, we can show that H^+ and Na^+ ions at the β'–a' interface are subject to the same equilibrium constant.

$$K = \frac{a_H{}^{a'} \, a_{Na}{}^{\beta'}}{a_H{}^{\beta'} \, a_{Na}{}^{a'}} \tag{3.45}$$

We turn now to the interfaces between the surfaces and interior of the membrane. The potential across the β–γ (or γ–β') interface results from the difference in the mobilities of the H^+ and Na^+ ions and should be given by the Henderson equation, eq (3.36). Since the mobilities of the anions inside the membrane are assumed to be zero, the prelogarithmic term simplifies, and we are left with

$$\Delta\Phi_{\beta\gamma} = \frac{RT}{F} \ln \frac{u_H a_H{}^\beta + u_{Na} a_{Na}{}^\beta}{u_{Na} a_{Na}{}^\gamma}$$

where we have assumed that the activity of H^+ ions in the interior of the membrane is zero and that the mobility of Na^+ ions is the same in the interior as at the surface of the membrane. Similarly,

$$\Delta\Phi_{\gamma\beta'} = \frac{RT}{F} \ln \frac{u_{Na} a_{Na}{}^\gamma}{u_H a_H{}^{\beta'} + u_{Na} a_{Na}{}^{\beta'}}$$

Thus the contribution to the membrane potential from the differential mobilities of the H^+ and Na^+ ions inside the membrane is

$$\Delta\Phi_{\beta\gamma} + \Delta\Phi_{\gamma\beta'} = \frac{RT}{F} \ln \frac{(u_{Na}/u_H) a_{Na}{}^\beta + a_H{}^\beta}{(u_{Na}/u_H) a_{Na}{}^{\beta'} + a_H{}^{\beta'}}$$

Adding this potential difference to the contributions from the solution–membrane surface interfaces, eq (3.42), we get the total membrane potential,

$$\Delta\Phi_{membrane} = \frac{RT}{F} \ln \frac{(u_{Na}/u_H)Ka_{Na}{}^a + a_H{}^a}{(u_{Na}/u_H)Ka_{Na}{}^{a'} + a_H{}^{a'}}, \qquad (3.46)$$

where we have used eqs (3.44) and (3.45) to eliminate the activities of the H^+ and Na^+ ions in the β- and β'-phases. We can define a parameter called the *potentiometric selectivity coefficient,*

$$k_{H,Na} = K(u_{Na}/u_H) \qquad (3.47)$$

which is a function of properties of the membrane—the Donnan equilibrium constant and the relative mobilities of Na^+ and H^+. If the inner electrolyte solution (the a'-phase) is constant in composition, the denominator of the log term in eq (3.46) makes a constant additive contribution to the membrane potential. Incorporating $k_{H,Na}$ to simplify the expression further, eq (3.46) becomes

$$\Delta\Phi_{membrane} = constant + \frac{RT}{F} \ln[a_H{}^a + k_{H,Na}a_{Na}{}^a] \quad (3.48)$$

If the potentiometric selectivity coefficient were zero, then the membrane would respond exclusively to the H^+ activity; the selectivity coefficient thus is a measure of the interference of Na^+ in the operation of the glass electrode. Although eq (3.48) was derived for the specific case of a glass membrane, models of other membrane systems lead to the same general result *(D11,D14),* which can be expressed by

$$\Delta\Phi_{membrane} = constant + \frac{RT}{F} \ln[a_i + \sum_j k_{ij}a_j] \qquad (3.49)$$

where species i is the principal diffusing species of interest and k_{ij} is the selectivity coefficient for j relative to i.

REFERENCES

(Reference numbers preceded by a letter, *e.g. (C3)*, refer to a book listed in the Bibliography.)

1. T. Shedlovsky and L. Shedlovsky in *Physical Methods of Chemistry*, Part IIA (*Techniques of Chemistry*, Vol. I), A. Weisberger and B. W. Rossiter, eds, New York: Wiley, 1971, p 163.
2. F. J. Holler and C. G. Enke in *Laboratory Techniques in Electroanalytical Chemistry*, P. T. Kissinger and W. R. Heineman, eds, New York: Marcel Dekker, 1984, p 235.
3. P. Erman, *Ann. Phys.* **1801,** *8,* 197; **1802,** *10,* 1.
4. F. Kohlrausch, *Wiss. Abh. Phys. Tech. Reichsanstalt* **1900,** *3,* 155.
5. M. Spiro in *Physical Methods of Chemistry*, Part IIA (*Techniques of Chemistry*, Vol. I), A. Weissberger and B. W. Rossiter, eds, New York: Wiley, 1971, p 205.
6. J. W. Hittorf, *Ann. Phys. Chem.* **1853,** *89, 177.*
7. R. D. Shannon, *Acta Crystallogr.* **1976,** *A32,* 751.
8. L. Onsager, *Phys. Z.* **1927,** *28,* 277; L. Onsager and R. M. Fuoss, *J. Phys. Chem.* **1932,** *36,* 2689.
9. P. Debye and H. Falkenhagen, *Phys. Z.* **1928,** *29,* 121, 401; *Z. Elektrochem.* **1928,** *23,* 562.
10. H. Sack, *Physik. Z.* **1928,** *29,* 627.
11. M. Wien, *Ann. Phys.* **1927,** *388,* 327; **1928,** *380,* 795; **1929,** *393,* 400.
12. P. Debye, *J. Chem. Phys.* **1933,** *1,* 13.
13. R. Zana and E. B. Yeager, *Modern Aspects of Electrochemistry* **1982,** *14,* 1.
14. F. Kohlrausch and A. Heydweiller, *Ann. Phys.* **1894,** *53,* 209.
15. J. W. Loveland, in *Treatise on Analytical Chemistry*, I. M. Kolthoff and P. J. Elving, eds, Part I, Vol. 4, New York: Interscience, 1963, p 2569.
16. P. Henderson, *Z. phys. Chem.* **1907,** *59,* 118.
17. F. G. Donnan, *Z. Elektrochem.* **1911,** *17,* 572; F. G. Donnan and E. A. Guggenheim, *Z. phys. Chem.* **1932,** *162,* 346.

PROBLEMS

3.1   (a) A conductance cell filled with 0.1 M KCl solution had a resistance of 24.96 Ω at 25°C. Calculate the cell constant (L/A) given that the conductivity of 0.1 M KCl is known to be 1.1639 S m^{-1} at this temperature.
(b) When the cell was filled with 0.01 M acetic acid, the resistance was 1982 Ω. Compute the molar conductivity of acetic acid at this concentration.
(c) Using the molar ionic conductivities of H$^+$ and CH$_3$CO$_2^-$ from Table A.7, compute $\Lambda°$ for acetic acid.
(d) Compute the degree of dissociation a for 0.01 M acetic acid and estimate the acid dissociation equilibrium constant.

3.2   The molar conductivity of AgNO$_3$, measured at 25°C for various concentrations, is as follows:

C/mM: 100Λ/S m^{2}mol^{-1}:	0.0276 1.329	0.0724 1.326	0.1071 1.325	0.3539 1.316	0.7538 1.308

Use these data to determine $\Lambda°$ for AgNO$_3$.

3.3   Compute the Onsager limiting law slope for AgNO$_3$ and compare it with the slope of the plot obtained in Problem 3.2.

3.4   Compute the Onsager limiting law slope for BaCl$_2$.

3.5   What fraction of the total current is carried by K$^+$ in a solution 0.10 M in KCl and 0.05 M in HCl?

3.6   The conductivity of a saturated silver iodate solution at 18°C is 13.0 × 10^{-4} S m^{-1} when water with a conductivity of 1.1 × 10^{-4} S m^{-1} was used to prepare the solution. The sum of the molar ionic conductances at 18°C is

$$\Lambda°(\text{Ag}^+) + \Lambda°(\text{IO}_3^-) = 87.3 \times 10^{-4} \text{ S m}^2\text{mol}^{-1}$$

What is the solubility of silver iodate at this temperature? You can assume that the species leading to the conductivity of the water used do not interfere with the dissolution of AgIO$_3$ and that the conductivity contributions are additive.

3.7   The conductivity of a 0.0384 M solution of Cl$_2$ was found to be

0.385 S m^{-1} at $0°C$. Given that the molar conductivity of HCl is 0.0250 S m^{2}mol^{-1} at $0°C$, what is the equilibrium constant for the disproportionation of Cl_2,

$$Cl_2(aq) + H_2O \rightleftarrows H^+(aq) + Cl^-(aq) + HOCl(aq)?$$

The ionization of HOCl is negligible under these conditions.

3.8   Compute the Stokes' law radii, the ionic mobilities, frictional coefficients, and diffusion coefficients of $Fe(CN)_6^{3-}$ and $Fe(CN)_6^{4-}$.

3.9   A potassium salt of an uncharacterized anion, A^{n-}, contains 12.50% potassium by weight. Conductance measurements on some solutions of varying composition gave the following data (aqueous solutions at $25°C$):

C/g L^{-1}	κ/S m^{-1}
0.0250	1.09
0.0500	2.17
0.1000	4.31
0.200	8.53
0.500	10.94

(a) Compute the "equivalent weight", MW/n, and use this value to convert the concentrations to equivalents per cubic meter $(nC/mol$ m$^{-3})$ and the conductances to "equivalent conductivities," Λ/n, with units of S m^{2}equiv^{-1}.
(b) Plot Λ/n vs. $(nC)^{\frac{1}{2}}$, determine the intercept and the slope. How does this slope differ from the slope of a plot of Λ vs. $\sqrt{C}$?
(c) Given the molar ionic conductivity of K^+, 73.5×10^{-4} S m^{2}mol^{-1}, determine the transference numbers, t_+ and t_-.
(d) Suppose that the salt is a 1:1 electrolyte (n = 1). Does the slope obtained in part (b) agree with theory? Suppose it is a 2:1 electrolyte? Is the fit better? A 3:1 or 4:1 electrolyte?

3.10 Verify that eq (3.22) is a solution to Fick's second law, eq (3.21).

3.11 Show that the diffusion coefficient of a spherical particle of radius r in a medium of viscosity η is

$$D = kT/6\pi\eta r$$

3.12 A simple calculation can show that the development of a central nervous system was essential to the evolution of large organisms. Suppose that a primitive fish swimming in a primeval sea at 25°C relied on the diffusion of information–carrying molecules to control body movement. Imagine that the information–carrying molecule was 10 nm in diameter and that diffusion was through an essentially aqueous medium, $\eta = 0.89 \times 10^{-3}$ kg m^{-1}s^{-1}. Estimate the diffusion coefficient (see Problem 3.11). Now suppose that a primitive predator bit the fish's tail. How long would it take for the information to reach the fish's head if the fish was 10 cm long?

3.13 A solute diffuses through a long cylindrical tube of 2.00 mm diameter such that during each second 0.025 nanomol crosses a plane where the concentration gradient is 5.0 M m^{-1}. Compute the diffusion coefficient.

3.14 A solution of $CuSO_4$ was electrolyzed in a Hittorf apparatus with Cu electrodes. The electrode processes were

$$\text{anode:} \quad Cu(s) \rightarrow Cu^{2+} + 2\,e^-$$

$$\text{cathode:} \quad Cu^{2+} + 2\,e^- \rightarrow Cu(s)$$

The initial concentration was 0.1473 M. After passage of 1.372×10^{-3} Faradays, the cathode compartment contained 30 mL of 0.1183 M $CuSO_4$. Calculate the transference number of Cu^{2+} in the $CuSO_4$ solution.

3.15 150 mL of a saturated solution of $Ba(OH)_2$ was titrated with 1.00 M H_2SO_4 and the resistance of the solution measured with a dip–type cell. The following data were obtained:

Vol/mL	R/Ω	Vol/mL	R/Ω
0.00	53	6.00	1080
1.00	63	6.30	2500
2.00	84	6.50	1200
3.00	100	7.00	350
4.00	140	7.50	210
5.00	250	8.00	150
5.50	400	8.50	110
5.75	560	9.00	90

Given that the cell constant is $L/A = 104 \text{ m}^{-1}$, compute the conductivity κ for each point and plot κ vs. volume of acid and locate the equivalence point. Compute the concentration of $Ba(OH)_2$ in the saturated solution. Knowing the concentration of $Ba(OH)_2$ and the conductivity of the initial solution, determine the molar conductivity of $Ba(OH)_2$.

3.16 Show that the Donnan membrane equilibrium condition for an electrolyte where the cations have charge z_1 and the anions have charge z_2 is

$$\left[\frac{C_1{}^a}{C_1{}^\beta} \right]^{|z_2|} = \left[\frac{C_2{}^\beta}{C_2{}^a} \right]^{z_1}$$

3.17 A 0.01 M solution of a colloidal electrolyte NaX is placed on one side of a membrane which is permeable to Na^+, but not to X^-. Calculate the equilibrium distribution of ions if on the other side of the membrane there is placed a 0.025 M solution of Na_2SO_4. Assume equal volumes in the two phases. Calculate the electric potential difference across the membrane at 298 K.

3.18 Sketch conductometric titration curves for the following systems. Ignore the contribution to the conductivity from weak electrolytes or insoluble precipitates.
(a) 0.01 M ammonia titrated with 1 M hydrochloric acid.
(b) 0.01 M ammonia titrated with 1 M acetic acid.
(c) 0.01 M silver nitrate titrated with 1 M hydrochloric acid.
(d) 0.01 M silver nitrate titrated with 1 M potassium chloride.

4 VOLTAMMETRY OF REVERSIBLE SYSTEMS

In Chapter 1, we considered galvanic cells in which chemical energy is converted to electrical energy. In this and the next two chapters, we will discuss the reverse process, where electricity is used to induce chemical change. Electrochemical cells through which current is passed from an external power source are called *electrolysis cells*.

In an electrolysis cell, reduction occurs at the cathode and oxidation at the anode just as in a galvanic cell. However, the cathode and anode are negative and positive, respectively, in an electrolysis cell, seemingly opposite to the convention adopted for galvanic cells. Consider the Daniell cell (discussed in Section 1.1)

$$Zn\,|\,Zn^{2+}(aq)\,\|\,Cu^{2+}(aq)\,|\,Cu$$

When operated as a galvanic cell, the cell reaction

$$Zn + Cu^{2+} \rightarrow Cu + Zn^{2+}$$

corresponds to oxidation of Zn and reduction of Cu^{2+}. The Zn electrode then is the anode and the Cu electrode is the cathode. When we measure the cell potential, we find that the Zn electrode is negative and the Cu electrode is positive. When an external potential source is attached and the direction of current flow is reversed, the relative polarities of the electrodes are unchanged—Cu is still positive and Zn is negative. However, the process taking place at the Zn electrode is now a reduction, so that this electrode is the cathode; Cu is oxidized, so that the Cu electrode is the anode. Strictly speaking, the designations anode and cathode refer to the direction of current flow. In order to avoid confusion, we will usually focus our attention on a single indicator or working electrode and will describe its role by saying that the current is anodic or cathodic, *i.e.*, the electrode process at the indicator electrode is an oxidation or a reduction. We adopt the following sign convention for current at the indicator electrode:[‡] *cathodic current is positive, anodic current is negative.*

When current flows through an electrochemical cell, it may be limited by any one of the three steps:

(1) conductance of the bulk solution;

(2) transport of reactants to the electrode or products away from the electrode; or

(3) electron transfer and associated chemical reactions.

Examples can be found where any one of these steps is rate determining. In Chapter 3, we were concerned with cells designed, so that step (1) was rate limiting. In this chapter, we consider cases where the mass transport step is slowest; we will assume that the solution conductivity is high and that electron transfer is fast and reversible and is not complicated by associated chemical reactions. We will consider cases where the current is limited by the rates of electron transfer or of coupled chemical reactions in Chapter 5. We will be concerned here with processes where the total charge passed is small, so that the bulk solution composition is essentially unchanged by electrolysis. In Chapter 6, we will consider experiments in which the bulk concentrations change during electrolysis.

We begin this chapter with a study of the general problem of diffusion–controlled current. In Section 4.2, we digress to some

---

[‡] Our sign convention for indicator electrode current is the usual one among electroanalytical chemists, but many other electrochemists employ (and the International Union of Pure and Applied Chemistry recommends) the opposite convention with positive anodic current.

practical matters, such as instrumentation and the choice of solvents. In Section 4.3, we survey the host of electroanalytical methods which rely on diffusion control. In Sections 4.4 and 4.5, we discuss polarography, historically one of the most important of the electroanalytical methods, and some of the many variations on the polarographic method. In Section 4.6, we examine the characteristics of the rotating disk electrode and finally, in Section 4.7, look at some of the applications of the various methods.

4.1 DIFFUSION–LIMITED CURRENT

There are several mechanisms for transport of ions or molecules to an electrode surface:

(1) *diffusion* (transport of ions or molecules through a chemical potential gradient);

(2) *electric migration* (transport of ions through an electric potential gradient);

(3) *convection* (transport of ions or molecules through mechanical motion of the solution). Convection may occur either through gravity operating on a density gradient (sometimes called *natural convection*) or through stirring of the solution or motion of the electrode (sometimes called *forced convection*).

In this section we will assume that diffusion is the only significant transport process and will restrict our attention to experiments in which precautions are taken to eliminate or reduce other transport mechanisms. Thus the solution should not be stirred and the cell should be free of vibration. In addition, we will require that the ratio of electrode area to solution volume be small and that experiments be relatively short in duration so that relatively little electrolysis takes place. This precaution will minimize local changes in concentration or local heating, which could lead to density gradients and convective mixing. The solution should contain a large excess of an inert electrolyte (called a *supporting electrolyte*) to ensure that the double layer is very compact and that the electric potential is nearly constant throughout the solution, thus minimizing electric migration effects.

Solution of the Diffusion Equation

Our goal is an equation giving the current as a function of electrode potential; the current will also be a function of time, the concentrations of electroactive species, the diffusion coefficients of these species, the electrode area, and the number of electrons transferred per molecule oxidized or reduced. We assume that the electrode process involves an oxidized species O and a reduced species R, both in solution, and that the process consists simply of the transfer of n electrons:

$$O + n e^- \rightleftarrows R$$

The species O and/or R may be charged but we will omit the charges for clarity. We will need to find the concentrations of O and R, $C_O(x,t)$ and $C_R(x,t)$, as functions of time and distance from the electrode. We define the problem by the following specifications:

(1) The electrode is planar and has surface area A. Only diffusion along the x–axis perpendicular to the electrode surface need be considered. The electrode is sufficiently large that edge effects can be neglected.

(2) The solution is initially homogeneous. Specifically, the initial concentrations of O and R are $C_O(x,0) = C_O{}^*$ and $C_R(x,0) = 0$ for all values of x.

(3) The electrolysis cell is sufficiently large that the bulk concentrations of O and R are unchanged from the initial values even after electrolysis has been under way for a while. In other words, $C_O(x,t) \rightarrow C_O{}^*$, $C_R(x,t) \rightarrow 0$ as $x \rightarrow \infty$.

(4) For every O molecule consumed, an R molecule is formed; in other words, the fluxes of O and R at the electrode surface are equal and opposite in sign: $J_O(0,t) = - J_R(0,t)$.

(5) The electron–transfer reaction is very fast so that O and R are always in equilibrium at the electrode surface with the concentration ratio given by the Nernst equation:‡

---

‡ Since the reactant and product concentrations at the electrode surface obey the Nernst equation, we sometimes refer to an

$$\theta = \frac{C_O(0,t)}{C_R(0,t)} = \exp \frac{nF(E - E°)}{RT} \tag{4.1}$$

Specification (1) tells us that we need to consider diffusion of O and R along the x–axis only. The diffusion problem is governed by Fick's second law, eq (3.21). Since we have two diffusing species, we must solve two diffusion equations:

$$\frac{\partial C_O}{\partial t} = D_O \frac{\partial^2 C_O}{\partial x^2}$$

$$\frac{\partial C_R}{\partial t} = D_R \frac{\partial^2 C_R}{\partial x^2}$$

Specification (2) represents the initial conditions imposed on the solutions to the differential equations, and specifications (3), (4), and (5) represent boundary conditions.

The solution of two coupled partial differential equations might seem to be a formidable task, but it is really quite easy using the method of Laplace transforms.[‡] The solutions to the differential equations are the concentrations of O and R as functions of x and t:

$$C_O(x,t) = C_O{}^* \frac{\xi\theta + \text{erf}(x/2D_O^{\frac{1}{2}}t^{\frac{1}{2}})}{1 + \xi\theta} \tag{4.2a}$$

$$C_R(x,t) = C_O{}^* \frac{\xi[1 - \text{erf}(x/2D_R^{\frac{1}{2}}t^{\frac{1}{2}})]}{1 + \xi\theta} \tag{4.2b}$$

---

electrochemically reversible system as nernstian. Strictly speaking, eq (4.1) should include the activity coefficient ratio, γ_O/γ_R, but we will ignore this complication; thus the standard potentials used here are really formal potentials—see Section 1.3.

[‡] The use of Laplace transform methods for the solution of differential equations is discussed in Appendix 4. See also Delahay (B1), Bard and Faulkner (B12), and MacDonald (F5).

where $\xi = (D_O/D_R)^{\frac{1}{2}}$ and erf(ψ) is the error function, defined by eq (3.23) and plotted *vs.* ψ in Figure 3.9. Since erf$(0) = 0$, the concentrations at the electrode surface $(x = 0)$ are given by

$$C_O(0,t) = C_O^* \; \frac{\xi \theta}{1 + \xi \theta} \tag{4.3a}$$

$$C_R(0,t) = C_O^* \; \frac{\xi}{1 + \xi \theta} \tag{4.3b}$$

so that the ratio of concentrations at the surface is equal to θ, as required by eq (4.1). The concentrations $C_O(x,t)$ and $C_R(x,t)$ are plotted *vs.* x in Figure 4.1.

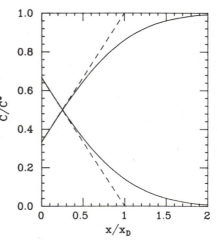

Figure 4.1 Concentration profiles for diffusion to a planar electrode with $\xi = 1$ and $\theta = 0.5$. The x–axis scale is in units of the diffusion layer thickness, $x_D = (\pi Dt)^{\frac{1}{2}}$.

Current–Potential Curves

The net current is proportional to the flux of O molecules at the electrode surface:

$$i = - nFAJ_O(0,t) \tag{4.4}$$

The flux is given by Fick's first law, eq (3.19),

$$J_O(x,t) = - D_O \; \frac{\partial C_O(x,t)}{\partial x}$$

and we can use eq (3.24) to differentiate eq (4.2a) to obtain

$$J_O(x,t) = -C_O \frac{(D_O/\pi t)^{\frac{1}{2}}}{1 + \xi\theta} \exp \frac{-x^2}{4D_O t}$$

Setting $x = 0$, we have the flux at the electrode surface

$$J_O(0,t) = -C_O^* \frac{(D_O/\pi t)^{\frac{1}{2}}}{1 + \xi\theta} \tag{4.5}$$

Substituting eq (4.5) into eq (4.4), we have

$$i = nFAC_O^* \frac{(D_O/\pi t)^{\frac{1}{2}}}{1 + \xi\theta} \tag{4.6}$$

Notice that the current is apparently infinite at zero time. This is an artifact which results from the inconsistent assumptions that $C_R(0,0) = 0$ and that $C_O(0,t)/C_R(0,t)$ is finite. We will derive a more accurate description of the current, including the finite rate of electron transfer, in Section 6.2. In practice, eq (4.6) is accurate for reversible systems when $t > 0$.

When the electrode potential is large and negative, $E \ll E°$, eq (4.1) tells us that θ and $C_O(0,t)$ are very small. In this case, every O molecule that arrives at the electrode will be reduced. Making the potential more negative cannot make the current any larger because it is limited by the rate of diffusion of O to the electrode. Setting $\theta = 0$ in eq (4.6) then gives the limiting *diffusion current:*

$$i_D = nFAC_O^*(D_O/\pi t)^{\frac{1}{2}} \tag{4.7}$$

Equation (4.7) was first derived by Cottrell[‡] in 1903 *(1)* and is commonly called the Cottrell equation.

[‡] Frederick G. Cottrell (1877–1948) is best known as the inventor of the electrostatic precipitator for removing particles from flue gases and as founder of the Research Corporation. His contributions to electrochemistry were made during his graduate work with Ostwald at the University of Leipzig.

Example 4.1 Compute the current for a one–electron reduction of a species with $C_O{}^* = 1$ mol m^{-3} (1 mM) and $D_O = 10^{-9}$ m^{2}s^{-1} at an electrode with $A = 10^{-6}$ m^2 (1 mm^2) at times 1, 10, and 100 s after application of the potential.

If we use SI units for all the quantities on the right of eq (4.7), we will get the current in amperes (A). In particular, A must be in square meters, $C_O{}^*$ must be in moles per cubic meter, D_O must be in meters squared per second, and t in seconds. Thus eq (4.7) gives

$$i_D = 1.72 \times 10^{-6}\, t^{-\frac{1}{2}} \text{ amperes}$$

so that $i_D = 1.72,\ 0.54,$ and $0.17\ \mu$A at t = 1, 10, and 100 s.

The potential dependence of the current is most clearly seen if we eliminate time and the other parameters of eq (4.6) by dividing by i_D:

$$\frac{i}{i_D} = \frac{1}{1 + \xi\theta}$$

It is important that we keep in mind that the way in which we derived this expression implies that i and i_D are measured at the same times in different experiments. Rearranging and substituting for ξ and θ, we obtain a form which bears a superficial resemblance to the Nernst equation:

$$E = E^\circ - \frac{RT}{nF} \ln \frac{D_O{}^{\frac{1}{2}}}{D_R{}^{\frac{1}{2}}} + \frac{RT}{nF} \ln \frac{i_D - i}{i} \tag{4.8}$$

A plot of i/i_D vs. E is shown in Figure 4.2. The shape of this curve led electrochemists to refer to such plots as waves. A common example of current–potential curves like that of Figure 4.2 are the polarographic waves discussed in Section 4.4, but similar curves arise in other experimental contexts. The first two terms on the right–hand side of eq (4.8) are usually lumped together to define a new parameter, called the *half–wave potential:*

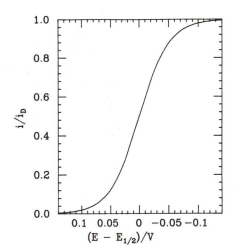

Figure 4.2 Current–potential curve for reversible electron transfer and diffusion–limited current ($n = 1$).

$$E_{\frac{1}{2}} = E^\circ - \frac{RT}{2nF} \ln \frac{D_O}{D_R} \qquad (4.9)$$

Equation (4.8) can then be written as

$$E = E_{\frac{1}{2}} + \frac{RT}{nF} \ln \frac{i_D - i}{i} \qquad (4.10)$$

Equation (4.10) was derived by Heyrovský[‡] and Ilkovič[#] in 1935 *(2)* and is called the Heyrovský–Ilkovič equation.

[‡] Jaroslav Heyrovský (1890–1967) was Professor of Chemistry at the Charles University in Prague. Heyrovský's interest in electrochemistry began during his graduate work with Donnan at University College, London. He invented polarography and was responsible for many of the developments and elaborations of the polarographic method; in recognition of this work, Heyrovský was awarded the Nobel Prize in Chemistry in 1959.

[#] Dionyz Ilkovič (1907–1980) was a student and colleague of Heyrovský in Prague and later was Professor of Physics at the Komenský University in Bratislava. Ilkovič is best known for his pioneering theoretical work in electrochemistry.

Because of the term involving the diffusion coefficients, half–wave potentials are not the same as standard (or formal) potentials, even for the electrochemically reversible processes we have been discussing. However, even if the diffusion coefficients differ by a factor of 2, $E_{\frac{1}{2}}$ will differ from $E°$ by less than 10 mV (for n = 1, 25°C) and for many practical purposes, half–wave potentials measured from current–potential curves can be used as approximations to standard potentials. Some examples of such applications are discussed in Section 4.7.

Example 4.2 Compute the electrode potential for $i = \frac{1}{3^2}\, i_D$ and the potential difference corresponding to $i = \frac{1}{4}\, i_D$ and $i = \frac{3}{4}\, i_D$.

When $i = \frac{1}{2}\, i_D$,

$$(i_D - i)/i = 1$$

so that $E = E_{\frac{1}{2}}$, hence the name "half–wave potential."

When $i = \frac{1}{4}\, i_D$ and $\frac{3}{4}\, i_D$,

$$(i_D - i)/i = 3 \quad \text{and} \quad (i_D - i)/i = \frac{1}{3}$$

Thus

$$E_{\frac{1}{4}} - E_{\frac{3}{4}} = \frac{RT}{nF}\, \ln 9$$

or, at 25°C,

$$E_{\frac{1}{4}} - E_{\frac{3}{4}} = 56.5/n \ \text{mV}$$

This property of a reversible wave was proposed in 1937 by Tomeš *(3)* as a quick indication of nernstian behavior and is called the Tomeš criterion of reversibility.

Diffusion Layer Thickness

It is sometimes convenient to think in terms of a layer near the electrode surface in which the solution is depleted of electroactive material. We define the thickness of this depleted layer, x_D, as the distance over which a linear concentration gradient would produce the same flux at the electrode surface as calculated from an exact solution to the diffusion equation.[‡] The flux through a linear concentration gradient is

$$J_O(x,t) = -D_O \frac{\partial C_O(x,t)}{\partial x}$$

$$J_O(0) = -D_O \frac{C_O{}^* - C_O(0,t)}{x_D} \qquad (4.11)$$

Substituting for $C_O(0,t)$ from eq (4.3a), we have

$$J_O(0) = -C_O{}^* \frac{D_O/x_D}{1 + \xi\theta}$$

Comparing this expression with the flux obtained from the exact solution, eq (4.5), we obtain

$$x_D = (\pi D_O t)^{\frac{1}{2}} \qquad (4.12)$$

Note that the thickness of the diffusion layer increases as $t^{\frac{1}{2}}$ for diffusion to a planar electrode in unstirred solution.

The abscissa of the concentration *vs.* distance plots of Figure 4.1 is in units of x_D. Notice that the concentrations are indeed linear in x/x_D for a short distance from the electrode surface. The hypothetical linear concentration gradients (the dashed lines in Figure 4.1) correspond to extrapolation of this linear region to $x = x_D$. We will find it convenient to use the concept of diffusion layer thickness and to assume a linear concentration gradient in several problems discussed below. In effect, we are then assuming that the concentration *vs.* distance profile follows the dashed lines of Figure 4.1 and that $C_O(x) = C_O{}^*$ and $C_R(x) = 0$ for $x > x_D$.

[‡] This approach to diffusion problems is due originally to Nernst and we sometimes refer to the *Nernst diffusion layer*.

The delivery of a reactant to the electrode may be thought of as a first–order heterogeneous process governed by a mass transport rate constant k_{DO}:

$$O^* \underset{k_{DO}}{\overset{k_{DO}}{\rightleftarrows}} O_0$$

where the net rate of delivery of O to the surface is

$$\text{Rate} = k_{DO}C_O{}^* - k_{DO}C_O(0)$$

The units of this rate must be $\text{mol s}^{-1}\text{m}^{-2}$. If the concentrations have units of mol m^{-3}, we see that the rate constant k_{DO} must have units of m s^{-1}, the SI unit for a first–order heterogeneous rate constant. To put it another way, the rate we are describing is just (the negative of) the flux of O at the surface,

$$J_O(0) = - k_{DO}[C_O{}^* - C_O(0)]$$

Comparing this with eq (4.11), we get the relation between the mass transport rate constant and the diffusion layer thickness

$$k_{DO} = D_O/x_{DO} \qquad (4.13)$$

It is sometimes convenient to write eq (4.7) in terms of x_{DO} or k_{DO}:

$$i_D = nFAC_O{}^*D_O/x_{DO} \qquad (4.14a)$$

$$i_D = nFAk_{DO}C_O{}^* \qquad (4.14b)$$

Example 4.3 Compute x_D and k_{DO} for $D_O = 10^{-9} \text{ m}^2\text{s}^{-1}$ at t = 1, 10, and 100 s.

Substituting D_O into eq (4.12), we have

$$x_D/m = 5.60 \times 10^{-5} \, t^{\frac{1}{2}}$$

so that x_D = 0.056, 0.177, and 0.560 mm at t = 1, 10, and 100 s. Substituting these values in eq (4.13), we find k_{DO} = 17.9, 5.6, and 1.8 μm s^{-1} at the same times.

General Current–Potential Curve

The generality of eq (4.10) is restricted by our assumption that the bulk concentration of R is zero. The general problem with both O and R present is easily solved using Laplace transforms. To illustrate the use of the diffusion layer thickness, we will follow a less rigorous route which nonetheless gives the correct result.

When the electrode potential is sufficiently positive that every R molecule which arrives at the electrode surface is immediately oxidized, the current is again diffusion limited and, by analogy with eq (4.14a), should be given by

$$i_{Da} = -nFAC_R{}^*D_R/x_{DR} \tag{4.15}$$

where x_{DR} is the diffusion layer thickness for R and the negative sign is required to preserve the sign convention. In general, the current is proportional to the fluxes $J_O(0,t)$ or $J_R(0,t)$,

$$i = -nFAJ_O(0,t) = nFAJ_R(0,t)$$

and the fluxes can be written in terms of the hypothetical linear concentration gradients:

$$J_O(0,t) = -D_O \frac{C_O{}^* - C_O(0,t)}{x_{DO}}$$

$$J_R(0,t) = -D_R \frac{C_R{}^* - C_R(0,t)}{x_{DR}}$$

Solving for $C_O(0,t)$ and for $C_R(0,t)$, we have

$$C_O(0,t) = C_O{}^* - (x_{DO}/D_O)(i/nFA)$$

$$C_R(0,t) = C_R{}^* + (x_{DR}/D_R)(i/nFA)$$

Solving for $C_O{}^*$ and $C_R{}^*$ from eqs (4.14a) and (4.15) and substituting into the expressions for the surface concentrations, we have

$$C_O(0,t) = \frac{x_{DO}}{D_O} \frac{i_{Dc} - i}{nFA}$$

$$C_R(0,t) = \frac{x_{DR}}{D_R} \frac{i - i_{Da}}{nFA}$$

If the Nernst equation is obeyed at the electrode surface, we have

$$E = E° + \frac{RT}{nF} \ln \frac{C_O(0,t)}{C_R(0,t)}$$

Substituting the expressions for the surface concentrations gives

$$E = E° + \frac{RT}{nF} \ln \frac{x_{DO}D_R}{x_{DR}D_O} + \frac{RT}{nF} \ln \frac{i_{Dc} - i}{i - i_{Da}}$$

or

$$E = E_{\frac{1}{2}} + \frac{RT}{nF} \ln \frac{i_{Dc} - i}{i - i_{Da}} \tag{4.16}$$

where $E_{\frac{1}{2}}$ is again given by eq (4.9). The current–potential curve represented by eq (4.16) is shown in Figure 4.3. Notice that this curve is identical in shape to that derived from eq (4.10) and shown in Figure 4.2. Indeed, the only difference is a displacement of the curve along the current axis.

When the solution contains two reducible species, O_1 and O_2, we can usually assume that they diffuse independently and that the electron–transfer processes do not interfere. If one of the components of a couple is adsorbed or deposited on the electrode surface, this may not be a good assumption, but ordinarily the contribution to the current from each process can be computed from eq (4.16) independently and added to get the total current. If the half-wave potentials are separated by more than about 150/n mV, the composite current–potential curve will show two resolved waves. Figure 4.4 shows some composite voltammetric curves computed for $C_{O1}{}^* = C_{O2}{}^*$, $C_{R1}{}^* = C_{R2}{}^* = 0$, $n_1 = n_2 = 1$, and $E_{\frac{1}{2}}(2) - E_{\frac{1}{2}}(1) = 100, 200, 300,$ and 400 mV.

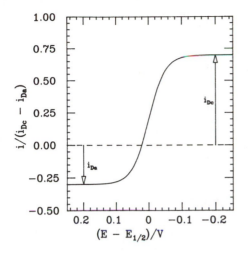

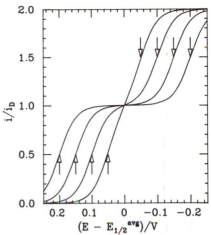

Figure 4.3 Current–potential curve for a one–electron process where $C_O{}^*/C_R{}^* = \frac{7}{3}$.

Figure 4.4 Current–potential curves for a two–component system showing the effect of the separation of half–wave potentials (indicated by arrows).

There is one notable case which is not covered by the general result, eq (4.16). When one member of the electrode couple is a solid coating the electrode, its activity at the electrode surface will be independent of potential; if the solid is pure, its activity is 1. The most obvious example of this situation is the reduction of a metal ion at an electrode of the same metal, *e.g.*, Hg^{2+} reduced at a mercury cathode. In this case the analog of eq (4.16) is

$$E = E° + \frac{RT}{nF}\ln(C_O{}^*) + \frac{RT}{nF}\ln\frac{i_D - i}{i_D} \qquad (4.17)$$

When $i = 0$, the last term of eq (4.17) vanishes and the Nernst equation is recovered. Since $C_O{}^*$ in this equation is an approximation to the activity, its units must be mol L^{-1}. The current–potential curve represented by eq (4.17) is shown in Figure 4.5. Notice that the anodic current increases without limit as the potential becomes positive.

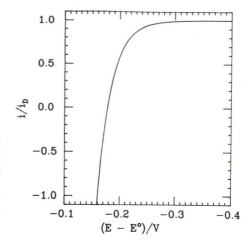

Figure 4.5 Current–potential curve for reversible electron transfer where transport of O is diffusion limited but R is a solid on the electrode surface, $C_O^* = 1$ mM.

4.2 EXPERIMENTAL TECHNIQUES

Measurement and Control of Potential

Before World War II, voltammetry experiments were done with two electrodes, an indicator electrode, often the dropping mercury electrode, and a reference electrode, often the saturated calomel electrode. In Heyrovský's polarography experiments (see Section 4.4), a d.c. ramp voltage was applied across the cell from a motor–driven potentiometer which was calibrated with a standard cell. The current was recorded as the voltage drop across a measuring resistor. Heyrovský recorded the current with a damped galvanometer[‡] which reflected a light beam onto a piece of photographic paper mounted on a drum which was rotated by the potentiometer motor. This device was later replaced by a strip–chart recorder. The two–electrode system works reasonably well for aqueous polarography where the solution resistance is low and the current is small. When nonaqueous systems are used or the experiment generates a larger current, iR drop in the solution and polarization of the reference electrode become serious

---

[‡] Because of the slow response time of the damped galvanometer, this instrument recorded the average polarographic current. Thus many early treatments of polarography discuss average currents rather than the current at drop fall.

problems.

Beginning in the 1950's, electrochemical experiments which require control of the indicator electrode potential have been done with three electrodes and instrumentation built around a *potentiostat*. A common design based on an operational amplifier is shown in Figure 4.6. The function of the reference electrode in a two–electrode system is divided in two. A true reference electrode is used to measure the potential and an auxiliary electrode is used to pass the bulk of the current. A voltage follower (see Section 1.4) is attached to the reference electrode so that very little current is drawn in this part of

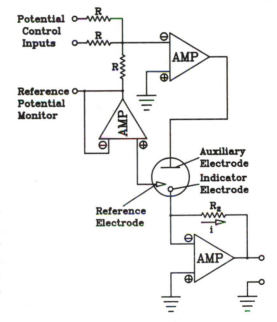

Figure 4.6 Potentiostat circuit for control of potential in a three-electrode cell. Cell current is monitored with a current-follower circuit.

the circuit. A control potential, equal and opposite to the desired potential of the reference electrode, is added to the actual reference electrode potential and the sum applied to the negative (inverting) input of the operational amplifier. The positive input is at ground potential (0 V) and, as we will see, the indicator electrode is also at 0 V. Suppose that we want the indicator electrode to have a potential –1.00 V relative to the reference. Since the indicator electrode is at ground potential, the reference electrode should be at +1.00 V; thus the control potential should be –1.00 V. If the actual reference electrode potential is 0.99 V, the potential at the inverting input will then be –0.01 V and the output will be large and positive. A large current is then passed through the cell, increasing the potential of the solution relative to the indicator electrode and thus increasing the reference electrode potential. As the reference potential approaches 1.00 V, the input to the amplifier approaches zero and the output becomes less positive. At equilibrium, the output voltage applied to the auxiliary electrode will produce enough current to polarize the cell such that the reference electrode potential exactly cancels the control potential.[‡]

The current is measured by means of another operational amplifier circuit attached to the indicator electrode lead. Since the input impedance of the operational amplifier is high, the cell current must flow through resistor R_2. However, the high gain of the amplifier produces an output voltage such that the potential at the inverting input is 0 V. Thus the indicator electrode is held at virtual ground and the output potential, $-iR_2$, is proportional to the cell current.

If the reference electrode is to be held at constant potential, a simple battery could be used to supply the control voltage. Many experiments require a potential which changes linearly with time. Such a linear ramp voltage is most commonly supplied by yet another operational amplifier circuit, which acts as a voltage integrator. This circuit, shown in Figure 4.7, again relies on a feedback loop which

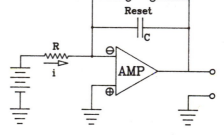

Figure 4.7 An operational amplifier integrator circuit provides a voltage ramp signal.

keeps the potential at the inverting input at virtual ground. Current from the battery flows through the resistor R. Because the amplifier input impedance is high, this current charges the capacitor C,

[‡] See Roe (4) for further details and more advanced designs.

$$i = \frac{dQ}{dt}$$

If the input voltage, Φ_{in}, is positive, the output will be negative and proportional to the charge on the capacitor:

$$\Phi_{out} = -Q/C$$

Thus if $Q = 0$ at $t = 0$,

$$\Phi_{out} = -\frac{1}{C} \int_{0}^{t} i\, dt$$

or, since $i = \Phi_{in}/R$,

$$\Phi_{out} = -\frac{\Phi_{in}t}{RC}$$

Thus for a positive input, the output voltage is a negative–going ramp. If, for example, $R = 10$ MΩ, $C = 10$ μF, and $\Phi_{in} = 100$ mV, then $\Phi_{in}/RC = 1$ mV s^{-1}. The output of the ramp generator is applied to one of the inputs of the potentiostat along with a constant potential source which sets the initial reference electrode potential.

Still other control potentials can be added to program the reference electrode potential as required. Thus a triangular wave potential is used for cyclic voltammetry (Section 4.3), a sinusoidal a.c. potential for a.c. polarography (Section 5.4), and various kinds of voltage pulses for pulse polarography (Section 4.5).

The three–electrode configuration avoids the problem of reference electrode polarization and somewhat reduces the measured iR drop in the solution. One way of further reducing solution iR drop is to locate the reference electrode junction as close as possible to the indicator electrode, often through use of a salt bridge with a long curved capillary tip as shown in Figure 4.11b (a Luggin probe).[‡] Even then, iR drop often remains a problem and most commercial electrochemical

––––––––––––

[‡] Named after the glassblower who constructed the prototype.

instruments contain a circuit which partially compensates for iR drop by electrically subtracting an adjustable voltage, proportional to to the current, from the measured cell potential.

Control of Current

In some experiments (*e.g.*, chronopotentiometry; see Section 4.3), a constant current is passed through the cell and the potential of the indicator electrode is monitored using a reference electrode. An operational amplifier circuit which performs this function (a *galvanostat* circuit) is shown in Figure 4.8. Since the operational amplifier input impedance is high, the cell current must flow through the resistor R. Feedback through the cell ensures that the potential at the inverting input is at virtual ground. Thus the potential drop across the resistor is equal to the battery potential and the cell current is

$$i = \Delta\Phi/R$$

With the battery polarity as shown in Figure 4.8, the indicator electrode is the cathode; reversal of polarity would change the direction of current flow. In this circuit, the ultimate source of current is the battery; the function of the operational amplifier is to adjust the potential of the auxiliary electrode to keep the current constant. In experiments where the current is to be controlled as a function of time, the battery can be replaced by a potential program circuit as described above.

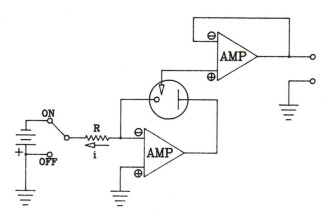

Figure 4.8 Galvanostat circuit for control of current through a cell.

Choice of Solvent

There is no single solvent which is ideal for all electrochemical work. The choice is often dictated by the solubility and reactivity of the materials to be studied. In organic and organometallic work, strongly basic anions or radical anions are often produced which are rapidly protonated by solvents like water or alcohols. The ubiquity of water as an impurity in other solvents makes the ease of solvent purification and drying an important consideration.[‡] Several common solvents are listed in Table A.8, together with some other parameters which merit attention:

Liquid range. Generally, the greater the liquid range, the greater the flexibility in experiments. It is often useful to examine electrochemical behavior at low temperature if follow–up chemical reactions are fast at room temperature or at higher temperature if electron transfer is slow.

Vapor pressure. Since oxygen interferes with most electrochemical experiments, the apparatus is usually flushed with nitrogen or argon. The purge gas may be presaturated with solvent, but when the vapor pressure is high (*e.g.*, CH_2Cl_2) it is often difficult to maintain constant composition and constant temperature because of solvent evaporation.

Dielectric constant. To obtain conducting electrolyte solutions, the solvent dielectric constant should be large. With a dielectric constant of only 2.2, p–dioxane is a poor electrochemical solvent and is usually used with 5–25% water added. DME, THF, and CH_2Cl_2 are only marginally acceptable; solution iR drop is usually a serious problem in these solvents.

Solvent viscosity. Solvents of low viscosity will generally give electrolyte solutions with greater conductivities. On the other hand, because diffusion is faster in such cases, diffusion–controlled chemical reactions will also be faster. Thus it is often found that an electrode process which is irreversible (because of a fast following chemical reaction) in a nonviscous solvent becomes reversible in a more viscous medium.

[‡] See Mann *(5)* or Fry and Britton *(6)* for further details.

Supporting Electrolytes

Voltammetry and related electrochemical techniques require an excess of inert electrolyte to make the solutions conducting and to reduce the electrode double–layer thickness. In principle, any strong electrolyte will satisfy these basic requirements, but there are other considerations (6).

In aqueous solutions, KCl and HCl are common choices since the liquid junction potentials with s.c.e. or Ag/AgCl reference electrodes can be eliminated. Aqueous solutions are usually pH–buffered and the components of the buffer system often act as the supporting electrolyte.

Clearly, the salt should not be easily oxidizable or reducible. The accessible potential range at Pt and Hg electrodes is given in Table A.9 for several electrolyte/solvent combinations.

In low–dielectric–constant solvents, tetraalkylammonium salts are more soluble and less easily reduced than alkali metal salts. Quaternary ammonium ions do not form tight ion pairs with anions. Thus R_4N^+ salts are by far the most common choice, although tetraphenylphosphonium salts and lithium salts are sometimes used in organic solvents. However, when tetraalkylammonium ions are reduced, they form surface–active polymers which coat electrodes, foul dropping mercury electrode capillaries, and generally raise havoc with experiments.

The choice of anion is less obvious. The simple halides, Cl^-, Br^-, and I^-, are relatively easily oxidized and often form tight ion pairs, so that they are frequently avoided. The most common choices—ClO_4^-, BF_4^-, PF_6^-, and BPh_4^-—have delocalized charge, so that their salts are often soluble in organic solvents. Ion pairing usually is not severe and these anions are not easily oxidized or reduced at electrodes.

Reference Electrodes

In aqueous solutions, the calomel and Ag/AgCl electrodes are well characterized and give reproducible potential readings with a minimum of experimental difficulty. Unfortunately, there is no universally accepted reference electrode for use with nonaqueous solvents. There are two commonly used approaches.

One school of thought is to use an aqueous s.c.e. with a salt bridge. The advantage is that the reference electrode is well understood and generally reproducible. There are some major disadvantages: an unknown liquid junction potential is introduced and water contamination through the salt bridge is difficult to avoid completely. Since the s.c.e. electrolyte is normally KCl, contamination by K^+ and Cl^- ions can sometimes lead to problems. For example, a KCl–filled salt bridge in contact with a solution containing ClO_4^- ions frequently leads to precipitation of $KClO_4$, which is quite insoluble in organic solvents and not very soluble in water. The precipitation usually takes place at the point of solution contact and often clogs the salt bridge, blocking current flow. One way of avoiding this particular problem is to replace the s.c.e. electrolyte with NaCl; $NaClO_4$ is much more soluble.

A second common approach is to use the Ag/Ag^+ couple as a reference, dissolving a suitable silver salt in the same solvent/supporting electrolyte system used in the experiment and placing it in contact with a silver wire. The advantages of this approach are that the liquid junction problems and solvent and electrolyte cross–contamination problems are minimized. The disadvantage is that solvent evaporation from the reference electrode or change in the silver surface with time may lead to nonreproducible potentials.

With many different reference electrodes in use, potential measurements from different laboratories are often difficult to compare. It is becoming standard practice in electrochemical studies of nonaqueous systems to use a standard reference couple of known (or at least commonly accepted) potential against which the reference electrode can be checked from time to time; potentials can then be reported relative to the standard couple or corrected to a common scale. Ferrocene, $(C_5H_5)_2Fe$, is reversibly oxidized to the ferrocenium ion, $(C_5H_5)_2Fe^+$, at $+0.08$ V $vs.$ Ag/Ag^+ in acetonitrile (7) and is the most commonly used potential standard. Ferrocene and ferrocenium salts are soluble in most nonaqueous solvents; in addition, the ferrocene/ferrocenium couple is relatively insensitive to solvation or ion–pairing effects and so provides an approximation to an absolute reference.

Indicator Electrodes

In Section 4.1, we assumed that the indicator electrode was planar and sufficiently large that we could ignore edge effects in solving the diffusion problem. Provided that the radius of curvature is large compared with the diffusion layer thickness, qualitatively similar results are obtained for cylindrical or spherical electrodes, indeed for any stationary electrode in an unstirred solution. Common designs for stationary electrodes include all three geometries:[‡] planar (metal disk electrodes), cylindrical (wires), and spherical (hanging mercury drop).

Two other indicator electrodes are commonly used in electroanalytical and mechanistic work. The dropping mercury electrode (d.m.e.), shown in Figure 4.9, consists of a fine–bore capillary through which mercury flows, forming a drop at the end of the capillary which grows until its weight exceeds the force of surface tension holding it to the capillary. Depending on the length and bore of the capillary, the pressure of mercury above the capillary, and the Hg-solution interfacial tension, the lifetime of a mercury drop can be anywhere from 1 to 10 s. The factors governing the drop time were discussed in Section 2.5. The current through a d.m.e. is time dependent, but because the solution is stirred when a drop falls, each new drop starts the experiment anew. If the current is measured just before the drop falls, experiments using a d.m.e. are essentially at constant time. We will discuss the operation of the d.m.e. in greater detail in Section 4.4.

The rotating disk electrode (r.d.e), shown in Figure 4.10, consists of a flat disk, usually 1–3 mm in diameter, mounted at the end of an insulating rod which is rotated rapidly in the solution. The rotational motion stirs the solution so that the diffusion problem is reduced to transport across a stagnant layer at the electrode surface. Since the rotation speed is constant and the stirring effect is reproducible, an experiment using the r.d.e. is carried out under steady–state (*i.e.*, time independent) conditions. We will discuss the operation of a r.d.e. in greater detail in Section 4.6.

Equation (4.16) applies to current–potential curves measured using a d.m.e. or r.d.e. indicator electrode. The diffusion currents for these electrodes, i_{Dc} and i_{Da}, can be expressed by eqs (4.14), although, because of the electrode motion, the diffusion layer thickness is not the same as that for a planar electrode. We will discuss these differences in Sections 4.4 and 4.6.

[‡] For reviews on electrodes, see Adams *(8,C3)*, Galus *(9)*, Dryhurst and McAllister *(10)*, or Winograd *(11)*.

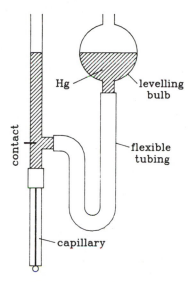

Figure 4.9 Dropping
mercury electrode.

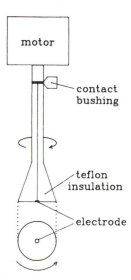

Figure 4.10 Rotating
disk electrode.

Cell Design

The detailed design of electrochemical cells for voltammetric experiments depends on the technique and on the requirements of the chemical system being studied *(12)*. Here we describe two common designs, shown in Figure 4.11, as representative examples.

The so–called H–cell, shown in Figure 4.11a, is a popular design for aqueous polarography. A separate compartment for the reference electrode prevents contamination of the test solution by KCl. Provision is made for purging the test solution with nitrogen (or argon) gas before the experiment and blanketing the solution with inert gas during the experiment. The indicator electrode (here a dropping mercury electrode) is mounted through a stopper with a small hole for escape of the purge gas. This cell is easily adapted to a three–electrode configuration by contacting the pool of waste mercury with a tungsten wire and using this as the auxiliary electrode. However, because the reference and indicator electrodes are widely separated, iR drop can be a major problem with this cell.

A somewhat more flexible design is shown in Figure 4.11b. Here a small glass cell body is clamped to a plastic top provided with ports

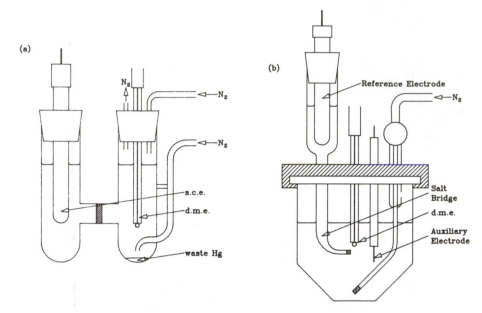

Figure 4.11 Examples of cells for voltammetric experiments: (a) H–cell   for   two–electrode   polarography;   (b)   three–electrode polarography cell.

for the electrodes and gas purge tube. The reference electrode is mounted in a salt–bridge tube which terminates in a Luggin probe situated as close as possible to the indicator electrode. The auxiliary electrode is typically a simple platinum wire and the purge tube is equipped with a two–way stopcock to permit gas to flow through or over the solution. The tip of the purge tube is sintered glass to disperse the gas into fine bubbles. Cells like this are quite well suited to most voltammetric techniques and work well provided that the solvent is not too volatile. The cell body can be fitted with a flow-through thermostat jacket for work away from room temperature and the entire experiment can be mounted in a glove box in the event that the analyte is very sensitive to air or moisture.

When solvent volatility is a problem or when the sample is exceedingly sensitive to air or moisture, cells are often designed to be filled on a vacuum line so that the purge gas is unnecessary.

4.3 A SURVEY OF ELECTROANALYTICAL METHODS

Because the diffusion–limited current in a voltammetric experiment is proportional to the concentration of the electroactive material, such an experiment is potentially adaptable to chemical analysis, and for many years electrochemists have been busy developing instrumental methods for measuring concentrations of reducible or oxidizable species in solution. There are so many methods now available that the neophyte is often quite overwhelmed. Indeed, the very wealth of electroanalytical techniques has acted sometimes as a barrier to their use by other chemists. In this section we will attempt to reduce the mystery of electroanalytical methods to manageable proportions.

Constant Time Experiments

If electron transfer is nernstian and the current is diffusion controlled, the current and potential are related by eq (4.16) to three parameters: the half–wave potential $E_{\frac{1}{2}}$ and the cathodic and anodic diffusion currents i_{Dc} and i_{Da}. The diffusion currents are related to the bulk solution concentrations of the participants in the electrode process, O and R, and to the diffusion layer thicknesses, x_{DO} and x_{DR}, by eqs (4.14) and (4.15). The diffusion layer thicknesses depend on the type of indicator electrode used but in general are functions of time. Thus an electrochemistry experiment has four variables which may be held constant, varied, or measured: current, potential, time, and solution composition.

We will consider first those electroanalytical methods for which time is not an explicit variable, either because the experiment is carried out under steady–state conditions (the r.d.e.), because measurements are always made at the same times in repetitive experiments (the d.m.e.), or because the current is zero. In these cases, the x_D's are constant and it is convenient to lump the various terms of eqs (4.14) and (4.15) together and write

$$i_{Dc} = \delta_O C_O^*$$

$$i_{Da} = -\delta_R C_R^*$$

where

$$\delta_O = nFAD_O/x_{DO}$$

$$\delta_R = nFAD_R/x_{DR}$$

It is also convenient to let X be the mole fraction of the electroactive material in the oxidized form and C^* be to the total concentration so that

$$C_O^* = XC^*$$

$$C_R^* = (1 - X)C^*$$

With these changes in parameters, eq (4.16) becomes

$$E = E_{\frac{1}{2}} + \frac{RT}{nF} \ln \frac{\delta_O XC^* - i}{i + \delta_R(1 - X)C^*} \tag{4.18}$$

Equation (4.18) describes the interrelationships of three variables: potential, current, and composition. These interrelationships are most easily visualized by thinking of a surface in three–dimensional space as shown in Figure 4.12.[‡] We can describe a number of electroanalytical techniques as excursions on the surface of Figure 4.12.

Potentiometric Titrations

When the current is equal to zero, eq (4.18) reduces to the Nernst equation, which we recall from Chapter 1 describes a potentiometric titration. In such a titration, the cell potential is measured as a function of solution composition at zero current. In Figure 4.12, the potentiometric titration curve corresponds to a path across the surface at zero current as shown in Figure 4.13. Of course, in a real case, the titrant and its reduced form (assuming that the titrant is an oxidizing agent) form another electrode couple, and beyond the endpoint it is this couple which determines the cell potential. Thus, to complete the titration curve, we would have to graft another similar surface onto Figure 4.12, so that when X approaches 1, the potential goes to some finite value rather than to infinity as implied by the single surface.

[‡] This way of describing electroanalytical experiments is due to Reilley, Cooke, and Furman *(13)*.

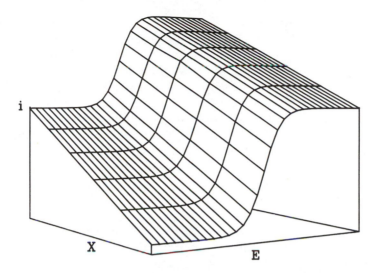

Figure 4.12 Surface showing the interrelationship of current, potential, and composition for a constant time experiment.

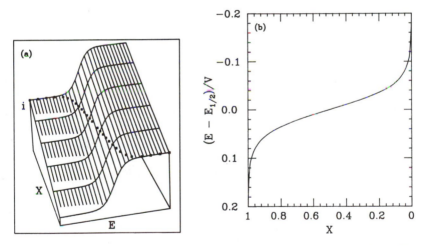

Figure 4.13 The $i = 0$ path on the three–dimensional surface (a) is the familiar potentiometric titration curve (b).

Polarography and Rotating Disk Voltammetry

Current measurements using a dropping mercury electrode (d.m.e.) are essentially constant time experiments. Since very little current is passed, little electrolysis takes place, so that the solution composition is constant. Thus a curve at constant X on the surface of Figure 4.12 is identical to the current–potential curve of Figure 4.2. When a d.m.e. is used, the voltammetric experiment is called *polarography* (or d.c. polarography to distinguish the technique from the variations discussed in Sections 4.5 and 5.4) and the current–potential curve is called a *polarogram*. The current on the diffusion-limited plateau, of course, is related to the bulk solution concentration. Polarography is one of the oldest electroanalytical techniques and, although it has been displaced by other methods for many analytical applications, it continues to be widely used. We will return to a more detailed discussion of the method in Section 4.4 and will give some examples of analytical applications in Section 4.7.

Currents through a rotating disk electrode (r.d.e.) are measured under steady–state conditions. Although currents through a r.d.e. are typically much larger than those through a d.m.e, the total current is small enough that little net electrolysis occurs. A rotating disk voltammogram thus resembles a polarogram and again is essentially the curve at constant X on the surface of Figure 4.12 or the current–potential curve shown in Figure 4.2.

Amperometric Titrations

Imagine an excursion upon the surface of Figure 4.12 where the current is measured as a function of composition, holding the potential constant. We expect to get a straight line terminating at X = 1. Such an excursion would be obtained if the current were measured during a titration of the electroactive material. In practice, such a titration would be carried out using a d.m.e. or r.d.e., so that again, the current would be measured at constant time (*i.e.*, the drop time) or under steady–state conditions. The potential could be set so that either the analyte or the titrant is reduced (or oxidized) at the electrode. If the analyte is reduced, then the titration curve corresponds to a straight line going to zero at the endpoint; if the titrant is reducible, then a linearly increasing current is obtained, commencing at the endpoint.

Amperometric endpoint detection is applicable to a wide variety of titration reactions *(14)*. The titration reaction does not have to be a redox reaction. Thus, for example, we could monitor the Pb^{2+} concentration in a titration of $Pb(NO_3)_2$ with Na_2SO_4 to produce insoluble $PbSO_4$; the current would decrease to zero at the endpoint.

Example 4.4 Show how the titration of Pb^{2+} with $Cr_2O_7^{2-}$ could be followed amperometrically.

The titration reaction is

$$2\ Pb^{2+} + Cr_2O_7^{2-} + H_2O \rightarrow 2\ PbCrO_4(s) + 2\ H^+$$

The polarograms of Pb^{2+} and $Cr_2O_7^{2-}$ are shown schematically in Figure 4.14. If we follow the titration with the potential set at –0.7 V, on the diffusion plateau of the Pb^{2+} polarogram, then the current will fall linearly during the course of the titration as the Pb^{2+} is used up. At the endpoint, the current should be near zero, but beyond the endpoint, the current should start to rise again as the excess dichromate is reduced. The resulting titration curve is also shown in Figure 4.14. An alternative arrangement for the same titration would be to set the potential at –0.2 V so that dichromate is reduced but Pb^{2+} is not. The resulting titration curve would then show nearly zero current up to the endpoint with current increasing beyond.

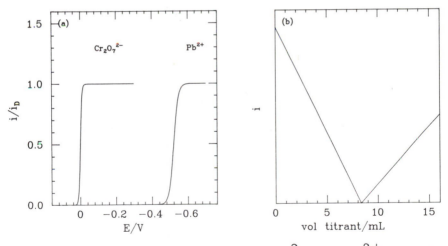

Figure 4.14 (a) Polarograms of $Cr_2O_7^{2-}$ and Pb^{2+} and (b) amperometric titration curve for the titration of 25 mL of 0.001 M Pb^{2+} with 0.0015 M $Cr_2O_7^{2-}$ at $E = -0.7$ V.

Constant Composition Experiments

We now consider experiments involving a stationary electrode where time is an explicit variable. We introduce the time dependence by writing the diffusion currents as

$$i_{Dc} = K_c t^{-\frac{1}{2}}$$

$$i_{Da} = -K_a t^{-\frac{1}{2}}$$

where for a planar electrode

$$K_c = nFA(D_O/\pi)^{\frac{1}{2}} C_O{}^*$$

$$K_a = nFA(D_R/\pi)^{\frac{1}{2}} C_R{}^*$$

Equation (4.16) then becomes

$$E = E_{\frac{1}{2}} + \frac{RT}{nF} \ln \frac{K_c t^{-\frac{1}{2}} - i}{i + K_a t^{-\frac{1}{2}}}$$

or

$$E = E_{\frac{1}{2}} + \frac{RT}{nF} \ln \frac{K_c - i t^{\frac{1}{2}}}{i t^{\frac{1}{2}} + K_a} \tag{4.19}$$

This equation also produces a three–dimensional surface representing the interrelationships of potential, current, and time (for fixed solution composition) which is shown in Figure 4.15. A number of additional techniques can be understood as excursions upon this surface.[‡] One which we have already found on the other surface—polarography—is found here as the current–potential curve at constant time.

Chronoamperometry

Perhaps the most obvious path to be followed on the surface is the constant potential curve where the current decays as $t^{-\frac{1}{2}}$. Since

---

[‡] The description of electroanalytical techniques in terms of an i–E–t surface is due to Reinmuth (15).

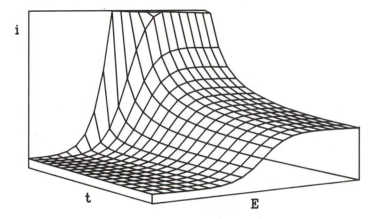

Figure 4.15 Surface showing the interrelationship of current, potential, and time for a constant composition experiment.

current is measured as a function of time, the experiment is called *chronoamperometry (16)*. In practice, a chronoamperometric experiment is done by stepping the potential applied to a cell from an initial value where negligible current flows to a final value well beyond the half–wave potential. The current–time response is then recorded.

An interesting variant on the technique is called *double potential step chronoamperometry*. Suppose that at time τ after the application of the potential step, the potential is stepped back to the initial value. Any R which remains near the electrode then will be oxidized back to O and an anodic current should be observed which decays to zero as the R molecules in the diffusion layer are used up. The current during the first stage of the experiment is given by the Cottrell equation,

$$i = nFA\,(D_O/\pi)^{\frac{1}{2}}C_O^* \, t^{-\frac{1}{2}} \qquad\qquad t < \tau \qquad\qquad (4.20a)$$

If we assume that the electron transfer is nernstian and that R is stable during the time of the experiment, it can be shown (see Appendix 4) that the current during the second stage is

$$i = -nFA\,(D_O/\pi)^{\frac{1}{2}}C_O^* \, [(t-\tau)^{-\frac{1}{2}} - t^{-\frac{1}{2}}] \qquad\qquad t > \tau \quad (4.20b)$$

Figure 4.16 shows the potentials and currents for such an experiment as functions of time. Suppose that the current is measured at time t_1 after the first potential step and again at time t_2 after the second step.

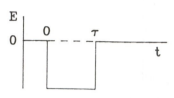

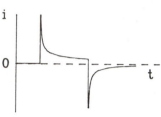

Figure 4.16 Applied potential and current response for a double potential step chronoamperometry experiment.

The ratio of the two currents will be

$$i_2/i_1 = [t_1/t_2]^{\frac{1}{2}} - [t_1/(t_2 - \tau)]^{\frac{1}{2}}$$

or if $t_1 = \tau$, $t_2 = 2\tau$,

$$i_2/i_1 = (2^{-\frac{1}{2}} - 1) = -0.293$$

Such a result would provide a good indication of an uncomplicated reversible electron transfer. If the product R was consumed by a chemical reaction, for example, the current after the second potential step would be less than expected from eq (4.20), $|i_2/i_1| < 0.293$ (see Example 5.8).

Chronopotentiometry

The curve followed on the surface of Figure 4.15 for constant (nonzero) current is shown in Figure 4.17. This experiment, where the potential is measured as a function of time at constant current, is called *chronopotentiometry* (16,17). In practice, a planar electrode is used and a constant current is switched on at zero time. The O molecules near the electrode are quickly reduced and the diffusion layer grows. The O/R ratio at the electrode surface decreases and the potential swings negative. Eventually, diffusion can no longer supply enough O to provide the required current and the potential heads toward $-\infty$. The time τ required for this potential swing is called the *transition time*,

$$\tau^{\frac{1}{2}} = K_c/i$$

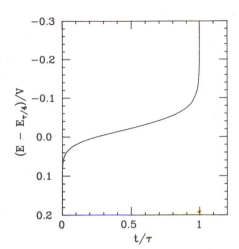

Figure 4.17 Chronopotentiometry follows a constant
current path on the i–t–E
surface.

In reality of course the potential doesn't go to $-\infty$ since there is always
something else in the solution reducible at a more negative potential
(solvent, supporting electrolyte, etc.).

When the diffusion equations are solved using the proper
boundary conditions for the experiment (see Appendix 4), it is found
that K_c differs by a factor of $\pi/2$ from its value in a constant potential
experiment, so that the transition time is given by

$$\tau^{\frac{1}{2}} = nFA(\pi D_O)^{\frac{1}{2}}C_O^*/2i \tag{4.21a}$$

This expression was first derived by Sand in 1900 *(18)* and is
sometimes called the Sand equation. The shape of the
chronopotentiogram for a nernstian process is given by eq (4.19),
which can be rewritten as

$$E = E_{\tau/4} + \frac{RT}{nF}\ln\frac{\tau^{\frac{1}{2}} - t^{\frac{1}{2}}}{t^{\frac{1}{2}}} \tag{4.21b}$$

where $E_{\tau/4}$ $(= E_{\frac{1}{2}})$ is the potential when $t = \tau/4$. Since the
concentration of the electroactive species is proportional to the square
root of the transition time, chronopotentiometry can be used
analytically.

A variation on the technique, called *current reversal
chronopotentiometry,* is analogous to double potential step
chronoamperometry. A cathodic current i is passed through the cell,

reducing O to R. At time t_1 (less than the cathodic transition time), the current is reversed and the R molecules remaining near the electrode are reoxidized. The potential then rises, reflecting the increase in the O/R ratio at the electrode surface. When the diffusion layer is depleted in R, the potential goes to $+\infty$. If R is stable in solution, the anodic transition time is exactly $\frac{1}{3}t_1$. Thus during the experiment, two thirds of the R molecules escape by diffusion and the remaining one third are reoxidized at the electrode during the current reversal phase. If R is consumed by a chemical reaction, of course, the second transition time will be shorter and in favorable cases the rate of the chemical step can be determined (see Example 5.9).

Linear Potential Sweep Voltammetry

If the potential is changed linearly with time, a current–potential curve is produced which corresponds to a diagonal excursion on the

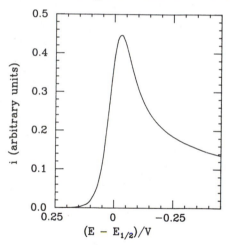

Figure 4.18 Linear potential sweep voltammogram.

surface of Figure 4.15. The resulting curve, shown in Figure 4.18, is called a *linear potential sweep voltammogram* or *peak polarogram*. The origin of the peak is easily understood. Consider a negative–going potential sweep; when the reduction potential of species O is reached, the current rises sharply as O near the electrode surface is reduced; the current then becomes diffusion controlled and begins to fall. When the potential is well past the half–wave potential, the current falls as $t^{-\frac{1}{2}}$ as in a chronoamperometry experiment.

Solution of the diffusion equations for the time–dependent boundary condition imposed by a rapid potential scan is complicated. The mathematics ends with an integral equation which must be solved numerically *(19)*. The peak current for a reversible process at a planar electrode is found to be

$$i_p = 0.446 \, nFA(nFv/RT)^{\frac{1}{2}} \, D_O^{\frac{1}{2}} C_O^* \qquad (4.22)$$

where v is the potential scan rate in V s^{-1} and the other parameters are in SI units. The current peak is somewhat displaced from $E_{\frac{1}{2}}$,

$$E_p = E_{\frac{1}{2}} - 1.11 \, RT/nF$$

so that at 25°C, the peak potential is 28.5/n mV past the half–wave potential. (The half–peak potential comes 28.0/n mV before $E_{\frac{1}{2}}$.) For a reversible process, the shape of the curve and the peak potential are independent of scan rate, but the peak current is proportional to $\sqrt{v}$.

Since the peak current is proportional to concentration, a linear potential sweep voltammogram can be used for analytical purposes. However, there are two experimental complications which reduce accuracy. Because of the changing potential, the capacitive charging current may be quite significant in a linear potential sweep experiment. Since the charge stored in the double–layer capacitance is

$$Q = C_d A \Delta \Phi$$

where C_d is the double–layer capacity in f m^{-2}, the charging current is

$$i_c = dQ/dt = C_d A v \qquad (4.23)$$

Example 4.5 Compute the faradaic and capacitive currents for a typical case with n = 1, $C_O^* = 1$ mol m^{-3} (1 mM), $D_O = 10^{-9}$ m^{2}s^{-1}, $C_d = 0.5$ F m^{-2}, $v = 1$ V s^{-1}, $A = 10^{-6}$ m^2 (1 mm^2), and T = 298 K.

Inserting T = 298 K and the physical constants, eq (4.22) becomes

$$i_p = (2.69 \times 10^5) \, n^{\frac{3}{2}} \, AD_O^{\frac{1}{2}} C_O^* v^{\frac{1}{2}}$$

Inserting the other variables, we get

$$i_p = 8.5 \, \mu A$$

Substituting C_d, A, and v in eq (4.23), we have

$$i_c = 0.5\mu A$$

For low concentrations and/or high scan rates, the capacitive current is a serious interference in linear potential sweep experiments.

Comparing the results of Examples 4.1 and 4.5, we see that linear potential scan peak currents can be quite large. Because of this, ohmic potential drop in the solution is often a problem. Since the cell potential includes the solution iR drop and the iR drop varies as the peak is traversed, the indicator electrode potential is not linear in the applied potential. Thus the observed peak potential becomes a function of scan rate. In practice, the reversible peak potential can usually be extracted from experimental data by measuring the peak potential at several scan rates and extrapolating to $v = 0$, but the absolute peak current is more difficult to correct.

Cyclic Voltammetry

The analog of double potential step chronoamperometry and current reversal chronopotentiometry is possible for linear potential sweep voltammetry. A triangular-wave potential is applied to the cell so that the indicator electrode potential is swept linearly through the voltammetric wave and then back again. On the forward scan, the current response is just the linear potential sweep voltammogram as O is reduced to R. On the reverse scan, the R molecules near the electrode are reoxidized to O and an anodic peak results. The resulting curve, shown in Figure 4.19, is called a *triangular-wave cyclic voltammogram*.

Detailed analysis of cyclic voltammetry also leads to integral equations which must be solved numerically (19). However, we can get a qualitative feeling for the experiment without a lot of mathematics. If the initial scan is carried far beyond the cathodic peak so that the diffusion layer is very thick and the cathodic current has decayed nearly to zero, then the concentration of R at the electrode surface is equal to C_O^* [within a factor of $(D_O/D_R)^{\frac{1}{2}}$]. Thus the amount of R available for oxidation on the reverse scan is the same as the O available on the forward scan and the current peak has the same shape and magnitude as on the forward scan, but reflected at $E = E_{\frac{1}{2}}$ and changed in sign.

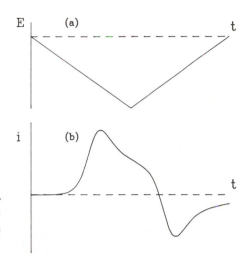

Figure 4.19 Cyclic volt–ammetry experiment. (a) Potential a function of time; (b) current as a function of time.

When the switching potential is less negative such that the cathodic current is still significant at the switching point, the diffusion layer is thinner, the R concentration falls to zero more rapidly with distance from the electrode, and the resulting anodic peak is smaller. However, it turns out that, if the anodic peak current is measured from a baseline equal to the cathodic current which would have flowed at the time of the anodic peak had the potential scan continued in the negative direction (rather than from the zero current baseline), then the anodic–to–cathodic peak current ratio is exactly 1 (for a reversible process uncomplicated by capacitive current and iR drop in solution). This result is shown schematically in Figure 4.20 with cyclic voltammograms computed using the numerical methods of Nicholson and Shain *(19)*.

Provided that the process is reversible and is uncomplicated by solution iR drop, the cathodic peak occurs 28.5/n mV more negative than the half-wave potential and the anodic peak is 28.5/n mV more positive (at 25°C). Thus $E_{\frac{1}{2}} = \frac{1}{2}(E_p^{\,c} + E_p^{\,a})$ and the anodic and cathodic peaks are separated by 57.0/n mV. A peak separation of 57.0/n mV is often used as a criterion for nernstian behavior. Since the current direction is different for the two peaks, the sign of iR is also different so that ohmic potential drop in the solution tends to increase the peak separation. Again, extrapolation to zero sweep rate allows corrections to be made. Since the sign of the potential sweep changes at the switching point, the sign of the capacitive current also changes. Thus a current discontinuity is usually observed at the switching point.

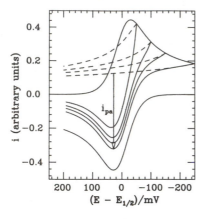

Figure 4.20 Cyclic voltammogram for a reversible one–electron process showing the effect of the switching potential. The bottom curve shows the expected anodic trace when the potential is negative long enough to completely polarize the electrode. The dashed (current–time) curves show the response which would have occurred had the potential scan continued in the negative direction.

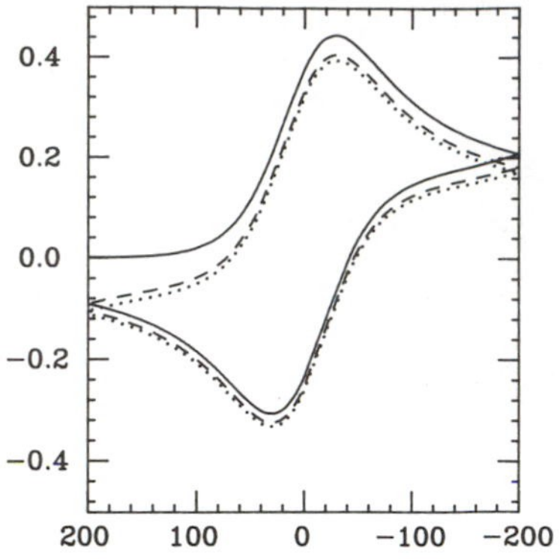

Figure 4.21 Cyclic voltammogram showing the effect of multiple scans. The first cycle is shown as a solid line, the second as a dashed line, and the third as a dotted line.

The response on the second and subsequent cycles of a cyclic voltammogram is qualitatively similar to that of the first cycle but with current peaks somewhat reduced in amplitude. This effect is shown in Figure 4.21. The reason for this decrease is that the concentration profiles of O and R do not return to initial conditions at the completion of a cycle. This is most easily seen by following these concentration *vs.* distance profiles over the course of the first cycle as shown in Figure 4.22. Initially, $C_O(x) = C_O^*$ and $C_R(x) = 0$. As the cathodic peak is traversed, O is depleted at the electrode surface and the R which is formed diffuses away from the electrode (curves a–c of Figure 4.22). On the reverse scan, R is oxidized to O but there remains a region where $C_O < C_O^*$ and $C_R > 0$. On subsequent cycles, the response is as if the bulk concentration of O were reduced. On multiple cycles, a

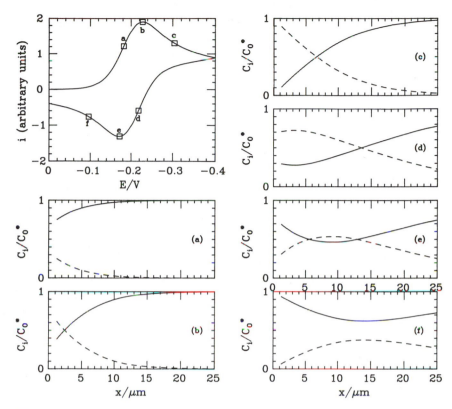

Figure 4.22 Concentration profiles for O (solid lines) and R (dashed lines) at various points on a cyclic voltammogram. Concentrations computed using a digital simulation technique (see Appendix 5) for $v = 1$ V s^{-1}, $D_O = D_R = 5 \times 10^{-10}$ m^{2}s^{-1}.

steady state is eventually reached with damped concentration waves propagating out into the solution.

Voltammetry at a Microelectrode

One way of dealing with the problem of solution iR drop in a voltammetric experiment is to decrease the electrode area, thus decreasing the size of the current. Indeed, it is possible to to decrease the electrode area to the point that solution iR drop is completely negligible, even in the absence of a supporting electrolyte (20). Not only does a microelectrode eliminate iR drop, it has the somewhat surprising consequence of also decreasing the relative importance of the double–layer capacitive charging current. Our approach to voltammetry has been through insights gained from the solution to the diffusion equation under conditions where diffusion is linear and edge effects can be ignored. Thus we have assumed implicitly that electrode radii are large compared with diffusion layer thicknesses. When very small electrodes are used, this assumption fails and we must consider radial diffusion as well as diffusion normal to the electrode surface.

The diffusion equation cannot be solved in closed form, but Heinze *(21)* has used digital simulation techniques to show that the Cottrell equation requires a correction for very small electrodes:

$$i_D = nFAC^*(D/\pi t)^{\frac{1}{2}}[1 + b(Dt)^{\frac{1}{2}}/r]$$

where r is the electrode radius and b $= 2.257$ when $Dt/r^2 \gg 1$. For an electrode of radius 10 μm and D $= 10^{-9}$ m^{2}s^{-1}, $D/r^2 = 10$ s^{-1}. Thus for t > 0.1 s, the radial diffusion correction term dominates and the current quickly becomes time independent. Indeed, for microelectrodes, the usual $t^{-\frac{1}{2}}$ dependence of the current amounts to a short–lived transient in the chronoamperometric response. Chronoamperometry is thus a very different kind of experiment with a microelectrode; in principle, it should be possible to use microelectrodes to obtain a time–independent current which is directly proportional to the concentration of the electroactive species.

The large radial diffusion contribution to the current has several consequences in cyclic voltammetry experiments. In the first place, the current density is larger with a microelectrode, perhaps by several powers of 10; the double–layer capacity, on the other hand, is substantially unchanged by the decreased electrode size so that the ratio of electrolysis current to capacitive current is greater. Thus cyclic voltammetry experiments can be done at higher scan rate, probing mechanistic details which are inaccessible with normal–size electrodes. Because diffusional mass transport is so much greater, the current peak which is normally observed in cyclic voltammograms is largely absent for slow potential scan rates (v on the order of 10 mV s^{-1}) and a current–potential wave is observed; similarly, because the electrode product escapes more rapidly, no reoxidation peak is observed on the reverse scan for slow scan rates. For faster scans ($v > 100$ mV s^{-1}) cyclic voltammograms have the "normal" shape although the quantitative details are slightly different.

Construction of microelectrodes with diameters on the order of tens of microns presents no particular experimental problem as long as

fine wire of the desired diameter is available. The principal barrier to
the use of microelectrodes is instrumental. Because the current is so
much smaller, signal–to–noise ratios are lower and greater care must
be exercised to avoid noise pickup. In order to realize the advantages
of the shorter time scale of microelectrode voltammetry, potentiostat
circuits with better high–frequency response must be used.

Microelectrodes show considerable promise as *in vivo* probes in
bioelectrochemistry as well as in extending electrochemical mechanistic
studies to shorter time scales. The development of microelectrodes is
quite recent; for further information, see the introductory review by
Wightman *(22)*.

4.4 POLAROGRAPHY

The Dropping Mercury Electrode

Shortly after World War I, Heyrovský discovered that the
dropping mercury electrode (d.m.e.) could be used to measure
reproducible current–potential curves. The technique, which he named
polarography, flowered in Heyrovský's hands and become the source of
many of the subsequent developments in electroanalytical chemistry
(23,D1,D4,D5,D12). In a classical d.m.e. such as that in Figure 4.9,
mercury flows through the capillary by gravity and a drop grows until
its weight exceeds the surface tension force holding it to the capillary.
Depending on the length and bore of the capillary, the height of the
mercury head above the capillary, and the mercury–solution interfacial
tension (see Section 2.5), drop lifetimes can range roughly from 1 to 10
seconds using this arrangement. Modern dropping mercury electrodes
are usually furnished with a drop dislodging mechanism and a timer so
that drop times are constant and reproducible. A more recent
innovation is the so–called *static mercury drop electrode*, where the
mercury flow rate is controlled by a pump; this device will be discussed
in Section 4.5.

When a drop falls off the capillary, it stirs the solution so that
each new drop begins life in contact with a nearly homogeneous
solution. In effect the experiment is repeated over and over again. If
the current through a d.m.e. is measured at the same time in the lives
of successive drops (at the time of drop fall, for instance), we can think
of the experiment being done at constant time. In ordinary
polarography, the electrode potential is scanned slowly with time so

that the potential is nearly constant during one drop life. Thus each successive drop measures the current at a slightly different potential and the current–potential curve, the polarogram, is generated. A major disadvantage of the d.m.e. is that mercury is much more easily oxidized than electrode materials like platinum or gold. The standard potential for the reduction of Hg_2^{2+} to the metal is $+0.8$ V, but with even as weak a ligand as Cl^-, the oxidation shifts to ca. 0.24 V. Thus polarography is largely limited to the negative end of the potential scale and thus has been used mostly to study reduction processes.[‡]

An exact solution of the diffusion problem for a d.m.e. is not easy, even for the simplest reversible electron–transfer process. If the electrode were simply a sphere, the diffusion equations could be recast in spherical polar coordinates with only the radial coordinate important. To this level of approximation, the mathematics is not appreciably harder than for linear diffusion. However, the electrode expands into the solution and this makes the problem considerably more complicated. Furthermore, the mercury drop is really not a perfect sphere but is slightly distorted and (more important) is shielded on one side by the capillary. Finally, the stirring which occurs when a drop falls is not complete and a rigorous model should take account of the resulting nonuniform concentration distribution.

A simple approach to the problem of diffusion to the d.m.e. is to use the results of the linear diffusion problem with two multiplicative correction terms. The first correction accounts for the changing electrode area, the second for the expansion of the electrode surface into the solution. If we assume that the drop is a sphere of radius r(t) and that mercury flows through the capillary at constant rate u (mass per unit time), then the mass of the drop at time t after the previous drop fell is m = ut and the volume is

$$V(t) = 4\pi r^3/3 = ut/d$$

where d is the density of mercury (13.6 g mL^{-1}). The drop area then is

---

[‡] The convention that cathodic current is taken as positive results from this feature; polarographic currents are usually cathodic and much of the theory of voltammetry was developed with polarography in mind.

$$A(t) = 4\pi r^2 = 4\pi (3ut/4\pi d)^{\frac{2}{3}} \qquad (4.24)$$

The effect of the electrode expanding into the solution is to stretch the diffusion layer over a larger area, in other words to make x_D smaller. The thinner diffusion layer results in a larger concentration gradient and thus in a larger flux and current. The net effect is as if the effective diffusion coefficient were larger than the real D by a factor of $\frac{7}{3}$.

Substituting eq (4.24) in eq (4.7) and replacing D_O by $\frac{7}{3} D_O$ gives

$$i_D = nF[4\pi (3ut/4\pi d)^{\frac{2}{3}}] C_O * (7D_O/3\pi t)^{\frac{1}{2}}$$

or, including numerical values for F and d (13.6×10^3 kg m^{-3}), we have

$$i_D = 706\, nD_O^{\frac{1}{2}} C_O * u^{\frac{2}{3}} t^{\frac{1}{6}} \qquad (4.25)$$

i_D will have units of amperes (A) when D_O is in m^{2}s^{-1}, C_O* is in mol m^{-3} (mM), u is in kg s^{-1}, and t in s. Equation (4.25) was first derived by Ilkovic in 1934 *(24)* and is usually called the Ilkovic equation. Ilkovic followed a somewhat more rigorous route than we did but did not deal with all the problems discussed above. Others have made serious attempts to solve the problem correctly but the complexity of the resulting equations is not wholly compensated by increased accuracy. The basic functional form of eq (4.25) remains unchanged in the more complete treatments, so we will not pursue the matter further. See, for example, Bard and Faulkner *(B12)* for further details.

Since the current at a d.m.e. is related to that at a planar electrode by the same scale factors for most electrode processes, expressions such as eqs (4.10), (4.16), and (4.17), which were derived for linear diffusion to a stationary planar electrode, are also applicable to the d.m.e.

One of the significant results of the Ilkovic equation is that, as shown in Figure 4.23, the current increases with time rather than decreasing as with a planar electrode. This means that the current at drop fall is the maximum current during the drop life. Furthermore, the time rate of change of the current, di/dt, decreases with increasing time and is at a minimum at drop fall. These considerations make the measurement of the current at or just before drop fall relatively easy

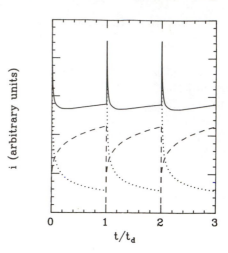

Figure 4.23 Faradaic current (dashed lines), capacitive current (dotted lines), and total current (solid lines) through a d.m.e. for $C_O^* = 10^{-4}$ M.

and accurate.

The favorable time dependence of the faradaic current for a d.m.e. has a price. Since the electrode area changes with time, the double-layer capacitance, which is proportional to the electrode area, is also time dependent. The charge stored in the capacitance is

$$Q = CA\Delta\Phi$$

where C is the double-layer capacity (capacitance per unit area). Since Q is time dependent, there must be a capacitive charging current

$$i_C = dQ/dt = C(dA/dt)\Delta\Phi$$

Differentiating eq (4.24), we obtain

$$i_C = C(8\pi/3)(3u/4\pi d)^{\frac{2}{3}} t^{-\frac{1}{3}} \Delta\Phi$$

$$i_C = 0.00566\, C\, u^{\frac{2}{3}} t^{-\frac{1}{3}} \Delta\Phi \tag{4.26}$$

The capacitive current is shown as a function of time in Figure 4.23, together with the total current.

Example 4.6 Compute the faradaic and capacitive currents at t = 0.1 and 5 s assuming that u = 1 mg s^{-1}, an electrode–solution potential difference of 1 V, and a double layer capacity of 0.1 F

m^{-2}. Assume a one–electron reduction with $D_O = 10^{-9}$ $m^2 s^{-1}$ and $C_O{}^* = 1$ mM (1 mol m^{-3}).

Substitution into eqs (4.25) and (4.26) gives

$$i_D = (2.23 \times 10^{-6}) \, t^{\frac{1}{6}} \ \text{amperes}$$

$$i_C = (4.66 \times 10^{-8}) \, t^{-\frac{1}{3}} \ \text{amperes}$$

Thus $i_D = 1.5$ and 2.9 μA at $t = 0.1$ and 5 s, respectively. At the same times, $i_C = 0.12$ and 0.03 μA.

For millimolar solutions of electroactive materials, the maximum current occurs at the end of the drop life,[‡] and this current is mostly faradaic. However, when the concentration of electroactive material is reduced, the capacitive current remains the same and the ratio of faradaic to capacitive current is smaller. The practical lower limit of concentration in polarography is about 10 μM since at that level the faradaic and capacitive currents are comparable in magnitude.

Example 4.7 Suppose that diffusion–controlled current flows at a dropping mercury electrode ($u = 1$ mg s^{-1}, $t = 5$ s) for 10 minutes and that O is reduced by one electron to R. If the cell contains 25 mL of 1 mM O ($D_O = 10^{-9}$ $m^2 s^{-1}$), what fraction of the O has been reduced?

The charge passed during one drop can be determined by integrating eq (4.25):

$$Q = \int_{0}^{t} i_D dt = \tfrac{6}{7} \times 706 \ nD_O{}^{\frac{1}{2}} C_O{}^* u^{\frac{2}{3}} t^{\frac{7}{6}}$$

Substituting the parameters, we have

[‡] The frequency response of the current–measuring circuitry is usually too slow to catch the short–lived transient in the capacitive current at drop fall.

$$Q = 1.25 \times 10^{-5} \text{ C}$$

In 10 minutes, 120 drops fall and so the total charge transferred is 1.50×10^{-3} C. Dividing by the Faraday constant, we have 1.56×10^{-8} mol, which is 0.062% of the total O originally present.

This result is typical of voltammetric techniques. The net amount of electrolysis is usually extremely small.

Effects of Adsorption on the Mercury Drop

An important advantage of the d.m.e compared with solid electrodes is that each successive drop presents a clean surface to the solution. Thus products of the electrolysis or subsequent chemical steps do not accumulate at the electrode surface beyond the amounts produced during one drop life. The effects of adsorption are thus more reproducible on a d.m.e. than on solid electrodes and are usually easier to study and to understand.

Suppose that the product of the electrode process is adsorbed on the mercury surface such that the first monolayer is strongly adsorbed but subsequent layers are either weakly held or not adsorbed at all. The electrode process then consists of two alternative reactions:

$$O + n e^- \rightleftharpoons R_{ad}$$

$$O + n e^- \rightleftharpoons R_{soln}$$

If the standard potential for the second reaction, where R remains in solution, is $E_2°$ and the free energy of adsorption is $\Delta G_{ad}°$, then the standard potential for the first reaction is

$$E_1° = E_2° - \Delta G_{ad}°/nF \tag{4.27}$$

Thus, if $\Delta G_{ad}° < 0$, the two reactions are expected to be distinguished by different half-wave potentials. If the bulk concentration of O is small enough that less than monolayer coverage by R is achieved during one drop life, we expect a wave at $E_1°$ with a limiting current given by eq (4.25). If the drop is completely covered before it falls, then we expect the current to be limited by the available adsorption sites. Excess O then remains unreduced at the electrode surface when the potential is in the vicinity of $E_1°$ and a second wave is observed at $E_2°$. The total diffusion current of the two waves is limited by diffusion and is given by eq (4.25). The second wave, of course, is the expected one for a process uncomplicated by adsorption. The first wave is called an *adsorption prewave*.

A very similar phenomenon occurs when the substrate O is adsorbed on the mercury drop (we assume that R is not then adsorbed). For low bulk concentrations of O, the electrode will be partially covered and the electrode process will be

$$O_{ad} + n\,e^- \rightleftarrows R$$

When $C_O{}^*$ is big enough that complete coverage is achieved, direct electron transfer may be possible:

$$O_{soln} + n\,e^- \rightleftarrows R$$

The potential of the first process will be more negative this time (if $\Delta G_{ad}° < 0$), but the more positive main wave will occur only at higher concentrations. The more negative wave is called an *adsorption post-wave*. The appearance of adsorption pre–waves and post–waves are shown in Figure 4.24.

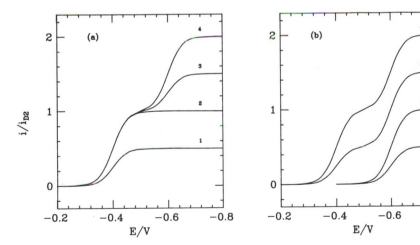

Figure 4.24 Polarograms for various substrate concentrations, $C_O{}^*$, where (a) the product R and (b) the reactant O are adsorbed on the mercury drop. The curves correspond to $C_O{}^*/C_O{}' = 0.5$, 1.0, 1.5, and 2.0, where $C_O{}'$ is the concentration leading to monolayer coverage at the time of drop fall.

Adsorption pre-waves and post-waves can be distinguished quite easily from ordinary electron-transfer processes. In addition to seeing the normal wave at higher substrate concentrations, heights of adsorption waves have a unique dependence on drop size and the variation of current during the drop life is different from the normal behavior. For further details, a specialized book on polarography should be consulted.

Polarographic Maxima

Throughout this chapter, we have assumed that diffusion is the only transport process involved in delivery of electroactive material to the electrode. There are several circumstances, encountered most frequently with the dropping mercury electrode, where convection also makes a significant contribution to mass transport. As we will see, convective mass transport is usually important in a limited potential range, and, since convection necessarily increases the current, the effect is a maximum in the current-potential curve.

Three kinds of polarographic maxima have been identified (25). Type I maxima are the most conspicuous and thus most commonly recognized. The polarographic current, instead of increasing smoothly with potential through the wave, seems to go "haywire," usually in the rising portion of the wave. Such erratic currents usually subside, often abruptly, on the diffusion-limited current plateau. Although the behavior appears erratic, it is usually quite reproducible. Type I maxima come about through variations in interfacial tension over the drop surface. These variations in turn reflect a variation in potential difference across the electrode-solution interface which arises through nonuniform current density over the surface of the mercury drop. The current density at the bottom of the drop is higher than at the "neck," where the drop is partially shielded by the capillary tube. Maxima of the first kind can be reduced by increasing the solution conductance (higher supporting electrolyte concentration), by decreasing the current (lower concentration of electroactive species), or by stabilizing the electrode-solution interface with a surfactant such as gelatin or Triton X-100 (polyethylene glycol p-isooctylphenyl ether).

A second kind of polarographic maximum can occur through motion within the mercury drop. If mercury emerges from the capillary fast enough, the new mercury will flow to the bottom of the drop and then up the sides. The drop may be thought of as growing longitudinally rather than radially. The circulation of mercury inside the drop will tend to drag the surface layer along, thus leading to convective mixing of the solution. Type II maxima are less dramatic in their effect; the convective contribution to mass transport is only weakly potential dependent, and the effect may pass unrecognized in many cases. Type II maxima depend on the mercury mass flow rate u and are most easily recognized by a departure from the dependence of current on flow rate predicted by the Ilkovič equation.

Type III maxima arise from nonuniform adsorption of surfactants which lead to variations in interfacial tension and thus to convective flow of solution in the vicinity of the mercury drop. Like Type I maxima, these effects can be reduced or eliminated by adding a competing surfactant.

Polarographic maxima are more common in aqueous solution polarography than when organic solvents are used, presumably because the mercury surface is hydrophobic and therefore susceptible to surfactants in aqueous media, whereas many organic solvents are themselves somewhat adsorbed on the mercury surface.

4.5 POLAROGRAPHIC VARIATIONS

Since the introduction of polarography in the 1920's, many instrumental modifications have been developed to improve the form of the polarogram and to increase sensitivity and/or resolution. We will discuss five of these improvements in this section.

Sampled D.C. Polarography

The simplest modification to polarography is called *sampled–d.c.* or *tast* polarography. In this method, a voltage ramp is applied to the cell exactly as in ordinary polarography, but instead of measuring the current continuously, it is measured for only a short time late in the drop life. A voltage proportional to the measured current is then applied to the recorder and held until the next current sample is obtained. The d.m.e. must be synchronized to the timing of the current–sampling circuitry so that the sample occurs at the same relative time during each drop life. Sampled d.c. polarography suppresses display of the current early in the drop life, when there is a large capacitive component, and focuses on the current late in the drop life, when the faradaic component is larger. This stratagem does nothing for sensitivity but does slightly improve the apparent signal–to–noise ratio, making polarograms somewhat easier to analyze. A typical sampled d.c. polarogram is shown in Figure 4.25.

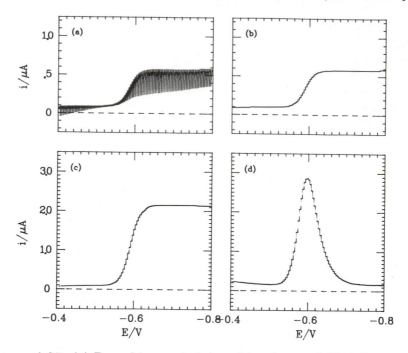

Figure 4.25 (a) D.c., (b) sampled d.c., (c) pulse, and (d) differential pulse polarograms of 0.1 mM $Cd(NO_3)_2$ in 0.10 M aqueous KCl. Potentials $vs.$ s.c.e., drop time 2 s, scan rate 2 mV s^{-1}, differential pulse height 10 mV.

Pulse Polarography

A weakness of both ordinary d.c. and sampled d.c. polarography is that these techniques allow faradaic current to flow during a time in the drop life when the capacitive current is large. A considerable improvement is possible with a technique called *pulse polarography*, introduced by Barker[‡] in 1960 *(26)*. A series of voltage pulses is applied to the cell as shown in Figure 4.26; the drop time is

---

[‡] Geoffrey C. Barker (1915–) was a member of the scientific staff of the Atomic Energy Research Establishment, Harwell (England). He is best known as an innovative inventor of electrochemical instrumental techniques. He is credited with square wave polarography (which led to the development of differential pulse polarography), normal pulse polarography, anodic stripping voltammetry, and several other techniques.

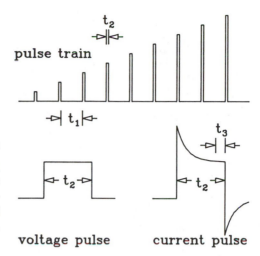

Figure 4.26 Pulse train for pulse polarography: t_1 is the drop time (0.5 – 5 s), t_2 is the pulse width (ca. 50 ms), and t_3 is the current sampling time (ca. 10 ms).

synchronized to the pulses which increase in amplitude at a rate comparable to the ramp voltage used in ordinary polarography. The total current response from such a pulse train still has a large capacitive component, but this is largely suppressed by sampling the current for a short time late in the pulse life after the capacitive current has mostly decayed. The sampled current is converted to a voltage which is held and applied to the recorder until the next sample. The appearance of a pulse polarogram thus is virtually indistinguishable from a sampled d.c. polarogram.

As we saw in the derivation of the Ilkovič equation, the current varies with time because of the time–dependent electrode area and because of the growth of the diffusion layer with time. In pulse polarography, essentially no current flows until late in the drop life, so that the diffusion layer is much thinner at the time the current is actually measured. In particular, we expect the measured current to depend on the times t_1 and t_2 (see Figure 4.26):

$$i \propto t_1^{\frac{2}{3}} t_2^{-\frac{1}{2}}$$

Time t_1 is the age of the drop when the current is measured (t_1 thus determines the drop area) and t_2 is the time between application of the voltage pulse and current measurement (t_2 thus determines the diffusion layer thickness). Suppose that the drop time is 5 s, that the pulse is applied in the last 50 ms of the drop life, and that the current is measured during the last 10 ms of the pulse. Then t_1 = 4.99 s, t_2 = 0.04 s and

$$t_1^{\frac{2}{3}} t_2^{-\frac{1}{2}} = 14.6$$

In sampled d.c. polarography, $t_1 = t_2 = 4.99$ s, so that

$$t_1^{\frac{2}{3}} t_2^{-\frac{1}{2}} = 1.31$$

Thus the measured current in a pulse polarogram is about 10 times bigger than in a sampled d.c. polarogram. The capacitive contribution to the current is about the same in the two techniques; thus the signal–to–noise ratio is improved by about a factor of 10 and detection limits are about 10 times lower.

Pulse polarography gives current–potential curves which are qualitatively similar to those from sampled d.c. polarography when the reduced half of the couple is absent in bulk solution (as we have assumed above). When $C_R^* \neq 0$, however, and anodic current flows at the initial potential, the situation is rather more complicated *(27)*. The complications can be turned to advantage, however, and Osteryoung and Kirowa–Eisner *(28)* have described a technique, called *reverse pulse polarography*, in which the pulse sequence is virtually a mirror image of that shown in Figure 4.26, *i.e.*, the "resting potential" is on the diffusion plateau of the wave and a series of positive–going pulses of linearly decreasing amplitude is applied.

Differential Pulse Polarography

Another pulse technique which is capable of good signal–to–noise ratios employs the pulse train shown in Figure 4.27. Small–amplitude voltage pulses are superimposed on a ramp voltage. Current is sampled twice during a drop life, once just before the pulse and again late in the pulse life. These currents are converted to analog voltages, subtracted, and applied to the recorder until the next drop. With this technique, called *differential pulse polarography*, a curve is produced which resembles the first derivative of the d.c. polarographic wave. In the limit that the pulse amplitude ΔE goes to zero, the curve is exactly the first derivative, but then the signal goes to zero as well. Pulse amplitudes of 5–50 mV are used in practice. If the differential current is plotted *vs.* the ramp voltage, the peak potential is displaced by $-\Delta E/2$ from $E_{\frac{1}{2}}$ (*i.e.*, a negative–going pulse ($\Delta E < 0$) shifts the peak to more positive potentials, a positive–going pulse shifts the peak to more negative potentials. A typical differential pulse polarogram is shown in Figure 4.25.

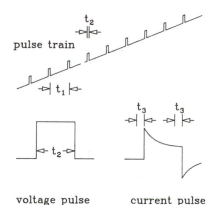

Figure 4.27 Pulse train, voltage and current pulses for differential pulse polarography.

The differential pulse peak current is proportional to the bulk concentration of the electroactive species, just as the diffusion current in d.c. polarography. However, detection limits are much lower, down to 0.1 μM or less in favorable cases. The improvement stems largely from suppression of the capacitive charging current. Because of the first derivative–like presentation of the output, the signal returns to the base line after a peak and successive current peaks for a multi-component system are somewhat better resolved than are d.c. polarographic waves. This effect is demonstrated in Figure 4.28, where differential pulse polarograms are plotted for a two–component system with $C_{O1}^* = C_{O2}^*$, $C_{R1}^* = C_{R2}^* = 0$, $n_1 = n_2 = 1$, and $\Delta E_{\frac{1}{2}} = 0.1, 0.2, 0.3,$ and 0.4 V. Comparison with Figure 4.4 leaves little doubt that the analytical accuracy of differential pulse polarography is considerably better than that of d.c. polarography when several components give closely spaced waves.

Square Wave Polarography

The time between pulses in differential pulse polarography (t_1 in Figure 4.27) is set equal to the d.m.e. drop time. Thus the pulse repetition rate is on the order of 0.1–1 Hz. If the pulse repetition rate is increased to 30 Hz, many pulses can be applied during one drop life and a scan over several hundred millivolts can be achieved during the lifetime of one drop. This approach is adopted in square wave polarography, where the positive– and negative–going pulses of Figure 4.27 are equal ($t_2 = \frac{1}{2}t_1$). In effect, the potential applied to the cell is a square wave–modulated potential scan. If the square wave

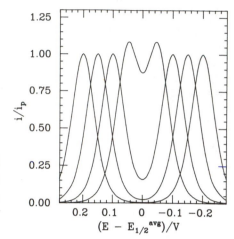

Figure 4.28 Computed differential pulse polarograms for a two–component system, showing the effect of half–wave potential separation.

frequency is 30 Hz and successive pulses advance the d.c. potential by 5 mV, the d.c. potential scan rate is 150 mV s^{-1}, comparable to typical scan rates in cyclic voltammetry. The current measurement strategy is the same as in the differential pulse technique: the current is measured at the end of successive positive–going and negative–going pulses, subtracted, and plotted as a function of the d.c. potential. The resulting square wave polarogram thus closely resembles a differential pulse polarogram *(29)*.

The instrumentation required for square wave polarography is rather more sophisticated; the potentiostat must have very good frequency response and the output must either be stored in a computer or recorded on an oscilloscope since an ordinary strip–chart recorder would not be able to keep up with such rapid data collection. In practice, a microprocessor is used to generate the wave form (the d.c. scan would then be a "staircase" rather than a ramp) and to collect the current signal. This arrangement allows the experiment to be repeated on several successive drops, adding the output signals to improve the signal–to–noise ratio.

Detection limits for square wave polarography are comparable to those of differential pulse polarography, on the order of 0.1 μM in favorable cases, but the data acquisition time is substantially shorter.

An advantage of the pulse techniques is that, for reversible couples, they work almost as well for stationary electrodes as they do for the d.m.e. In ordinary pulse polarography, the electrode spends most of its time at an initial potential where the current is very small;

thus the total current passed is small and time is allowed for relaxation of the diffusion layer between pulses. Square wave polarography treats the d.m.e. as if it were a stationary electrode and so clearly will work just as well on a real stationary electrode. A short rest period between scans would, of course, be required to allow the diffusion layer to dissipate.

The Static Mercury Drop Electrode

An interesting recent technical advance in the design of dropping mercury electrodes is the so-called *static mercury drop electrode* (s.m.d.e.). In this device a larger bore capillary is used and mercury flow is controlled with a pump. It can be used either to produce a stationary hanging drop or the pump can be pulsed on and off to simulate a d.m.e. Drop times can be as short as 0.1 s and as long as desired. Because of the larger bore capillary, drop growth occurs in about 50 ms, so that the drop is stationary and constant in area thereafter. Thus when a s.m.d.e. is used in sampled d.c. polarography, the current is measured on a stationary electrode, essentially eliminating the capacitive charging current. Furthermore, the current can be measured shortly after drop formation before the diffusion layer has been depleted. Bond and Jones *(30)* found that for sampled d.c. polarography using a s.m.d.e., detection limits are on the order of 0.1 μM for Cd^{2+} and Ni^{2+}, comparable to the performance of differential pulse polarography with a normal d.m.e. The pulse techniques provide no further enhancement in sensitivity with the s.m.d.e.

4.6 THE ROTATING DISK ELECTRODE

The dropping mercury electrode has many significant advantages but we have seen that it has some limitations: the electrode material is restricted to mercury, which is relatively easily oxidized; and the time dependence of the electrode area leads to complications in theoretical descriptions of the experiment. One way around these limitations is to construct the electrode from an inert metal such as platinum and to use the electrode to stir the solution in some reproducible way such that the experiment is done under steady-state conditions. Several electrode designs have been tried in attempts to implement this strategy. For example, the so-called *rotating platinum electrode* has a small Pt wire is attached to the side of a glass rod; the rod is rotated so that the electrode is swept through the solution. Reproducible steady-state current-voltage curves are obtained, but the theoretical hydrodynamic problem proves to be very difficult. Other similar approaches have included vibrating wires, rotating loops, and a variety of other devices.

In recent years, the strategy of a moving electrode has focused in the rotating disk electrode (r.d.e.). The r.d.e., which is shown in Figure 4.10, may be of virtually any material, but platinum, gold, and glassy carbon are particularly common choices.

Hydrodynamics of the Rotating Disk

As the electrode rotates, adjacent solution is pulled along by viscous drag and is thrown away from the axis of rotation by the centrifugal force. The expelled solution is replaced by flow normal to the electrode surface. The rotating disk thus acts as a pump which moves solution up from below and then out away from the electrode as shown schematically in Figure 4.29.

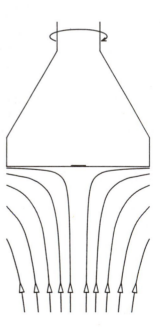

Figure 4.29 Schematic representation of solution flow at a rotating disk electrode.

The hydrodynamic problem can be solved to obtain the solution velocity. In cylindrical coordinates—r, ϕ, and x—the components of the velocity are expressed as a power series in x:

$$v_r = \omega r[a(x/x_H) - \tfrac{1}{2}(x/x_H)^2 - \tfrac{1}{3}b(x/x_H)^3 + ...] \qquad (4.28a)$$

$$v_\phi = \omega r[1 + b(x/x_H) + \tfrac{1}{3}a(x/x_H)^3 + ...] \qquad (4.28b)$$

$$v_x = -(\omega\nu)^{\frac{1}{2}}[a(x/x_H)^2 - \tfrac{1}{3}(x/x_H)^3 - \tfrac{1}{6}b(x/x_H)^4 + ...] \qquad (4.28c)$$

where $a = 0.510$, $b = -0.616$, ω is the angular speed in radians per second (2π times the rotation speed in hertz), $\nu = \eta/d$ is the kinematic viscosity of the solution,[‡] and

$$x_H = (\nu/\omega)^{\frac{1}{2}}$$

The x–component of the solution velocity is negative (motion is toward the electrode) and increases in magnitude with increasing x/x_H, approaching the limiting value, $v_x(\infty) = -0.884(\omega\nu)^{\frac{1}{2}}$. When $x = 3.6$ x_H, $v_x = 0.8\, v_x(\infty)$ and this value of x is taken as a measure of the thickness of the hydrodynamic layer, which tends to move with the electrode. For an aqueous solution with $\nu = 10^{-6}$ m^{2}s^{-1}, $x_H = 0.063$ mm when $\omega = 250$ rad s^{-1}. The hydrodynamic layer is about 0.23 mm thick under these conditions. Notice that this thickness decreases with increasing rotation speed; the solution can't keep up when the electrode rotates very rapidly.

Equations (4.28) were obtained assuming that the radius of the disk is large compared with the thickness of the hydrodynamic layer. In a typical r.d.e. (see Figure 4.10), the electrode is 2 mm in diameter but the disk as a whole (including insulation) is about 2 cm in diameter. As a rule of thumb, we can accept hydrodynamic layer thicknesses up to about one tenth the disk radius, i.e., up to a millimeter or so. This limitation imposes a lower limit on the rotation speed of about 10 rad s^{-1} (ca 100 rpm). In practice there is an upper limit on the rotation speed as well. The hydrodynamic problem was solved assuming laminar flow over the electrode surface. For very rapid rotation speeds, solution turbulence sets in and the equations fail. In practice turbulence usually becomes a problem with $\omega > 1000$ rad s^{-1} (10,000 rpm).

[‡] The units of the coefficient of viscosity η are kg m^{-1}s^{-1} and the units of density d are kg m^{-3}. The kinematic viscosity ν thus has units of m^{2}s^{-1} (the same units as diffusion coefficients).

A somewhat surprising, but important feature of the velocity is that the x-component is independent of r and ϕ. Thus the solution velocity perpendicular to the electrode is uniform over the electrode surface. At the surface only the ϕ-component is nonzero, reflecting the fact that solution on the surface moves with the rotating disk at angular speed ω.

Mass Transport to the Rotating Disk

In an experiment using a rotating disk electrode, material is transported to the electrode surface by a combination of diffusion and forced convection. The mass transport equation is obtained from Fick's second law by adding a term representing forced convection:

$$\frac{\partial C}{\partial t} = D \frac{\partial^2 C}{\partial x^2} - v_x \frac{\partial C}{\partial x} \tag{4.29}$$

Because v_x is independent of r and we are only interested in transport to the electrode (along the x-axis), we need only a one-dimensional equation. One of the great advantages of the r.d.e. is that the experiment is done under steady-state conditions. Thus we can set $\partial C/\partial t = 0$ and solve the simpler differential equation

$$D \frac{\partial^2 C}{\partial x^2} = v_x \frac{\partial C}{\partial x}$$

We take $v_x \cong -a(\omega \nu)^{\frac{1}{2}}(x/x_H)^2$, an acceptable approximation for $x/x_H < 0.2$, and use the boundary condition $C \to C^*$ as $x \to \infty$. Defining the dimensionless variable

$$u = (a/D)^{\frac{1}{3}} \omega^{\frac{1}{2}} \nu^{-\frac{1}{6}} x$$

the differential equation can be written as

$$\frac{d^2 C}{du^2} = -u^2 \frac{dC}{du}$$

Setting $C' = dC/du$, we can integrate

$$\int_{C_0'}^{C'} dC'/C' = -\int_0^u u^2 du$$

$$\ln(C'/C_0') = -\tfrac{1}{3} u^3$$

or

$$dC/du = (dC/du)_0 \exp(-u^3/3)$$

Integrating a second time, we have

$$C(x) = C(0) + (dC/du)_0 \int_0^{u(x)} \exp(-u^3/3)\, du \qquad (4.30)$$

We will need the derivative $(dC/du)_0$ in order to compute the flux at the electrode surface and the current. To evaluate it, we extend the integral limits to infinity and apply the boundary condition $C(x) \rightarrow C^*$ as $x \rightarrow \infty$. The definite integral

$$3^{-\frac{1}{3}} \int_0^{\infty} \exp(-u^3/3)\, du = \Gamma(\tfrac{4}{3}) = 0.893$$

is called a gamma function. Thus

$$C^* - C(0) = 3^{\frac{1}{3}} \Gamma(\tfrac{4}{3}) (dC/du)_0$$

or

$$C^* - C(0) = 1.288\, (D/a)^{\frac{1}{3}} \omega^{-\frac{1}{2}} \nu^{-\frac{1}{6}} (dC/dx)_0$$

For small x, we can assume a linear concentration gradient

$$(dC/dx)_0 = [C^* - C(0)]/x_D$$

where x_D is the diffusion layer thickness. Combining the last two equations, we can solve for x_D:

$$x_D = 1.288 \ (D/a)^{\frac{1}{3}} \omega^{-\frac{1}{2}} \nu^{\frac{1}{6}}$$

or, inserting a = 0.510,

$$x_D = 1.61 \ D^{\frac{1}{3}} \omega^{-\frac{1}{2}} \nu^{\frac{1}{6}} \tag{4.31}$$

The ratio of x_D to x_H then is

$$x_D/x_H = 1.61 \ (D/\nu)^{\frac{1}{3}}$$

For an aqueous solution ($D \cong 10^{-9} \ m^2 s^{-1}$, $\nu \cong 10^{-6} \ m^2 s^{-1}$) $x_D/x_H \cong$ 0.16, consistent with the assumption that the linear form of eq (4.28c) could be used in eq (4.29). This result affords a considerable simplification in practical calculations on experiments using the r.d.e. The same result could have been obtained had we neglected the forced convection term in eq (4.29) and solved the simple equation

$$D \ \frac{d^2C}{dx^2} = 0$$

We will have frequent recourse in Chapter 5 to this simplification.

Consider now our standard electrode process

$$O + n \ e^- \rightarrow R$$

The flux of O at the electrode surface is given by

$$J_O(0) = - D_O \ (dC/dx)_0$$

or

$$J_O(0) = - (D_O/x_D)[C_O{}^* - C_O(0)]$$

As usual, the electrode current is

$$i = - nFAJ_O(0)$$

$$i = nFA(D_O/x_D)[C_O{}^* - C_O(0)]$$

When the electrode is sufficiently negative that $C_O(0) = 0$, we have the limiting current,

$$i_L = nFAD_O C_O{}^*/x_D \qquad (4.32)$$

or

$$i_L = 0.62\, nFAC_O{}^* D_O{}^{\frac{2}{3}} \nu^{-\frac{1}{6}} \omega^{\frac{1}{2}} \qquad (4.33)$$

Equation (4.33) was first derived by Levich[‡] in 1942 *(31)* and is usually called the Levich equation. Notice that eq (4.32) is identical to eq (4.14). The analogy with the linear diffusion problem is still more complete, however, since we recall that we used eqs (4.14) and (4.15), together with the Nernst equation, to derive the Heyrovský–Ilkovič equation, eq (4.16), and this result is also valid for a reversible process at the rotating disk electrode. Since the flux of R can be written

$$J_R(0) = (D_R/x_{DR})C_R(0)$$

the current can be expressed in terms of $C_R(0)$:

$$i = nFA(D_R/x_{DR})C_R(0)$$

Thus we can write

$$\frac{i_L - i}{i} = \frac{(D_O/x_{DO})C_O(0)}{(D_R/x_{DR})C_R(0)}$$

or, taking logs, using eq (4.31) to evaluate x_{DO} and x_{DR}, and assuming nernstian behavior, we obtain the Heyrovský–Ilkovič equation:

$$E = E_{\frac{1}{2}} + \frac{RT}{nF} \ln \frac{i_L - i}{i}$$

[‡] Veniamin G. Levich (1917–) was Chairman of the Theoretical Department of the Institute of Electrochemistry of the Soviet Academy of Sciences. His contributions were mostly in the area of applications of hydrodynamics to problems in physical chemistry.

where

$$E_{\frac{1}{2}} = E° - \frac{2RT}{3nF} \ln \frac{D_O}{D_R} \qquad (4.34)$$

Current–potential curves for the r.d.e. thus look exactly like those for a planar electrode (Figure 4.2) at constant time, but half–wave potentials are shifted slightly.

Concentration profiles for a r.d.e. can be computed by numerical integration of eq (4.30) and are shown in Figure 4.30. The concentration gradient actually is quite linear over some distance from the electrode. Comparison with Figure 4.1 shows that the assumption of a linear concentration gradient is rather closer to the truth for the r.d.e. than for a stationary planar electrode. This of course means that results derived from a simplified linear diffusion model will be more accurate for the r.d.e. than for the d.m.e. or a planar electrode.

Further details on the rotating disk electrode can be found in the review by Riddiford *(32)* or in books by Albery *(A6)*, Levich *(C2)*, and Pleskov and Filinovskii *(C7)*.

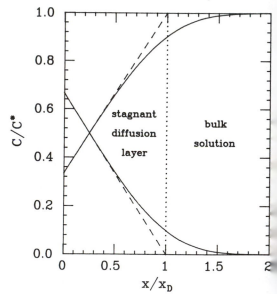

Figure 4.30 Concentration profiles for diffusion to a rotating disk electrode for $D_O = D_R$, $E = E° - 17.8$ mV.

Example 4.8 Estimate the range in limiting current for a one–electron reduction of O at a rotating disk electrode for $D_O = 10^{-9}$ m^2s^{-1}, $\nu = 10^{-6}$ m^2s^{-1}.

Substituting in eq (4.31), we have the diffusion layer thickness

$$x_D/m = 1.61 \times 10^{-4} \, \omega^{-\frac{1}{2}}$$

so that when ω ranges from 10 to 1000 rad s^{-1}, x_D goes from 51 μm to 5.1 μm. The limiting current for an electrode of 1 mm diameter and $C_O{}^* = 1$ mM is, from eq (4.33),

$$i_L/A = (3.03 \times 10^{-10})/x_D$$

Thus over the dynamic range of the r.d.e. ($10 < \omega < 1000$ rad s^{-1}) the limiting current ranges from 6 to 60 μA.

Notice that this range lies above the magnitude of currents expected for a d.m.e. This has advantages and disadvantages. Everything else being equal, the larger current implies a bigger signal–to–noise ratio. However, it also implies a larger iR drop in the solution. Rotating disk voltammetry generally is more susceptible to iR drop problems than is polarography.

The Rotating Ring–Disk Electrode

The hydrodynamics of the rotating disk electrode can be turned to advantage by adding another electrode as a ring surrounding the disk electrode as shown in Figure 4.31. As shown in Figure 4.29, solution which flows to the central disk electrode is then flung outward by centrifugal force and flows past the concentric ring electrode. Thus the solution can be sampled electrochemically at the ring shortly after undergoing an electron–transfer process. Solution of the convective diffusion problem is messy and we will not go into the details. The key parameter is the *collection efficiency*, i.e., the fraction of disk electrode product which can be sampled at the ring. This parameter depends on the electrode geometry only, independent of rotation speed, diffusion coefficients, etc., and is typically on the order of 0.5. Thus if a species is reduced at the disk, about half the product can be reoxidized at the ring. If the disk product is involved in a chemical reaction, the collection efficiency will be less than theoretical and will then depend on rotation speed, approaching theory at faster speeds.

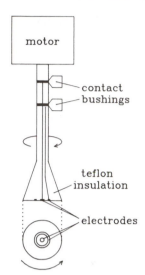

Figure 4.31 The rotating ring–disk electrode.

The method is comparable in some respects to cyclic voltammetry (see Section 4.3) in that the products of electron–transfer reactions can be studied, but, because the transport equations can be solved more exactly (it is a steady–state experiment) and because the ring electrode potential can be controlled independently of the disk, the method is capable of greater flexibility. The main disadvantage of the method is economic: ring–disk electrodes are difficult to make and thus are expensive. Two indicator electrode potentials must be controlled using a so–called bipotentiostat, which is about twice as complicated (and four times as expensive) as an ordinary potentiostat. For further details, see the book by Albery and Hitchman (*F2*).

4.7 APPLICATIONS

Applications of the voltammetric techniques discussed in this chapter have been extremely varied. For the purposes of discussion, however, we can roughly divide the applications into three groups:

(1) Quasi–thermodynamic applications, where a half–wave potential, taken as an approximation to the standard potential, is used in correlations with theory or with other kinds of experimental data or in computation of thermodynamic quantities.

(2) Analytical applications. Historically, analytical applications have been the most important and have provided the motivation for much of the development of voltammetric techniques.

(3) Mechanistic studies, in which electrochemical data are used to deduce the mechanism of the electrode process or the reactions which occur following oxidation or reduction of a substrate.

In this section, we will discuss some examples of the first two kinds of applications. Mechanistic applications will be discussed in Chapter 5.

"Standard" Potential/Electronic Structure Correlations

The free energy change associated with the one–electron reduction of O to R can be written

$$\Delta G^\circ = \Delta G^\circ(\text{gas}) + \Delta\Delta G^\circ(\text{solv})$$

where $\Delta G^\circ(\text{gas})$ is the free energy change for the gas–phase electron attachment process, and

$$\Delta\Delta G^\circ(\text{solv}) = \Delta G_R^\circ(\text{solv}) - \Delta G_O^\circ(\text{solv})$$

is the difference in solvation free energies of R and O. The free energy change is related as usual to the standard reduction potential,

$$\Delta G^\circ = -F(E^\circ + \text{const})$$

where the constant depends on the zero of the potential scale. If the gas–phase process involves no entropy change, then $\Delta G^\circ(\text{gas}) = -A$, the electron affinity. The half–wave potential measured in a polarographic experiment can be written, using eq (4.9):

$$E_{\frac{1}{2}} = \frac{A}{F} - \frac{\Delta\Delta G^\circ(\text{solv})}{F} - \frac{RT}{2F} \ln \frac{D_O}{D_R} - \text{const} \qquad (4.35)$$

If half–wave potentials are measured for a series of compounds using the same indicator and reference electrodes, the same solvent and supporting electrolyte, and the compounds are sufficiently similar that $\Delta\Delta G^\circ(\text{solv})$ and D_O/D_R may be assumed to be nearly constant, we might expect that $E_{\frac{1}{2}}$ will correlate well with changes in the electron affinity of the compounds. The electron affinity is expected to be related to the energy of the lowest unoccupied molecular orbital (the LUMO). Following a similar line of reasoning, we expect the oxidation potential to correlate with the energy of the highest occupied molecular orbital (the HOMO).

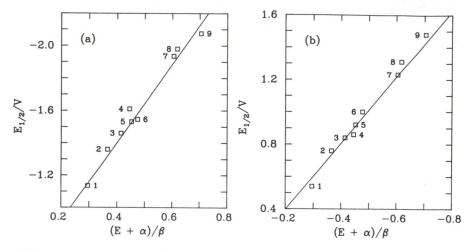

Figure 4.32 Correlation of half-wave potentials with Hückel MO theory energies for (a) polarographic reduction of aromatic hydrocarbons in 2-methoxyethanol solution (potentials measured *vs.* the mercury pool anode) and (b) oxidation of aromatic hydrocarbons at a rotating platinum electrode in acetonitrile solution (potentials measured *vs.* s.c.e.). The numbers correspond to: (1) tetracene, (2) 1,2–benzpyrene, (3) anthracene, (4) pyrene, (5) 1,2–benzanthracene, (6) 1,2,5,6–dibenzanthracene, (7) phenanthrene, (8) naphthalene, and (9) biphenyl. Data from Streitwieser *(34)*.

The original work along these lines was done in the 1950's by Hoijtink.[‡] Hoijtink *(33)* found that an excellent correlation is obtained between polarographic half-wave reduction potentials for aromatic hydrocarbons and LUMO energies obtained from Hückel molecular

---

[‡] G. Jan Hoijtink (1925–1978) was Professor of Physical Chemistry at the Free University, Amsterdam and later at the University of Amsterdam and the University of Sheffield. Hoijtink was a pioneer in mechanistic organic electrochemistry and was one of the first to apply ideas from molecular orbital theory to electrochemistry and spectroscopy.

orbital theory. Similar correlations are obtained between half–wave potentials for oxidations of aromatic hydrocarbons, measured with a rotating platinum electrode, and the HOMO energies obtained from Hückel molecular orbital theory. Some examples of such correlations are shown in Figure 4.32. (The energy scale of Figure 4.32 is in units of β, the "resonance integral" parameter of Hückel MO theory; the zero is relative to α, the "coulomb integral" parameter.)

This approach has been used extensively in organic electrochemistry not only for aromatic hydrocarbons but for compounds with many different functional groups. In the first instance, the motivation was to test molecular orbital theory calculations and the insights obtained from the calculations, but once the correlation is established, half–wave potentials for reductions or oxidations can be used to predict other chemical or physical properties which have been found to correlate with molecular orbital theory energies. A good review of this kind of work is given by Streitwieser *(34)*.

More recently, a similar approach has been used with organometallic systems. For example, it is well known that metal carbonyl complexes are stabilized by electron back donation from the metal into empty π^* orbitals on the CO ligand. Thus when CO is replaced by a less strongly π–acidic ligand, a phosphite for example, the electron density on the metal is expected to increase and the molecule should be easier to oxidize and harder to reduce. Since π–back donation to a CO ligand puts electron density in a CO π^* orbital, the infrared C–O stretching frequency is often taken as a measure of electron density on the metal. Thus with two physical parameters allegedly measuring the same thing, the obvious thing to do is to see if they correlate.

Example 4.9 de Montauzon and Poilblanc *(35)* have studied the reduction potentials of a series of nickel carbonyl complexes, $Ni(CO)_{4-n}[P(OMe)_3]_n$, where $n = 0, 1, 2$ and 3, in THF solution using a d.m.e. and a Ag/Ag^+ reference electrode. In this series, the reduction potential becomes more negative with increasing n, as expected. Similarly, the C–O stretching frequency decreases with increasing n, again as expected. The reduction potential is plotted *vs.* CO stretching frequency in Figure 4.33; a good (but somewhat nonlinear) correlation is obtained.

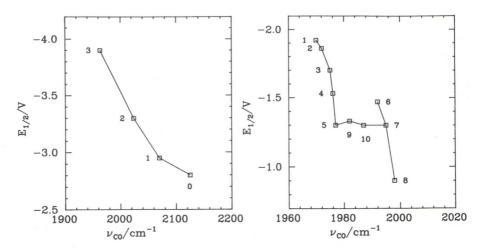

Figure 4.33 Correlation of polarographic half-wave potentials with C–O stretching frequencies for $Ni(CO)_{4-n}L_n$, where L = $P(OMe)_3$ and n = 0 – 3.

Figure 4.34 Correlation of polarographic half-wave potentials with C–O stretching frequencies for $[Co(CO)_3L]_2$ for various ligands L: (1) PMe_3, (2) PEt_3, (3) PMe_2Ph, (4) $PMePh_2$, (5) PPh_3, (6) $P(OEt)_3$, (7) $P(OMe)_3$, (8) $P(OPh)_3$, (9) Ph_2POMe, and (10) $PhP(OMe)_2$.

de Montauzon and Poilblanc also studied the correlation between $E_{\frac{1}{2}}$ and ν_{CO} for $[Co(CO)_3L]_2$ where L is a phosphine or phosphite ligand. The results are shown in Figure 4.34. Taking either the phosphines alone or the phosphites alone, there is a good correlation between $E_{\frac{1}{2}}$ and ν_{CO}, similar to that shown in Figure 4.33. However, the reduction potentials for L = PPh_3 and $P(OMe)_3$ are identical, whereas the C–O stretching frequencies differ by 18 cm^{-1}. Furthermore, the mixed ligands, $PPh_2[P(OMe)_3]$ and $PPh[P(OMe)_3]_2$, show equal values of $E_{\frac{1}{2}}$ but C–O stretching frequencies intermediate between the PPh_3 and $P(OMe)_3$ derivatives. The reason for this difference appears to be that both PPh_3 and $P(OMe)_3$ are strong σ–donors but $P(OMe)_3$ is a moderately good π–acid. Thus $P(OMe)_3$, which feeds considerable charge density into the metal acceptor orbitals, also removes charge density from the metal π–donor orbitals. The C–O stretching frequency is particularly sensitive to π–electron density, and C–O stretching frequency shifts probably reflect mostly change in metal π–electron density rather than overall charge density. Thus half–wave potentials may be a better measure of gross charge density than C–O stretching frequencies.

Stability Constants of Coordination Complexes

The voltammetric behavior of metal coordination complexes is determined in part by the kinetic stability or lability of the complex. For substitution–inert species such as Co(III) or Cr(III) complexes, electron transfer is often followed by rapid chemical steps. In more labile systems, it is often necessary to consider equilibria among several different species prior to electron transfer, with only the most easily reduced species participating in the electron–transfer step. We will discuss the general case in Section 5.6. If the system is sufficiently labile that the complexation steps can be assumed to be always at equilibrium, even at the electrode surface, we can deal with the expected behavior using the techniques we have developed in this chapter.

Consider an aquo metal ion, $M(H_2O)_m^{n+}$, which is reversibly reduced to the metal

$$M(H_2O)_m^{n+} + n\ e^- \rightleftarrows M(Hg) + m\ H_2O$$

and suppose that the metal ion forms a series of coordination complexes with the electroinactive ligand L:

$$M(H_2O)_m^{n+} + j\ L \rightleftarrows M(H_2O)_{m-j}L_j^{n+} + j\ H_2O$$

with overall formation constants

$$\beta_j = \frac{[M(H_2O)_{m-j}L_j^{n+}]}{[M(H_2O)_m^{n+}][L]^j}$$

To keep the notation relatively compact, we will label the species 0, 1, ... j ... m, so that C_0 is the concentration of $M(H_2O)_m^{n+}$ and C_j is the concentration of $M(H_2O)_{m-j}L_j^{n+}$. We suppose that these equilibria are very fast, so that the relative concentrations at the electrode surface are the same as in bulk solution. Since the various complex species are rapidly interconverted, the electrode current is related to the total flux of all species at the electrode surface:

$$i = -nFA \sum_{j=0}^{n} J_j(0)$$

The fluxes of the various species are determined by the concentration gradients. If k_{Dj} is the mass transport rate constant for delivery of species j to the electrode, we can write

$$J_j(0) = -k_{Dj} [C_j{}^* - C_j(0)]$$

so that the current is

$$i = nFA \sum_j k_{Dj} [C_j{}^* - C_j(0)]$$

When the potential is sufficiently negative that reduction of the most easily reduced species is nearly complete, then all $C_j(0) = 0$ and the current is diffusion limited,

$$i_D = nFA \sum_j k_{Dj} C_j{}^*$$

Subtracting i from i_D, we have

$$i_D - i = nFA \sum_j k_{Dj} C_j(0)$$

If the complexation steps are fast enough that they are at equilibrium even at the electrode surface, and if the free ligand concentration is sufficiently high that C_L is uniform throughout the solution, then we have

$$C_j(0) = \beta_j C_0(0) C_L{}^j$$

where $\beta_0 = 1$. Thus

$$i_D - i = nFAC_0(0) \sum_j k_{Dj} \beta_j C_L{}^j \tag{4.36}$$

Reduction of a metal complex at the d.m.e. produces metal atoms which dissolve in the mercury and diffuse into the interior of the drop with a flux

$$J_M(0) = k_{DM} [C_M{}^* - C_M(0)]$$

where k_{DM} is the rate constant for diffusion of M in the amalgam; we assume that the concentration of metal in the interior of the drop, C_M^*, is negligible. This flux is also related to the electrode current,

$$i = -nFAJ_M(0) = nFAk_{DM}C_M(0) \qquad (4.37)$$

If we assume that the electron–transfer process is reversible, the concentration ratio $C_0(0)/C_M(0)$ must be given by the Nernst equation

$$\frac{C_0(0)}{C_M(0)} = \exp \frac{nF(E - E^\circ)}{RT}$$

Using eqs (4.36) and (4.37), this ratio can be written

$$\frac{C_0(0)}{C_M(0)} = \frac{i_D - i}{\sum k_{Dj}\beta_j C_L^j} \cdot \frac{k_{DM}}{i}$$

Combining the two concentration ratio expressions, we have

$$E = E^\circ + \frac{RT}{nF} \ln \frac{k_{DM}}{\sum k_{Dj}\beta_j C_L^j} + \frac{RT}{nF} \ln \frac{i_D - i}{i} \qquad (4.38)$$

which is just the Heyrovský–Ilkovič equation with

$$E_{\frac{1}{2}} = E^\circ + \frac{RT}{nF} \ln \frac{k_{DM}}{k_D} - \frac{RT}{nF} \ln \sum_j \beta_j C_L^j \qquad (4.39)$$

where we have assumed that the complex species all have nearly the same diffusion coefficient so that $k_{Dj} = k_D$. Since

$$k_{DM}/k_D = (D_M/D_0)^{\frac{1}{2}}$$

the first two terms are just the half–wave potential in the absence of ligand, eq (4.9). The last term in eq (4.39) then represents the shift in half–wave potential as a function of ligand concentration. When the complexation steps are sufficiently well separated that, for a particular range of free ligand concentrations, only two species have significant concentrations in the bulk solution, then one term, $\beta_j C_L^j$, will dominate in the sum and we can simplify eq (4.39) to

$$E_{\frac{1}{2}} = E^{\circ} + \frac{RT}{2nF} \ln \frac{D_M}{D_0} - \frac{RT}{nF} \ln \beta_j C_L{}^j$$

or

$$\Delta E_{\frac{1}{2}} = -\frac{RT}{nF} \ln \beta_j - \frac{jRT}{nF} \ln C_L \qquad (4.40)$$

Equation (4.40) was first used by Lingane around 1940 and is sometimes called the Lingane equation.

Example 4.10 Lingane *(36)* used eq (4.40) to analyze polarographic half-wave potential data obtained for Pb(II) in 0.01–1.0 M NaOH solutions (of constant 1.0 M ionic strength). Figure 4.35 shows a plot of $E_{\frac{1}{2}}$ *vs.* pOH. The slope of the straight line is 0.0799 ± 0.0015, which corresponds to j = 2.70 ± 0.05. Lingane interpreted this result as indicating that Pb(II) in basic solution is $HPbO_2^-$ with the equilibrium

$$Pb^{2+} + 3\ OH^- \rightleftharpoons HPbO_2^- + H_2O$$

preceding electron transfer. The deviation of j from the integral value of 3 was explained as arising from the use of OH^- concentrations rather than activities in the plot.

Only a relatively small number of complexation equilibria can be characterized using Lingane's approach. Even when equilibria are fast enough, it is not often possible to focus on just one step. When several equilibria are simultaneously important, the more complete expression, eq (4.39), must be used. Several graphical methods have been developed to fit half–wave potential data to an equilibrium model. One way of dealing with the problem is to rearrange eq (4.39) to

$$\Delta E_{\frac{1}{2}} = -\frac{RT}{nF} \ln \sum_j \beta_j C_L{}^j$$

and define the function $F_0(C_L)$,

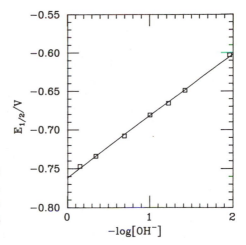

Figure 4.35 Polarographic half–wave potentials (*vs.* s.c.e.) for Pb(II) in KNO_3/NaOH solutions at $25°C$.

$$F_0(C_L) = \sum_i \beta_j C_L^{\,j} = \exp \frac{-nF\Delta E_{\frac{1}{2}}}{RT}$$

Since $\Delta E_{\frac{1}{2}}$ is measured, the function $F_0(C_L)$ can be computed for each experimental point. The function then is decomposed to

$$F_1(C_L) = [F_0(C_L) - 1]/C_L = \beta_1 + \beta_2 C_L + \beta_3 C_L^2 + \cdots$$

A plot of $F_1(C_L)$ *vs.* C_L should give a curve with an intercept β_1 at $C_L = 0$. Similarly defining $F_2(C_L)$,

$$F_2(C_L) = [F_1(C_L) - \beta_1]/C_L = \beta_2 + \beta_3 C_L + \cdots$$

we see that a plot of $F_2(C_L)$ *vs.* C_L allows the determination of β_2 by extrapolation to $C_L = 0$. Graphical methods such as this are inherently inaccurate but have the advantage that the experimenter gets a good "feel" for the data, making it less likely that the data will be fitted to a model which they really do not support. However, a nonlinear least–squares refinement of the data should follow the rough graphical analyses. See Crow *(37)* for further details and examples.

Analytical Applications

Polarography was the first partially automated instrumental analytical technique and the first technique capable of routine analysis at the submillimolar level; as such polarography had a tremendous impact on analytical chemistry and was widely used from the 1930's. With the development of competing instrumental methods with even greater sensitivity and specificity (*e.g.*, atomic absorption spectrophotometry), analytical applications of polarography went into a decline in the 1960's. Classical d.c. polarography is limited to analyte concentrations in the $10 \, \mu M - 1 \, mM$ range, but the detection limits are lowered considerably with the pulse techniques (or the a.c. methods to be discussed in Section 5.4), which suppress the interference of the capacitive charging current. The development of these techniques brought about a renaissance of analytical voltammetry in the 1970's *(38)*. Electrochemical methods are now competitive with or preferable to spectroscopic methods for many analyses. Developments currently under way—*e.g.*, the static mercury drop electrode and digital techniques to harness the power of microcomputers to electrochemical experiments—promise to keep analytical voltammetry competitive in the years to come *(39,D12)*.

Polarographic methods have been developed for most of the common metal ions, including the alkali metals. Indeed, Heyrovský's first paper on polarography *(40)* reported the reduction of the alkali metal ions. Advantage is taken of the large hydrogen overpotential on mercury (see Section 5.2) so that well–defined waves are seen for the alkali metals in weakly acidic or basic aqueous solutions. Unfortunately, however, Na^+, K^+, Rb^+, and Cs^+ all reduce at nearly the same potential so that polarography is unable to distinguish among them. (Lithium is sufficiently different that it can be determined in the presence of the other alkali metals.)

Although the polarographic diffusion current is proportional to concentration, the proportionality constant includes the diffusion coefficient and the dropping mercury electrode parameters (mercury flow rate and drop time). Polarographic methods reported in the analytical literature usually include the so–called diffusion current constant,

$$I_D = i_D/u^{\frac{2}{3}} t^{\frac{1}{6}} C$$

so that a reported procedure in principle can be used without calibration. As a practical matter, however, careful work should include the preparation of a calibration curve determined under the actual experimental conditions used in the analysis. Because of metal ion complexation and variations in activity coefficients, the half–wave potential and diffusion current constant are often strongly medium dependent. These parameters are given for Cu(II), Pb(II), Cd(II), and

TABLE 4.1 Polarographic Data for Some Metal Ions

Medium	$E_{\frac{1}{2}} (I_D)^{a,b}$			
	Cu(II)	Pb(II)	Cd(II)	Zn(II)
0.1 M KCl	0.04(3.23)	−0.40(3.80)	−0.60(3.51)	−1.00(3.42)
1 M KCl	0.04 −0.22(3.39)^c	−0.44(3.86)	−0.64	−1.02
1 M HNO$_3$	−0.01(3.24)	−0.40(3.67)	−0.59(3.06)	d
0.5 M H$_2$SO$_4$	0.00(2.12)	e	−0.59(2.6)^d	d
1 M NH$_3$, 1 M NH$_4$Cl	−0.24 −0.50(3.75)^c		−0.81(3.68)	−1.33(3.82)
1 M NaOH	−0.42(2.91)	−0.76(3.39)	e	−1.53(3.14)
pH 4.5 buffer^f	−0.09(2.37)	−0.48(2.37)	−0.64(2.34)	−1.23^d
pH 9.0 buffer^f	−0.12(2.24)	−0.50(2.30)	−0.64(2.34)	−1.15(2.30)

[a] From Kolthoff and Lingane (D1). Much more extensive tables have been published by Meites (41).

[b] Diffusion current constant in units of $\mu A\ mM^{-1}(mg\ s^{-1})^{-\frac{2}{3}} s^{-\frac{1}{6}}$ Half-wave potential in volts vs. s.c.e.

[c] Total diffusion current for double wave.

[d] Wave masked by hydrogen reduction.

[e] Insoluble.

[f] 0.5 M sodium tartrate buffer.

Zn(II) in several aqueous electrolyte solutions in Table 4.1. Copper(II), for example, gives polarographic waves ranging over about half a volt, depending on the medium. In the presence of high chloride ion concentrations, which stabilize Cu(I), two waves are observed which correspond to the stepwise reduction of Cu(II).

Classical polarographic analytical methods are often transferable with better sensitivity and/or better resolution to more modern techniques such as r.d.e. voltammetry and the various pulse techniques.

One of the advantages of polarography and the other voltammetric methods is the ability to distinguish among and analyze simultaneously for several species. Resolution is particularly good with the differential pulse technique (see Figure 4.28). Since half–wave potentials are medium dependent, the solvent or aqueous buffer system used in a given analysis is chosen to optimize resolution of the expected waves. In actual analytical work, it is usually necessary to separate the analytes from a complex mixture, leaving behind interfering species. Ideally, this initial separation step should leave the analytes in a form that will give a well–resolved voltammogram. One general approach to this problem is to extract metal ions as chelate complexes soluble in a polar aprotic solvent.

Example 4.11 A weakness of differential pulse polarography is that a single peak is obtained in analysis of solutions where the analyte is present in both oxidized and reduced forms. Thus, for example, a solution containing Fe(II) and Fe(III) as aquo ions (or other complexes related by simple electron transfer) gives a single differential pulse polarographic peak with peak current proportional to the sum of the Fe(II) and Fe(III) concentrations. In order to achieve the higher sensitivity of the differential pulse method while at the same time resolving Fe(II) and Fe(III) into separate peaks, Leon and Sawyer *(42)* extracted iron into propylene carbonate, Fe(II) as the *tris*–4,7–diphenyl–1,10–phenanthroline complex and Fe(III) as the *tris*–8–quinolinato complex. These complexing agents are highly selective for Fe(II) and Fe(III), respectively. The differential pulse polarogram then shows well–separated reduction peaks for Fe(III) and Fe(II) at –0.55 and –1.25 V *vs.* s.c.e., respectively, with peak currents linear in concentration in the 2 – 200 μM range.

In the presence of ligands which form complexes with mercury, *e.g.*, halides and pseudohalides (Cl^-, SCN^-, etc.), an anodic wave is observed in polarography which corresponds to the oxidation of mercury:

$$Hg + p\,X^- \rightarrow HgX_p^{\,2-p} + 2\,e^-$$

If the formation constant of the mercury complex is large enough, the wave is shifted to a more negative potential and is resolved from the normal mercury oxidation process. Since the process relies on the diffusion of X^- to the electrode surface, the limiting current is proportional to the bulk concentration of X^-. Using sensitive techniques such as pulse polarography or sampled d.c. polarography on a static mercury drop electrode, this wave provides a good way of determining a wide range of anions and other complexing agents. In addition, the wave shape and half-wave potential can be used to determine the stoichiometry of the process and to estimate the formation constant of the mercury complex (see Problems).

Example 4.12 Kirowa–Eisner and Osteryoung *(43)* have reported a polarographic method for the determination of small concentrations of hydroxide ion which takes advantage of the mercury oxidation wave. The electrode process is

$$Hg + 2\,OH^- \rightarrow Hg(OH)_2 + 2\,e^-$$

where $Hg(OH)_2$ is a solution species at the very low concentrations formed in polarography. The half-wave potential is linear in pOH, varying from 0.150 to 0.117 V (*vs.* s.c.e.) as [OH^-] increases from 10 μM to 0.1 mM. With the static mercury drop electrode, the technique affords a nearly unique method for the determination of hydroxide ion in the micromolar concentration range; the detection limit is 0.3 μM and a linear response is obtained up to 0.4 mM. Other species which form mercury complexes interfere, of course; thus Cl^-, for example, must be at least five times lower in concentration than OH^-.

Example 4.13 Polarograms of basic aqueous solutions of thiols show anodic waves corresponding to the formation of mercury thiolates,

$$Hg + 2\,RS^- \rightarrow Hg(SR)_2 + 2\,e^-$$

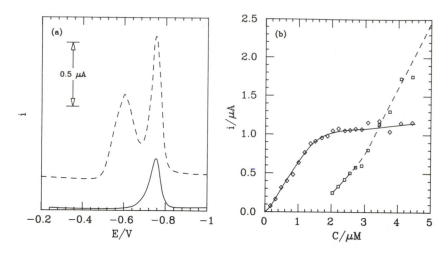

Figure 4.36 (a) Differential pulse polarograms of 0.05 mM (solid line) and 0.29 mM (dashed line) benzyl mercaptan in 0.2 M NaOH for 50 mV pulses and a negative-going potential scan. (b) Differential pulse peak currents for the –0.75 V peak (solid line) and for the –0.60 V peak (dashed line) as functions of concentration for benzyl mercaptan.

Birke and Mazorra *(44)* have shown that a variety of thiols give differential polarographic peak currents linear in thiol concentration over the range 0.1 μM to 0.1 mM. Above 0.1 mM, two peaks are observed as shown in Figure 4.36a, with the sum of peak currents approximately continuing the calibration curve (Figure 4.36b). This behavior is reminiscent of adsorption effects in d.c. polarography (see Section 4.4) and Birke and Mazorra suggest that the insoluble mercury thiolate product deposits on the electrode surface. At low concentrations, less than monolayer coverage occurs, but above about 0.1 mM, a multilayer structure is formed in two distinct steps. The method shows good sensitivity when the medium is 0.1 M NaOH and the thiols are ionized. The sensitivity decreases markedly when the pH is less than the pK_a of the thiol.

Voltammetry is by no means confined to inorganic analysis. Voltammetric methods have been developed for many organic functional groups *(45)*, for organometallics *(46)*, for pharmaceuticals *(47–49)*, and for compounds of biological interest *(50,51,D9)*. The advent of microelectrodes has been particularly important for the development of methods for *in vivo* determinations of electroactive species in biological systems *(52)*.

Example 4.14 Albery and co–workers *(53)* have performed linear potential sweep voltammetric experiments with a carbon paste microelectrode, a Ag/AgCl reference electrode, and a stainless steel auxiliary electrode implanted in the brain of a rat. It was hoped that oxidation peaks corresponding to catecholamines could be observed. (The catecholamines are intimately involved in the function of nerve cells.) Several oxidation peaks are observed as shown in Figure 4.37, and one of the goals of this research was the

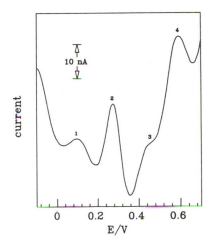

Figure 4.37 Voltammogram obtained with a carbon paste electrode in the striatum of a rat. The current peaks have been sharpened by electronic "semidiffer- entiation".

identification of the peaks. *In vitro* experiments showed that both catecholamines and ascorbic acid are oxidized at about the potential of peak 1 in the voltammogram. However, careful experiments in which catecholamines were injected into the rat's brain showed small shifts in the peak potential, whereas addition of ascorbic acid simply increased the peak size; thus peak 1 was assigned to ascorbic acid. Peak 2 was identified, after similar experiments, as corresponding to the oxidation of indoles (probably 5–hydroxyindoleacetic acid) and/or glutathione. Peaks 3 and 4 were not identified with certainty. Comparison of the *in vivo* and *in vitro* results suggest an ascorbic acid concentration on the order of 100 μM, and the resulting peak thus obscures the oxidation of much smaller amounts of catecholamines which might be present.

REFERENCES

(Reference numbers preceded by a letter, *e.g. (D4)*, refer to a book listed in the Bibliography.)

1. F. G. Cottrell, *Z. phys. Chem.* **1903**, *42*, 385.
2. J. Heyrovský and D. Ilkovič, *Coll. Czech. Chem. Commun.* **1935**, 7, 198.
3. J. Tomeš, *Coll. Czech. Chem. Commun.* **1937**, *9*, 12, 81, 150.
4. D. K. Roe in *Laboratory Techniques in Electroanalytical Chemistry*, P. T. Kissinger and W. R. Heineman, eds, New York: Marcel Dekker, 1984, p 193.
5. C. K. Mann, *Electroanalytical Chemistry* **1969**, *3*, 57.
6. A. J. Fry and W. E. Britton in *Laboratory Techniques in Electroanalytical Chemistry*, P. T. Kissinger and W. R. Heineman, eds, New York: Marcel Dekker, 1984, p 367.
7. G. Gritzer, *Inorg. Chim. Acta* **1977**, *24*, 5; G. Gritzer and J. Kuta, *Electrochim. Acta* **1984**, *29*, 869.
8. R. N. Adams in *Treatise on Analytical Chemistry*, I. M. Kolthoff and P. J. Elving, eds, Part I, Vol. 4, New York: Interscience, 1963, p 2381.
9. Z. Galus in *Laboratory Techniques in Electroanalytical Chemistry*, P. T. Kissinger and W. R. Heineman, eds, New York: Marcel Dekker, 1984, p 267.
10. G. Dryhurst and D. L. McAllister in *Laboratory Techniques in Electroanalytical Chemistry*, P. T. Kissinger and W. R. Heineman, eds, New York: Marcel Dekker, 1984, p 289.
11. N. Winograd in *Laboratory Techniques in Electroanalytical Chemistry*, P. T. Kissinger and W. R. Heineman, eds, New York: Marcel Dekker, 1984, p 321.
12. F. M. Hawkridge in *Laboratory Techniques in Electroanalytical Chemistry*, P. T. Kissinger and W. R. Heineman, eds, New York: Marcel Dekker, 1984, p 337.
13. C. N. Reilley, W. D. Cooke, and N. H. Furman, *Anal. Chem.* **1951**, *23*, 1226.
14. J. T. Stock, *Amperometric Titrations*, New York: Interscience, 1965.
15. W. H. Reinmuth, *Anal. Chem.* **1960**, *32*, 1509.
16. P. Delahay in *Treatise on Analytical Chemistry*, I. M. Kolthoff and P. J. Elving, eds, Part I, Vol. 4, New York: Interscience, 1963, p 2233.
17. D. G. Davis, *Electroanalytical Chemistry* **1966**, *1*, 157.
18. H. T. S. Sand, *Phil. Mag.* **1900**, *1*, 45; *Z. phys. Chem.* **1900**, *35*,

641.

19. R. S. Nicholson and I. Shain, *Anal. Chem.* **1964**, *36*, 706.
20. A. M. Bond, M. Fleischmann, and J. Robinson, in *Electrochemistry: The Interfacing Science,* D. A. J. Rand and A. M. Bond, eds, Amsterdam: Elsevier, 1984.
21. J. Heinze, *J. Electroanal. Chem.* **1981**, *124*, 73.
22. R. M. Wightman, *Anal. Chem.* **1981**, *53*, 1125A.
23. L. Meites in *Treatise on Analytical Chemistry,* I. M. Kolthoff and P. J. Elving, eds, Part I, Vol. 4, New York: Interscience, 1963, p 2303.
24. D. Ilković, *Coll. Czech. Chem. Commun.* **1934**, *6*, 498.
25. H. H. Bauer, *Electroanalytical Chemistry* **1975**, *8*, 170.
26. G. C. Barker and A. W. Gardner, *Z. anal. Chem.* **1960**, *173*, 79.
27. J. R. Morris, Jr. and L. R. Faulkner, *Anal. Chem.* **1977**, *49*, 489.
28. J. Osteryoung and E. Kirowa-Eisner, *Anal. Chem.* **1980**, *52*, 62.
29. J. H. Christie, J. A. Turner, and R. A. Osteryoung, *Anal. Chem.* **1977**, *49*, 1899; J. A. Turner, J. H. Christie, M. Vukovic, and R. A. Osteryoung, *Anal. Chem.* **1977**, *49*, 1904.
30. A. M. Bond and R. D. Jones, *Anal. Chim. Acta* **1980**, *121*, 1.
31. V. G. Levich, *Acta Physicochim. URSS* **1942**, *17, 257.*
32. A. C. Riddiford, *Adv. Electrochem. Electrochem. Engin.* **1966**, *4*, 47.
33. G. J. Hoijtink, *Rec. trav. chim.* **1955**, *74*, 1525; *Adv. Electrochem. Electrochem. Engin.* **1970**, *7*, 221.
34. A. Streitwieser, *Molecular Orbital Theory for Organic Chemists,* New York: John Wiley, 1961.
35. D. de Montauzon and R. Poilblanc, *J. Organometal. Chem.* **1976**, *104*, 99.
36. J. J. Lingane, *Chem. Rev.* **1941**, *29*, 1.
37. D. R. Crow, *Polarography of Metal Complexes,* New York: Academic Press, 1969.
38. J. B. Flato, *Anal. Chem.* **1972**, *44* (11), 75A.
39. A. M. Bond, *J. Electroanal. Chem.* **1981**, *118*, 381.
40. J. Heyrovský, *Chem. Listy* **1922**, *16*, 256; *Phil. Mag.* **1923**, *45*, 303.
41. L. Meites (and co-workers), *Anal. Chim. Acta* **1956**, *14*, 390, 482; **1959**, *20*, 397; **1978**, *98*, 163.
42. L. E. Leon and D. T. Sawyer, *Anal. Chem.* **1981**, *53*, 706.
43. E. Kirowa-Eisner and J. Osteryoung, *Anal. Chem.* **1978**, *50*, 1062.
44. R. L. Birke and M. Mazorra, *Anal. Chim. Acta* **1980**, *118*, 257.
45. P. Zuman, *Talanta* **1965**, *12*, 1337.
46. M. D. Harris, *Electroanalytical Chemistry* **1974**, *7*, 79.
47. G. J. Patriarche, M. Chateau-Gosselin, J. L Vandenbalck, and P.

Zuman, *Electroanalytical Chemistry* **1979**, *11*, 141.

48. H. Siegerman, *Electroanalytical Chemistry* **1979**, *11*, 291.

49. M. A. Brooks in *Laboratory Techniques in Electroanalytical Chemistry*, P. T. Kissinger and W. R. Heineman, eds, New York: Marcel Dekker, 1984, p 569.

50. P. Zuman, in *Experimental Methods in Biophysical Chemistry*, C. Nicolau, ed, New York: John Wiley, 1973.

51. A. L. Underwood and R. W. Burnett, *Electroanalytical Chemistry* **1973**, *6*, 1.

52. J. Koryta, M. Březina, J. Praděč, and J. Praděčová, *Electroanalytical Chemistry* **1979**, *11*, 85.

53. R. D. O'Neill, R. A. Grünewald, M. Fillenz, and W. J. Albery, *Neuroscience* **1982**, *7*, 1945.

PROBLEMS

4.1 Show that eqs (4.2) are indeed solutions to the diffusion equations and that they satisfy the boundary conditions.

4.2 Derive eq (4.17) assuming diffusion of O across a linear concentration gradient of thickness x_D.

4.3 Show that inclusion of the activity coefficient ratio in eq (4.1) leads to another term in the expression for $E_{\frac{1}{2}}$:

$$E_{\frac{1}{2}} = E^\circ - \frac{RT}{2nF} \ln \frac{D_O}{D_R} + \frac{RT}{nF} \ln \frac{\gamma_O}{\gamma_R}$$

4.4 Use the standard potential for the Fe^{3+}/Fe^{2+} couple, diffusion coefficients, estimated from the molar ionic conductivities of Fe^{3+} and Fe^{2+}, and activity coefficients, estimated using the Debye–Hückel limiting law, to compute the half–wave potential for the reduction of Fe^{3+} to Fe^{2+} at 25°C in 0.1 M $KClO_4$ at 25°C.

4.5 A triangular wave signal (needed for potential control in cyclic voltammetry) can be generated using a voltage integrator such as shown in Figure 4.7. What kind of input signal would be required?

4.6 An alternative design for a potentiostat connects the reference electrode directly to the positive operational amplifier input, eliminating the need for a voltage follower. The potential control inputs are summed and applied to the negative input. Sketch the circuit diagram. How would the indicator electrode potential (relative to the reference) vary for a negative–going ramp as the potential control input?

4.7 The analysis of the performance of the operational amplifier circuits in Section 4.2 assumed infinite gain and infinite input impedance. In real circuits, the gain and input impedance are finite, typically 10^4–10^6 and 1–10^5 MΩ, respectively. Consider the effect of finite gain and input impedance for each of the following circuits:

(a) Current follower (the current measuring circuit of Figure 4.6). Assume that R_2 = 10 kΩ and estimate the minimum amplifier gain required to keep the indicator electrode within 0.1 mV of ground potential when the cell current is 0.1 mA.

(b) Potentiostat (Figure 4.6). Assuming that R = 100 kΩ, that the amplifier input impedance is 100 MΩ, the control potential is 1.00 V, and the auxiliary electrode potential must be –2.00 V, what gain is required if the magnitude of the reference electrode potential differs from the control potential by no more than 0.1 mV.

(c) Voltage ramp generator (Figure 4.7). Assuming that R = 10 MΩ, C = 10 μF, Φ_{in} = 1.0 V, an input impedance of 10 MΩ and a gain of 10^6, estimate the nonlinearity of the ramp; *i.e.*, by how much does the potential scan rate differ when the output voltage is 0 V and when it is –1 V?

(d) Galvanostat (Figure 4.8). Assuming the operational amplifier parameters of part (c), a control voltage of –1.00 V, resistance R = 10 kΩ, and a maximum output voltage from the amplifier of 10 V, what is the maximum effective cell resistance? By how much would the current then differ from the nominal value?

4.8 Assuming that reduced species R forms a monolayer on the electrode surface, what is the critical concentration below which only a polarographic adsorption pre–wave is observed if the drop time is 4 s and the mercury flow rate is 1.78 mg s^{-1}, D_O = 10^{-9} m^{2}s^{-1}, and each adsorbed R molecule occupies a surface area of 0.25 nm^2?

4.9 In the polarogram of O, which is reduced to R, the main reduction wave is preceded by an adsorption pre–wave. On the plateau following the pre–wave, the current is limited by the available surface area on the electrode. Assuming that the concentration of O is large, that only a monolayer of R is adsorbed, and that the potential is on the pre–wave plateau, prepare a plot of the current variation with time during the life of one drop; compare with Figure 4.23.

4.10 In a potential step chronoamperometry experiment, the observed current includes a contribution due to charging of the double-layer capacitance. Consider an experiment in which species O is reduced by one electron at a disk electrode 2 mm in diameter with a double–layer capacity $C_d = 0.5$ f m^{-2} in contact with a solution of effective resistance $R = 1500$ Ω. Assume $C_O{}^* = 1$ mM, $D_O = 10^{-9}$ m^{2}s^{-1} and a potential step from $E = E_{\frac{1}{2}} + 0.25$ V to $E_{\frac{1}{2}} - 0.25$ V.
(a) Compute and plot the faradaic current over the time range 0–20 ms.
(b) On the same graph plot the capacitive charging current and the total current. Hint: see Problem 2.8.
(c) In practice, how could the interference of capacitive charging current be avoided?

4.11 The technique of chronoamperometry can be improved for analytical purposes by electronically integrating the current and recording charge *vs.* time curves. The experiment is then called *chronocoulometry*.
(a) Show that the chronocoulometric response for a reversible electrode process is

$$Q = 2nFAC_O{}^*(D_O t/\pi)^{\frac{1}{2}}$$

(b) One advantage of chronocoulometry over chronoamperometry is that the signal proportional to concentration grows with time rathen than decaying. However, the current contribution from double–layer charging is also integrated and thus contributes to the signal long after the capacitive current has decayed to zero. Use the data of Problem 4.10 to compute the faradaic and capacitve charges and the total charge and plot these *vs.* t and $\sqrt{t}$.
(c) In practice, it is not difficult to separate the faradaic and capacitive contributions to Q. Describe how this would be done in reference to the data plotted in part (b).

4.12 What current should be used to obtain a chronopotentiometric transition time of 10 s if the concentration of electroactive material is 0.1 mM, the electrode is a disk of 1 cm diameter, the diffusion coefficient is 5×10^{-10} m^2 s^{-1} and the electrode process is a one–electron reduction?

4.13 The assumption of linear diffusion breaks down when the dimensions of the electrode are comparable to the thickness of the diffusion layer. Suppose that a chronoamperometry experiment lasts for one second, and that the electroactive material has a diffusion coefficient $D_O = 10^{-9}$ m^2 s^{-1}. What is the minimum diameter of the electrode if we require that the diameter be more than ten times the maximum diffusion layer thickness?

4.14 The diffusion–controlled current at a microelectrode does not decay to zero but reaches a time independent value. In a potential step chronoamperometry experiment, what electrode radius is required if the current is to fall to within 10% of its limiting value in 0.1 s. Assume a diffusion coefficient of 10^{-9} $m^2 s^{-1}$.

4.15 A series of rotating disk voltammograms at varying rotation speeds gave the following half–wave potentials:

ω/rad s^{-1}	$E_{\frac{1}{2}}$/V
20	−0.699
50	−0.703
100	−0.707
200	−0.713
500	−0.724

(a) Assuming that the electrode process is reversible and that the shift in half–wave potential is due to solution iR drop, determine the "true" value of $E_{\frac{1}{2}}$.
(b) If the limiting current for $\omega = 500$ rad s^{-1} is 42.0 μA, what is the effective solution resistance?

4.16 The pulse sequence for differential pulse polarography shown in Figure 4.27 gives current–potential curves like those of Figure 4.28. Suppose that the pulse sequence is rearranged so that positive–going pulses are superimposed on a negative–going ramp. Sketch the resulting current–potential curve. How is the peak potential related to the half–wave potential?

4.17 Hückel molecular orbital theory gives energies in units of a parameter β, the C–C bond resonance integral. Use the half-wave potential correlations shown in Figure 4.32 to obtain an estimate for β in units of eV.

4.18 An anodic wave is observed for solutions containing halide ions or other species which coordinate Hg(II). The wave corresponds to the process

$$Hg + p\,X^- \rightarrow HgX_p^{2-p} + 2\,e^-$$

Show that the wave has the diffusion current, half-wave potential, and i–E shape given by

$$i_D = -2FAk_D C_X{}^*/p$$

$$E_{\frac{1}{2}} = E_{Hg}{}^\circ - \frac{RT}{2F}\,\ln[p\beta_p(C_X{}^*/2)^{p-1}]$$

$$E = E_{\frac{1}{2}} - \frac{RT}{2F}\,\ln(i_D/2i)[2(i_D - i)/i_D]^p$$

where β_p is the overall formation constant for HgX_p^{2-p}, k_D is the mass transport rate constant (all diffusion coefficients are assumed equal), and $C_X{}^*$ is the bulk concentration of X^-. Hint: Start with the Nernst equation for the Hg^{2+}/Hg couple and expressions for the current in terms of the flux of X^- toward the electrode and the complex away from the electrode.

4.19 Plot i vs. E for mercury oxidation waves in the presence of a coordinating ligand X^- assuming that $p = 2$ and $p = 4$; compare with the wave shape expected in the absence of ligands, Figure 4.5, and with a normal reversible two–electron wave. What are the differences?

4.20 Kirowa–Eisner, Talmor, and Osteryoung [Anal. Chem. **1981**, 53, 581] studied the polarographic oxidation of mercury in the presence of cyanide ion using pulse and sampled d.c. polarography on a static mercury drop electrode. A wave for 50.8 μM cyanide is shown in Figure 4.38. Potentials were measured vs. a Ag/AgCl (saturated KCl) reference electrode, $E = 0.205$ V, and a few points are given in the following table:

E/V	i/μ A	E/V	i/μ A
−0.299	−0.0112	−0.250	−0.0963
−0.282	−0.0321	−0.241	−0.1124
−0.273	−0.0481	−0.228	−0.1284
−0.266	−0.0642	−0.209	−0.1444
−0.255	−0.0802	−0.000	−0.1605

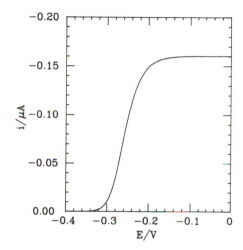

Figure 4.38 Sampled d.c. polarogram at a s.m.d.e. for 50.8 μ M cyanide.

(a) Using the equations given in Problem 4.17, determine p and $E_{\frac{1}{2}}$ by plotting $\ln(i_D/2i)[2(i_D - i)/i_D]^p$ vs. E for several values of p.

(b) Use the half−wave potential found in part (a) to estimate the formation constant for the mercury(II) cyanide complex.

(c) Polarograms for higher concentrations of cyanide had shapes which suggested formation of complexes with larger values of p, at least in the leading part of the wave. Measured diffusion currents, however, were linear in the bulk cyanide concentration over the range 0.003–10 mM. Explain why the diffusion current would correspond to a constant value of p when higher complexes are apparently formed in some instances.

4.21 Describe an experimental arrangement for an amperometric acid–base titration of hydroxide ion in aqueous solution, taking advantage of the anodic wave described in Example 4.12. What titrant should be used? At what potential should the current be monitored? Sketch the titration curve.

4.22 The diffusion current constants given in Table 4.1 are uniformly smaller for polarograms in tartrate buffer solutions than for the other media listed. Give a qualitative explanation for this phenomenon.

4.23 Using the data of Figure 4.35, estimate the equilibrium constant for

$$Pb^{2+} + 3 OH^- \rightleftharpoons HPbO_2^- + H_2O$$

$E_{\frac{1}{2}}$ for the polarographic reduction of Pb^{2+} in acid solution is -0.388 V $vs.$ s.c.e.

5 ELECTRODE KINETICS

In Chapter 4, we considered experiments in which the current was controlled by the rate of transport of electroactive material to the electrode surface. In this chapter, we turn to voltammetric experiments where the current is influenced by the rate of electron transfer at the electrode or by the rates of coupled homogeneous chemical reactions. We begin with a general discussion of electrode kinetics and then consider experiments under limiting conditions where transport is fast and electron tranfer is rate limiting. In Section 5.3, we consider voltammetric experiments where the rates of transport and electron transfer are both important. In Section 5.4, we turn to alternating–current experiments, which are best understood in terms of the faradaic impedance, and in Sections 5.5–5.9, we deal with voltammetric experiments where the electrode process is complicated by associated chemical reactions.

5.1 KINETICS OF ELECTRON TRANSFER

In order to focus solely on the electron–transfer process, we will assume that the solution is well stirred and that transport of electro- active material to the electrode is fast. Initially we will also assume that the electrolyte concentration is sufficiently high that the diffuse part of the double layer is very thin and that the potential just outside the Helmholtz layer is essentially zero, *i.e.*, equal to the potential of the bulk solution. This is not always a good assumption and we will have to correct it later. To the extent that our assumptions are correct, the concentrations of the electroactive species, O and R, will be essentially the same at the outer edge of the Helmholtz layer as in the bulk solution. In other words, we are assuming that transport of reactants and products to or from the electrode is very fast and that variations in potential as they approach the electrode can be neglected.

Consider an electron–transfer process consisting of the single elementary step

$$O + e^- \rightleftharpoons R$$

Suppose that the rate of the forward reaction (the cathodic or reduction rate) is first–order in O:

$$\text{cathodic rate} = k_c C_O(0,t)$$

and that the reverse rate (anodic or oxidation rate) is first–order in R:

$$\text{anodic rate} = k_a C_R(0,t)$$

where k_c and k_a are heterogeneous rate constants with units of m s^{-1}. If the concentrations have units of mol m^{-3} (mM), then the rates have units of mol m^{-2}s^{-1}, the same units as flux. Indeed, the difference between the cathodic and anodic rates is the net flux of O at the electrode surface and thus is proportional to the net current:

$$i = FA[k_c C_O(0,t) - k_a C_R(0,t)] \qquad (5.1)$$

We assume that the temperature dependence of the rate constants k_c and k_a is given by the following expressions from transition state theory:

$$k_c = Z \exp \frac{-\Delta G_c^{\ddagger}}{RT} \qquad (5.2a)$$

$$k_a = Z \exp \frac{-\Delta G_a^{\ddagger}}{RT} \qquad (5.2b)$$

where the $\Delta G^{\ddagger}$'s are free energies of activation. The pre–exponential factors Z can be estimated theoretically (1) but we will be content with noting that the Z's in eqs (5.2a) and (5.2b) must be equal to satisfy the principle of microscopic reversibility; otherwise, we will treat Z simply as an empirical parameter with units of m s^{-1}.

Consider now the special case of equilibrium at the electrode. The *net* current must be zero and the surface and bulk concentrations should be equal. Substituting eqs (5.2) into eq (5.1), taking logs and rearranging, we have

$$\ln \frac{C_O(0,t)}{C_R(0,t)} = \ln \frac{k_{a,0}}{k_{c,0}} = \frac{\Delta G_c^{\ddagger} - \Delta G_a^{\ddagger}}{RT} \tag{5.3}$$

where $k_{c,0}$ and $k_{a,0}$ are the rate constants at zero current. The ratio of the reactant and product concentrations at equilibrium is given by the Nernst equation,

$$\ln \frac{C_O}{C_R} = \frac{F}{RT} (E_e - E°) \tag{5.4}$$

where E_e and $E°$ are the equilibrium and standard half–cell potentials, respectively. Combining eqs (5.3) and (5.4), we have

$$\Delta G_c^{\ddagger} - \Delta G_a^{\ddagger} = F(E_e - E°)$$

This expression is a statement of a familiar idea from chemical kinetics: as shown graphically in Figure 5.1, the difference between the forward and reverse activation free energies is equal to the standard free energy change for the reaction. The rates of the cathodic and anodic processes depend on the electrode potential. By varying the potential, we can change the free energy of an electron in the electrode and in general the free energies of O and R at the electrode surface as well. Thus the activation free energies must depend on the potential. The details of this dependence might be rather complicated, but for the moment we will take a phenomenological approach and assume a simple linear relationship:

$$\Delta G_c^{\ddagger} = \Delta G_0^{\ddagger} + a F(E_e - E°) \tag{5.5a}$$

$$\Delta G_a^{\ddagger} = \Delta G_0^{\ddagger} - \beta F(E_e - E°) \tag{5.5b}$$

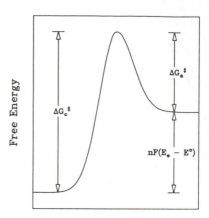

Figure 5.1 Free energy – reaction coordinate diagram for an electron–transfer process.

Reaction Coordinate

where the parameters a and $\beta = 1 - a$ are, respectively, the *cathodic* and *anodic transfer coefficients*, and $\Delta G_0^{\ddagger}$ is the activation free energy for the process when $E_e = E°$. We will assume that $\Delta G_0^{\ddagger}$ is a constant at constant temperature, characteristic of the electrode process. Substitution into eqs (5.2) then gives

$$k_{c,0} = k_0 \exp \frac{-a\, F(E_e - E°)}{RT} \tag{5.6a}$$

$$k_{a,0} = k_0 \exp \frac{\beta\, F(E_e - E°)}{RT} \tag{5.6b}$$

where

$$k_0 = Z \exp \frac{-\Delta G_0^{\ddagger}}{RT} \tag{5.7}$$

When a net current flows through the electrochemical cell, the cell is not at equilibrium. The deviation of the half–cell potential from the equilibrium value is called the *overpotential* or *overvoltage*, η.

$$E = E_e + \eta \tag{5.8}$$

We assume that the form of eqs (5.5) is retained under nonequilibrium conditions so that the cathodic activation energy changes by an amount

$aF\eta$ and the anodic activation free energy by an amount $-\beta F\eta$. The rate constants then differ from the equilibrium values according to the relations

$$k_c = k_{c,0} \exp \frac{-aF\eta}{RT} \tag{5.9a}$$

$$k_a = k_{a,0} \exp \frac{\beta F\eta}{RT} \tag{5.9b}$$

Substituting these into eq (5.1) gives

$$i = FA \left[k_{c,0}C_O(0,t) \exp \frac{-aF\eta}{RT} - k_{a,0}C_R(0,t) \exp \frac{\beta F\eta}{RT} \right]$$

or

$$i = i_0 \left[\frac{C_O(0,t)}{C_O^*} \exp \frac{-aF\eta}{RT} - \frac{C_R(0,t)}{C_R^*} \exp \frac{\beta F\eta}{RT} \right] \tag{5.10}$$

where i_0, the *exchange current*, is equal to the cathodic current (and to the negative of the anodic current) at equilibrium:

$$i_0 = i_{c,e} = FAk_0C_O^* \exp \frac{-aF(E_e - E^\circ)}{RT} \tag{5.11a}$$

$$i_0 = -i_{a,e} = FAk_0C_R^* \exp \frac{\beta F(E_e - E^\circ)}{RT} \tag{5.11b}$$

If transport is really fast so that $C_O(0,t) = C_O^*$ and $C_R(0,t) = C_R^*$, then eq (5.10) reduces to

$$i = i_0 \left[\exp \frac{-aF\eta}{RT} - \exp \frac{\beta F\eta}{RT} \right] \tag{5.12}$$

which is called the Butler–Volmer [,] equation after the two electrochemists who, in 1924[‡] and 1930,[#] respectively, contributed to its formulation *(2)* and experimental test *(3)*. The current–potential curve predicted by eq (5.12) for a = 0.4 is shown in Figure 5.2.

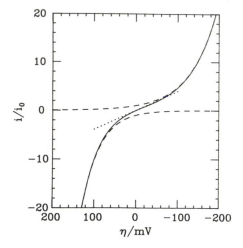

Figure 5.2 Current–potential curve predicted by the Butler–Volmer equation for a = 0.4. The dashed lines represent the anodic and cathodic components of the current. The dotted line represents the linear (ohmic) region for small overpotentials.

The Transfer Coefficient

Before going further, let us consider the significance of the transfer coefficients a and β (= $1 - a$), which were introduced in order to separate the potential contributions to $\Delta G_c^{\ddagger}$ and $\Delta G_a^{\ddagger}$. These parameters have taken an apparently significant role in eqs (5.9): a and β are seen to reflect the dependence of k_c and k_a on the overvoltage η. Thus if a = 0, β = 1, the cathodic free energy of activation (and thus the cathodic current) would be independent of overpotential and all the variation of net current with potential would be through the anodic component. This situation is shown schematically in Figure 5.3a. In the opposite extreme, a = 1, β = 0, shown in Figure 5.3b, the anodic activation free energy and anodic

---

[‡] John A. V. Butler (1899–1977) was a lecturer at the University of Edinburgh during the 1920's and 1930's. His interests shifted to biophysical chemistry after World War II.

[#] Max Volmer (1885–1965), a student of Nernst, was Professor at the Technical High School in Berlin and a leading figure in German electrochemistry before World War II.

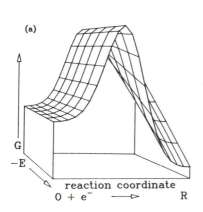

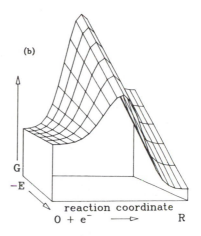

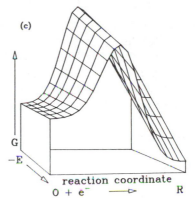

Figure 5.3 Hypothetical free energy – reaction coordinate – potential surfaces for (a) $a = 1$ (b) $a = 0$, and (c) $a = 0.5$.

current would be independent of overpotential. Ordinarily, the anodic and cathodic transfer coefficients are equal, $a = \beta = \frac{1}{2}$, and the anodic and cathodic currents respond symmetrically to changes in overpotential as shown in Figure 5.3c.[‡]

In order to get a physical feeling for the significance of the transfer coefficients, let us consider an electron–transfer process at the molecular level, outlining a theory developed by Marcus *(1)*. An electron transfer is like a charge–transfer transition in electronic spectroscopy. It occurs in a very short time and the various nuclei can

---

[‡] Since a and β govern the symmetry of a current–potential curve, some authors refer to transfer coefficients as "symmetry factors."

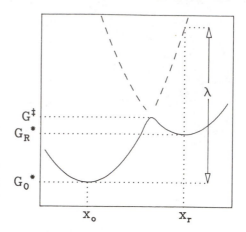

Figure 5.4 Free energy – internal coordinate diagram for an electron–transfer process, $O + e^- \rightarrow R$.

be thought of as fixed in position during the transition (the Franck–Condon principle). Since O and R generally have somewhat different structures, there is a barrier to electron transfer associated with changes in bond lengths and bond angles. This is shown schematically in Figure 5.4, where the free energies of O and R are plotted as functions of a generalized internal coordinate x which represents a composite of bond lengths and bond angles which change on going from O to R. The energy variation with x is assumed to be harmonic, so that we have

$$G_O = G_O{}^* + \tfrac{1}{2} k(x - x_o)^2$$

$$G_R = G_R{}^* + \tfrac{1}{2} k(x - x_r)^2$$

where k is a composite force constant and $G_O{}^*$ and $G_R{}^*$ are the free energies of O and R in the conformations, represented by x_o and x_r, just prior to and just after electron transfer. The transition state corresponds to the crossing of the G_O and G_R curves. Thus the value of x at the transition state is found from

$$G^{\ddagger} = G_O{}^* + \tfrac{1}{2} k(x - x_o)^2 = G_R{}^* + \tfrac{1}{2} k(x - x_r)^2$$

$$G_O{}^* - G_R{}^* + \tfrac{1}{2} k(x_o{}^2 - x_r{}^2) = k(x_o - x_r)x$$

$$x = \tfrac{1}{2}(x_o + x_r) + \frac{G_O{}^* - G_R{}^*}{k(x_o - x_r)}$$

Substituting this value of x back into the expression for $G^{\ddagger}$, we have

$$G^{\ddagger} = \tfrac{1}{8} k(x_0 - x_r)^2 + \tfrac{1}{2}(G_O^* + G_R^*) + \frac{(G_O^* - G_R^*)^2}{2k(x_0 - x_r)^2} \qquad (5.13)$$

The lead term is seen to be proportional to the energy required to reorganize O to the conformation of R,

$$\lambda = \tfrac{1}{2} k(x_0 - x_r)^2 \qquad (5.14)$$

If we take the free energy of O in bulk solution as zero, then G_O^* is the work required to bring O from bulk solution to the electrode and, if necessary, to affect any gross changes in conformation or coordination such that the small distortion represented by Figure 5.4 will suffice for electron transfer. Thus we have

$$G_O^* = w_0$$

Similarly, there will be work w_r required to bring R from bulk solution to the electrode and affect any gross changes required before electron transfer. With the zero of free energy defined as above, the bulk solution free energy of R is $F(E - E^\circ)$. Thus

$$G_R^* = w_r + F(E - E^\circ)$$

Substituting G_O^*, G_R^*, and λ in eq (5.13), we have

$$\Delta G_c^{\ddagger} = G^{\ddagger} - 0$$

$$\Delta G_c^{\ddagger} = \tfrac{1}{4}\lambda + \tfrac{1}{2} F(E - E^\circ) + \tfrac{1}{2}(w_0 + w_r) + \frac{[F(E - E^\circ) - (w_0 - w_r)]^2}{4\lambda}$$

$$(5.15a)$$

$$\Delta G_a^{\ddagger} = G^{\ddagger} - F(E - E^\circ)$$

$$\Delta G_a^{\ddagger} = \tfrac{1}{4}\lambda - \tfrac{1}{2} F(E - E^\circ) + \tfrac{1}{2}(w_0 + w_r) + \frac{[F(E - E^\circ) - (w_0 - w_r)]^2}{4\lambda}$$

$$(5.15b)$$

Thus when $E = E^\circ$, the activation free energy is

$$\Delta G_0^{\ddagger} = \tfrac{1}{4}\lambda + \tfrac{1}{2}(w_0 + w_r) + \frac{(w_0 - w_r)^2}{4\lambda} \tag{5.16}$$

According to eqs (5.5), the transfer coefficients are

$$a = \frac{\partial \Delta G_c^{\ddagger}}{\partial (FE)} = \tfrac{1}{2} + \frac{[F(E - E^\circ) - (w_0 - w_r)]}{2\lambda} \tag{5.17a}$$

$$\beta = -\frac{\partial \Delta G_a^{\ddagger}}{\partial (FE)} = \tfrac{1}{2} - \frac{[F(E - E^\circ) - (w_0 - w_r)]}{2\lambda} \tag{5.17b}$$

Thus the transfer coefficients, a and β, are expected to be exactly $\tfrac{1}{2}$ when $F(E - E^\circ)$ and $(w_0 - w_r)$ are small compared with the reorganization energy λ. Notice, however, that the transfer coefficients are not expected to be constant for large overpotentials. Several of the general references listed in the Bibliography *(A1,A6,B1,B6,B7,B8,B12)* give useful discussions of transfer coefficients and insights derived from somewhat different perspectives.

Example 5.1 Cyclooctatetraene, C_8H_8, is a nonaromatic cyclic polyene with a nonplanar "tub" conformation. The radical anion, $C_8H_8^-$, however, is planar or nearly so. The rate of electrochemical reduction in N,N–dimethylformamide solution was studied by Allendoerfer and Rieger *(4)* and by Heubert and Smith *(5)* using a.c. polarography (see Section 5.4). It was found the cathodic transfer coefficient is $a = 0.40$. Since the transfer coefficient is measured for $E \cong E^\circ$ in a.c. polarography, eq (5.17a) suggests that $w_O \gg w_R$. This is consistent with the notion that the electron is transferred only after the ring is flattened and that the activation barrier is mostly due to the conformation change. Indeed, the enthalpy of activation for the electron–transfer process (32 kJ mol^{-1}, determined by a.c. polarographic measurements of the electron–transfer rate as a function of temperature) is comparable to the estimated enthalpy of activation for bond isomerization, a process which goes through a planar transition state.

Double–Layer Effects

We have assumed in the discussion above that the potential drop between the electrode and solution occurs entirely within the immobile Helmholtz layer and that a molecule or ion experiences no variation in potential as it approaches the electrode. This ideal situation is rarely, if ever, obtained and a correction is usually required.

We showed in Section 2.1 that the concentrations of O and R at the outer surface of the Helmholtz layer $(x = a)$ are

$$C_O(a) = C_O{}^* \exp \frac{-z_O F \Phi_a}{RT}$$

$$C_R(a) = C_R{}^* \exp \frac{-z_r F \Phi_a}{RT}$$

where z_O and z_r are the charges on O and R and Φ_a is the potential at $x = a$ (relative to the bulk solution). If Φ_a is significantly different from zero, then we should use $C_O(a)$ and $C_R(a)$ in eqs (5.11) for the exchange current rather than the bulk concentrations $C_O{}^*$ and $C_R{}^*$. Furthermore, the potential difference contributing to the activation free energy needs to be corrected by subtracting Φ_a from E_e. Thus eqs (5.11) become

$$i_0 = FAk_0 C_O{}^* \exp \frac{-z_O F \Phi_a}{RT} \exp \frac{-\alpha F(E_e - E^\circ - \Phi_a)}{RT}$$

$$i_0 = FAk_0 C_R{}^* \exp \frac{-z_r F \Phi_a}{RT} \exp \frac{\beta F(E_e - E^\circ - \Phi_a)}{RT}$$

The apparent exchange current given by eqs (5.11) then is related to the true exchange current by

$$(i_0)_{app} = i_0 \exp \frac{-(\beta z_O + \alpha z_r) F \Phi_a}{RT}$$

or, since $z_O - z_r = 1$,

$$(i_0)_{app} = i_0 \exp \frac{(a - z_0)F\Phi_a}{RT} \tag{5.18a}$$

$$(i_0)_{app} = i_0 \exp \frac{(\beta + z_r)F\Phi_a}{RT} \tag{5.18b}$$

Alternatively, the apparent electron–transfer rate constant determined using eqs (5.11) can be corrected by

$$k_0 = (k_0)_{app} \exp \frac{(\beta z_0 + a z_r)F\Phi_a}{RT} \tag{5.19}$$

Equation (5.19) can also be obtained from eqs (5.15) by noting that the work required to bring O and R up to that point in the double layer where the potential is Φ_a is

$$w_0 + w_r = (z_0 + z_r)F\Phi_a$$

Thus the correction to the activation free energy is

$$\Delta G_{corr}{}^{\ddagger} = \tfrac{1}{2}(z_0 + z_r)F\Phi_a$$

Since $z_0 - z_r = 1$, the difference in the work terms due to the double–layer effect is

$$w_0 - w_r = F\Phi_a$$

so that there may be a contribution to the transfer coefficients. For $E \cong E°$, eq (5.17a) gives

$$a = \tfrac{1}{2} - F\Phi_a/4\lambda \tag{5.20}$$

The double–layer contributions to the exchange current and transfer coefficients depend on the sign and magnitude of the surface potential Φ_a. Φ_a is expected to go through zero at the potential of zero charge. Thus, for example, the potential of zero charge for mercury is about –0.43 V (*vs.* s.c.e.) so that Φ_a is positive for an electrode potential $E > -0.43$ V. As we saw in Section 2.5, the potential of zero charge can be determined, at least approximately, by the maximum in the electrocapillarity curve (for liquid electrodes) or by the minimum in the double–layer capacitance *vs.* potential curve.

The influence of the electric double layer on electron–transfer rates was discovered and explained theoretically by Frumkin[‡] (6) and the effect is usually referred to with his name. See also a review by Parsons (7).

5.2 CURRENT–OVERPOTENTIAL CURVES

The Butler–Volmer equation, eq (5.12), is often used in a strictly empirical way to analyze electrode kinetic data. The kinetic information contained in the Butler–Volmer equation is expressed in two parameters, the exchange current i_0 or exchange current density, $j_0 = i_0/A$, and the transfer coefficient a. There are basically two ways of extracting these parameters from experimental data on well-stirred solutions. Both involve the analysis of current–overpotential curves.

The Tafel Equation

When the overpotential η is sufficiently large, $|F\eta/RT| \gg 1$, one of the exponential terms in eq (5.12) will be negligible compared with the other. For example, when the overpotential is large and negative, the anodic component of the current is negligible and we have

$$i = i_0 \exp \frac{-a\,F\eta}{RT}$$

$$\ln i = \ln i_0 - \frac{a\,F}{RT}\,\eta \qquad (5.21)$$

or, at 25°C,

$$\log i = \log i_0 - 16.90\,a\eta$$

_____

[‡] Aleksandr N. Frumkin (1895–1976) was Director of the Institute of Electrochemistry of the Soviet Academy of Sciences. Frumkin was a leading figure in twentieth–century electrochemistry; his contributions to the understanding of surface effects were particularly important.

Equation (5.21) is known as the Tafel equation. This logarithmic current–potential relationship was discovered empirically by Tafel[‡] in 1905 *(8)*, some years before the theory of electrode kinetics was developed.

The Tafel equation suggests the means by which the exchange current and the transfer coefficient may be determined. If, for an equilibrium mixture of O and R, the current is measured as a function of overpotential and then plotted as log i vs. η, a linear region should be found. Extrapolation of the linear portions of the plot to zero overpotential yields the log of the exchange current as the intercept; the slopes should be –16.90 α and 16.90 β. Such a plot is shown in Figure 5.5.

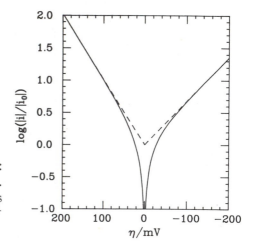

Figure   5.5   Tafel   plot: $\log(|\,i\,|\,/i_0)$ *vs.* η for α = 0.4. The slopes of the dashed lines are $-\alpha F/2.3RT = -6.76$ V^{-1} and $\beta F/2.3RT = 10.14$ V^{-1}.

The Tafel equation and Figure 5.5 suggest that the current increases exponentially with increasing overpotential. There must be a point at which the current becomes limited by the rate of transport and log i vs. η plots begin to flatten out. The point at which this happens depends on the efficiency of stirring and on the diffusion coefficients of the electroactive species but more critically on the electron–transfer rate. For slow electron–transfer processes such as the reduction of H^+(aq) at a mercury cathode, the exchange current is so small that

[‡]   Julius Tafel (1862–1918) was an organic chemist, a student of Emil Fischer, and a professor at the University of Würzburg. Tafel was a pioneer in the application of electrochemistry to organic synthesis.

the rate of transport is rarely a problem. For faster processes, however, the Tafel plot approach may be impractical because, with increasing overpotential, transport limitation sets in before the linear Tafel region has been established.

The Charge–Transfer Resistance

When the electron–transfer rate is too fast to be measured by the Tafel approach, all is not lost; the current–overpotential relationship for small η yields the same information. Returning to the Butler–Volmer equation, consider the limiting case of a small overpotential. If $|F\eta/RT| \ll 1$, the exponential terms in eq (5.12) can be expanded in a power series

$$i = i_0 \left[1 - \frac{\alpha F\eta}{RT} + \ldots - 1 - \frac{\beta F\eta}{RT} - \ldots \right]$$

Retaining only the first nonvanishing term, we have

$$i = - \frac{Fi_0}{RT} \eta \tag{5.22}$$

Thus for small departures from equilibrium, the current is approximately linear in η and the electrode behaves as an ohmic resistance. The quantity

$$R_{ct} = RT/Fi_0$$

is called the *charge–transfer resistance*. The charge–transfer resistance is best determined by a.c. impedance measurements (see Section 5.4), but d.c. measurements can also be used. Determination of the exchange current from the charge–transfer resistance does not directly yield the transfer coefficient; however, according to eqs (5.11a) or (5.11b), the exchange current depends on one of the concentrations, C_O or C_R, and on the equilibrium potential. Furthermore, it is easy to show, by differentiation of eq (5.11a) or (5.11b), using eq (5.4), that the exchange current varies with the equilibrium potential according to the partial derivatives:

$$\left[\frac{\partial \ln i_0}{\partial E_e} \right]_{C_R} = \frac{F}{RT} \beta \tag{5.23a}$$

$$\left[\frac{\partial \ln i_0}{\partial E_e} \right]_{C_O} = - \frac{F}{RT} a \qquad (5.23b)$$

Thus measurement of R_{ct} (and thus i_0) for a series of solutions with constant $C_O{}^*$ and variable $C_R{}^*$ (and thus variable E_e) allows the evaluation of a.

Example 5.2 Vetter and Manecke (9) studied the reduction of Mn^{3+} to Mn^{2+} at a platinum electrode in a well–stirred 7.5 M sulfuric acid solution. A Tafel plot—log of current density, $\log j$ ($j = i/A$), vs. E—is shown in Figure 5.6 for $[Mn^{2+}]$ = 0.01 and 0.001 M and $[Mn^{3+}]$ = 0.01 and 0.001 M. At large overpotentials, two curves are found corresponding to the different Mn^{3+} concentrations but independent of $[Mn^{2+}]$. The curves branch at small overpotentials; the equilibrium potential depends on the Mn^{3+}/Mn^{2+} concentration ratio, of course, so that the point of zero overpotential differs for the different curves. At large overpotentials, the current levels out to a transport–limited value. The linear regions of the two curves are separated by exactly one

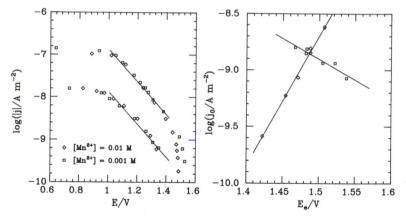

Figure 5.6 Tafel plot (cathodic branch) for the Mn^{3+}/Mn^{2+} couple in 7.5 M H_2SO_4 solution at 25°C. The upper points correspond to $[Mn^{3+}]$ = 0.01 M, the lower points to $[Mn^{3+}]$ = 0.001 M.

Figure 5.7 Exchange current density for the Mn^{3+}/Mn^{2+} couple in 7.5 M H_2SO_4 solution at 25°C.

base 10 log unit, as expected if the rate of the cathodic process is first–order in $[Mn^{3+}]$. The slope of the linear regions corresponds to a cathodic transfer coefficient $a = 0.28$.

Vetter and Manecke also determined exchange current densities by measurement of the charge–transfer resistance near equilibrium. A plot of their data, log j_0 vs. E_e, is shown in Figure 5.7. The two lines shown correspond to constant Mn^{3+} (squares) and constant Mn^{2+} (diamonds) concentrations. The slopes of the lines lead to independent estimates of the transfer coefficient using eqs (5.23), obtaining $a = 0.29$ and 0.23, respectively, in satisfactory agreement with the results of the Tafel plot method.

Multistep Mechanisms

We have thus far assumed that the electrode process is simply the addition of an electron to a single molecule of O to produce a single molecule of R. Consider a process with the stoichiometric half–cell reaction

$$\nu_o\, O + n\ e^- \rightleftharpoons \nu_r\, R$$

The concerted addition of two or more electrons in a single step is highly unlikely (see further discussion of this point in Section 5.5). In general, we expect that any electrode process involving two or more electrons necessarily involves two or more elementary steps. If one of these steps is rate limiting, the rate laws should have the relatively simple forms:

$$\text{anodic rate} = k_a C_R{}^{n_{ar}} C_O{}^{n_{ao}}$$

$$\text{cathodic rate} = k_c C_R{}^{n_{cr}} C_O{}^{n_{co}}$$

where the n's are the orders of the reactions. Equations (5.11a) and (5.11b) are then

$$i_0 = nFAk_0 C_R{}^{n_{cr}} C_O{}^{n_{co}} \exp \frac{-a\, F(E_e - E^\circ)}{RT} \tag{5.24a}$$

$$i_0 = nFAk_0 C_R{}^{n_{ar}} C_O{}^{n_{ao}} \exp \frac{\beta F(E_e - E^\circ)}{RT} \tag{5.24b}$$

The net current is given by eq (5.10). Since the equilibrium potential depends on C_O and C_R through the Nernst equation

$$E_e = E^\circ + \frac{RT}{nF} \ln \frac{C_O{}^{\nu_o}}{C_R{}^{\nu_r}}$$

i_0 is really a function of only two independent variables, the two concentrations or E_e and one of the concentrations. Differentiating eq (5.24a) or (5.24b), we have

$$\left[\frac{\partial \ln i_0}{\partial E_e} \right]_{C_R} = n_{co} \left[\frac{\partial \ln C_O}{\partial E_e} \right]_{C_R} - \frac{aF}{RT}$$

or

$$\left[\frac{\partial \ln i_0}{\partial E_e} \right]_{C_R} = n_{ao} \left[\frac{\partial \ln C_O}{\partial E_e} \right]_{C_R} + \frac{\beta F}{RT}$$

From the Nernst equation, we have

$$\left[\frac{\partial \ln C_O}{\partial E_e} \right]_{C_R} = \frac{nF}{\nu_o RT}$$

Combining the partial derivatives, we obtain

$$\left[\frac{\partial \ln i_0}{\partial E_e} \right]_{C_R} = \frac{F}{RT} \left[\frac{n n_{co}}{\nu_o} - a \right] \tag{5.25a}$$

or

$$\left[\frac{\partial \ln i_0}{\partial E_e} \right]_{C_R} = \frac{F}{RT} \left[\frac{n n_{ao}}{\nu_o} + \beta \right] \tag{5.25b}$$

Similarly, the partial derivatives for constant C_O are:

$$\left[\frac{\partial \ln i_0}{\partial E_e} \right]_{C_O} = - \frac{F}{RT} \left[\frac{n n_{cr}}{\nu_r} + a \right] \tag{5.25c}$$

or

$$\left[\frac{\partial \ln i_0}{\partial E_e} \right]_{C_O} = - \frac{F}{RT} \left[\frac{n n_{ar}}{\nu_r} - \beta \right] \tag{5.25d}$$

Comparing eqs (5.25a) and (5.25b), we obtain a relation between the kinetic orders in O and the stoichiometric coefficients

$$n_{co} - n_{ao} = \nu_o / n \tag{5.26a}$$

Equations (5.25c) and (5.25d) give a similar relation for the kinetic orders in R:

$$n_{ar} - n_{cr} = \nu_r / n \tag{5.26b}$$

For a simple electron–transfer process with $n_{co} = n_{ar} = 1$, $n_{cr} = n_{ao} = 0$, $\nu_o = \nu_r = 1$, and $n = 1$, eqs (5.25) reduce to eqs (5.23). The stoichiometric coefficients are generally known. Thus if the exchange current is determined as a function of the equilibrium potential, holding one of the concentrations constant, the number of independent unknowns can be further reduced from four to two and, with luck, there may be only one set of parameters which fit the constraints and make chemical sense.[‡]

Example 5.3 Vetter *(10)* studied the kinetics of the electrode couple

$$I_3^- + 2 e^- \rightleftharpoons 3 I^-$$

using an a.c. bridge technique to measure the faradaic impedance. An a.c. potential (50–100 Hz, 10 mV peak–to–peak) was applied to the cell, which contained a small platinum indicator electrode ($A = 0.3$ cm^2), a large Pt counter electrode ($A = 16$ cm^2), and a Hg/Hg$_2$SO$_4$ reference electrode. The electrolyte was 0.5 M H$_2$SO$_4$

[‡] For further details and examples, see Vetter *(B7)*.

with variable amounts of iodine and potassium iodide. The exchange current density was computed from the measured faradaic impedance (see Section 5.4). The equilibrium d.c. potential was a function of the iodide and triiodide ion concentrations as given by the Nernst equation:

$$E = E° + \frac{RT}{2F} \ln \frac{[I_3^-]}{[I^-]^3}$$

Figure 5.8 shows a plot of $(RT/F)\ln(j_0)$ vs. the equilibrium potential for two sets of data, one with $[I^-] = 0.01$ M and variable I_3^- concentration (diamonds), the other for $[I_3^-] = 0.0086$ M and variable I^- (squares).

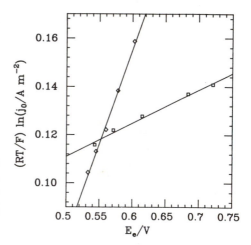

Figure 5.8 Exchange current density as a function of equilibrium potential for the I_3^-/I^- couple in 0.5 M H_2SO_4 solution at 25°C.

The slopes of the two lines are +0.13 and +0.78 for the constant I_3^- and I^- concentrations, respectively. With $n = 2$, $\nu_0 = 1$, and $\nu_r = 3$, eqs (5.25) and (5.26) give

$$0.13 = -\tfrac{2}{3}\, n_{cr} - a$$

$$n_{ar} - n_{cr} = \tfrac{3}{2}$$

$$0.78 = 2\, n_{co} - a$$

$$n_{co} - n_{ao} = \tfrac{1}{2}$$

There are five parameters and only four relations, but if we insist that the orders be integral or half–integral (or even third–integral), the only consistent set of parameters is

$$n_{ar} = 1 \qquad n_{ao} = 0$$

$$n_{cr} = -\tfrac{1}{2} \qquad n_{co} = +\tfrac{1}{2}$$

$$a = 0.21$$

Thus the rates of the anodic and cathodic processes are:

$$\text{anodic rate} = k_a[I^-]$$

$$\text{cathodic rate} = k_c[I_3^-]^{\frac{1}{2}}[I^-]^{-\frac{1}{2}}$$

A mechanism consistent with these rate laws is

$$I_3^- \rightleftharpoons I_2 + I^-$$

$$I_2 \rightleftharpoons 2\,I$$

$$2\,(I + e^- \rightleftharpoons I^-)$$

where the last step is rate limiting. The small transfer coefficient is not surprising; eq (5.17a) suggests that a should be less than $\tfrac{1}{2}$ when a lot of work is required to get O ready for electron transfer. The surface potential on platinum is probably positive in this experiment, so that eq (5.20) would also suggest a decrease in a from $\tfrac{1}{2}$.

Consider now an electrode process which consists of two electron–transfer steps, one of which is slow and rate limiting, the other of which is fast and can be assumed to be essentially at equilibrium:

$$O + e^- \rightleftharpoons I \qquad E_1^\circ$$

$$I + e^- \rightleftharpoons R \qquad E_2^\circ$$

The standard potential for the overall two–electron process is

$$E_{12}^\circ = \tfrac{1}{2}(E_1^\circ + E_2^\circ)$$

and the Nernst equation for the overall process gives the equilibrium
R/O concentration ratio

$$\frac{C_R}{C_O} = \exp \frac{-2F(E_e - E_{12}°)}{RT} = \theta_1 \theta_2$$

where

$$\theta_1 = \frac{C_I}{C_O} = \exp \frac{-F(E_e - E_1°)}{RT}$$

$$\theta_2 = \frac{C_R}{C_I} = \exp \frac{-F(E_e - E_2°)}{RT}$$

If we assume that $E_2° > E_1°$, so that the intermediate is never
thermodynamically stable, then two electrons will be transferred for
each occurrence of the rate–limiting step. Suppose that the first step is
rate limiting; the total current is twice the contribution of this step,

$$i = 2FA(k_c C_O - k_a C_I)$$

with

$$k_c = k_0 \exp \frac{-a F(E - E_1°)}{RT} = k_0 \theta_1{}^a \exp \frac{-a F\eta}{RT}$$

$$k_a = k_0 \exp \frac{\beta F(E - E_1°)}{RT} = k_0 \theta_1{}^{-\beta} \exp \frac{\beta F\eta}{RT}$$

where a and β ($= 1 - a$) are the transfer coefficients and k_0 is the
standard electron–transfer rate constant. Since the second step is
assumed to be in equilibrium at the electrode potential $E = E_e + \eta$,
we have

$$\frac{C_R}{C_I} = \theta_2 \exp \frac{-F\eta}{RT}$$

Substituting C_I, k_c, and k_a into the expression for the current, we have

$$i = 2FAk_0 \left[C_O\theta_1{}^a \exp \frac{-aF\eta}{RT} - \frac{C_R}{\theta_1{}^\beta\theta_2} \exp \frac{\beta F\eta}{RT} \exp \frac{F\eta}{RT} \right]$$

When $\eta = 0$, the net current is zero and the cathodic and anodic currents are equal to the exchange current i_0:

$$i_0 = 2FAk_0 C_O\theta_1{}^a \tag{5.27a}$$

$$i_0 = 2FAk_0 C_R/\theta_1{}^\beta\theta_2 \tag{5.27b}$$

For $\eta \neq 0$, the net current is given by

$$i = i_0 \left[\exp \frac{-aF\eta}{RT} - \exp \frac{(1+\beta)F\eta}{RT} \right] \tag{5.28}$$

which has the form of the Butler–Volmer equation, but with $a_{app} = a$, $\beta_{app} = 1 + \beta$. We thus have the interesting result: For a two–electron process where the first step is rate limiting, the apparent cathodic and anodic transfer coefficients are expected to be approximately $\frac{1}{2}$ and $\frac{3}{2}$, respectively.[‡] Notice that the exchange current in this case is a complicated function of the electron–transfer rate constant k_0, the transfer coefficient a, and the standard potentials for the two steps, $E_1{}^\circ$ and $E_2{}^\circ$. The individual standard potentials are usually unknown, so that k_0 cannot be determined directly.

The situation would be just reversed if the second step were rate limiting; the apparent cathodic and anodic transfer coefficients would be $\frac{3}{2}$ and $\frac{1}{2}$, respectively.

Example 5.4 Bockris, Drazic, and Despic *(11)* studied the rate of anodic dissolution of iron in 0.5 M $FeSO_4$, 0.5 M Na_2SO_4. Figure 5.9 shows Tafel plots for the process

$$Fe(s) \rightarrow Fe^{2+} + 2e^-$$

[‡] Some authors use the term "transfer coefficient" to refer to these empirical "apparent transfer coefficients," reserving the term "symmetry factor" for the transfer coefficient of an elementary process.

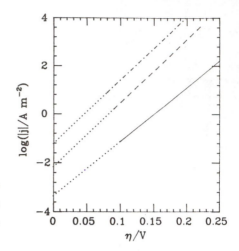

Figure 5.9 Tafel plots for the anodic dissolution of iron at pH 1.9 (solid), 3.1 (dashes), and 4.0 (dot–dash).

at pH 1.9, 3.1, and 4.0. The slopes of the three lines are approximately equal and give

$$\frac{\beta_{app}F}{2.303RT} = 25 \pm 3 \text{ V}^{-1}$$

so that β_{app} = 1.48 ± 0.18. Furthermore, the displacement of the lines at different pH values gives

$$\frac{\partial \log|j|}{\partial \text{pH}} = 1.01 \pm 0.02$$

suggesting that the process is first order in OH^- (or –1 order in H^+). The cathodic branch Tafel plot is more difficult to come by because of the interference of the reduction of H^+. However, the data available suggested that $\alpha_{app} \cong 0.5$. These results are thus consistent with the mechanism

$$\text{Fe(s)} + \text{H}_2\text{O} \rightleftarrows \text{FeOH(surface)} + \text{H}^+\text{(aq)} + \text{e}^-$$

$$\text{FeOH(surface)} \rightleftarrows \text{FeOH}^+\text{(aq)} + \text{e}^-$$

$$\text{FeOH}^+\text{(aq)} + \text{H}^+\text{(aq)} \rightleftarrows \text{Fe}^{2+}\text{(aq)} + \text{H}_2\text{O}$$

with the second step rate limiting. Since $FeOH^+$ is in equilibrium with Fe^{2+}, C_O in eq (5.27a) must be replaced by

$$C_O = [FeOH^+] = K_h[Fe^{2+}]/[H^+]$$

where K_h is the equilibrium constant for the hydrolysis of Fe^{2+}. Thus the exchange current is

$$i_0 = FAk_0 K_h \theta_1{}^a [Fe^{2+}]/[H^+]$$

consistent with the observed pH dependence.

Kinetics of Hydrogen Evolution

Probably the most thoroughly studied problem in electrode kinetics is the reduction of hydrogen ions, H^+ (aq), to form hydrogen gas. The mechanism has been the subject of some controversy over the years and the details are still imperfectly understood *(B7,B8)*. Some representative results are shown in Table 5.1. One of the most striking features of the experimental results is the enormous range in the exchange current density with electrode material. Hydrogen evolution at a platinum electrode has $j_0 = 10$ A m^{-2}, whereas at a lead electrode, $j_0 = 2 \times 10^{-9}$ A m^{-2}. Furthermore, the apparent cathode transfer coefficients vary from less than 0.5 to about 2.0. These features can be understood, at least qualitatively, in terms of a simple model.

Consider the electrode process

$$2\,H^+\,(aq) + 2\,e^- \rightarrow H_2(g)$$

The mechanism is likely to involve reduction of hydrogen ions to give hydrogen atoms adsorbed on the electrode surface:

$$M + H^+\,(aq) + e^- \rightarrow M\text{–}H \tag{i}$$

The hydrogen atoms then migrate on the surface and form hydrogen molecules:

$$2\,M\text{–}H \rightarrow M \cdot H_2 \tag{iia}$$

Alternatively, if the mobility of hydrogen atoms on the surface is slow, H–H bond formation may be concerted with reduction of a second hydrogen ion near an adsorbed atom:

TABLE 5.1 Kinetic Parameters for Hydrogen Evolution

Metal[a]	$\log(j_0/\text{A m}^{-2})$	a_{app}
Pt (smooth)	1.0	1.5
Pd	0.3	2.0
Ni	−1.4	0.5
Au	−2.0	1.3
Fe	−2.0	0.4
Mo	−2.0	1.5
W	−2.0	1.5
Cu	−2.7	0.5
Cd	−3.0	0.3
Sn	−4.0	0.4
Hg	−7.7	0.5
Pb	−8.7	0.5

[a] Metal in contact with 1.0 M HCl at 20°C. Data from Conway *(H2)*.

$$\text{M–H} + \text{H}^+(\text{aq}) + \text{e}^- \rightarrow \text{M} \cdot \text{H}_2 \tag{iib}$$

Finally, hydrogen molecules are desorbed from the surface:

$$\text{M} \cdot \text{H}_2 \rightarrow \text{M} + \text{H}_2(\text{g}) \tag{iii}$$

If step (i) is rate limiting, the mechanism is analogous to the reduction of iodine with two parallel and identical rate–limiting electron–transfer steps. In the absence of double–layer effects, the anodic and cathodic transfer coefficients should be near $\frac{1}{2}$. The data given in Table 5.1 are consistent (though they are hardly conclusive) with step (i) being rate limiting for Pb, Hg, Zn, Sn, Cd, Cu, Fe, and Ni.

If step (iib) is rate limiting, the mechanism is analogous to the oxidation of iron and we expect that the cathodic transfer coefficient should be approximately $\frac{3}{2}$ and the anodic transfer coefficient should be about $\frac{1}{2}$. Step (iib) is probably rate limiting for Pt, Au, Mo, and W.

This leaves the apparent transfer coefficient of 2 for Pd as an anomaly. Palladium behaves anomalously with hydrogen in other kinds of experiments. The mobility of hydrogen atoms in bulk palladium is known to be unusually high so it may be that the reaction goes via step (iia) which is rate limiting. If this is the case, the expected rate law is somewhat different. If k_c is the forward rate constant for step (iia) and C_H is the surface concentration of adsorbed hydrogen atoms, the net current for a cathodic overpotential should be given by

$$i = 2FAk_c C_H{}^2$$

For small overpotential and small surface concentration of H atoms, the electron–transfer step is nearly at equilibrium, so that the rates of the cathodic and anodic steps are nearly equal:

$$k_{ic}[H^+] = k_{ia}C_H$$

or

$$k_0[H^+] \exp \frac{-\alpha F(E - E^\circ)}{RT} \cong k_0 C_H \exp \frac{\beta F(E - E^\circ)}{RT}$$

Thus the surface concentration is

$$C_H \cong [H^+] \exp \frac{F(E - E^\circ)}{RT}$$

and the net current is given by

$$i = 2FAk_c[H^+]^2 \exp \frac{-2F(E - E^\circ)}{RT}$$

or

$$i = i_0 \exp \frac{-2F\eta}{RT}$$

Thus the apparent cathodic transfer coefficient for this mechanism is expected to be about 2, consistent with the results for Pd.

A strong M–H bond is expected to make step (i) more favorable but hinder step (iia) or (iib). Conversely, a weak M–H bond would make step (i) slow and (iia) or (iib) fast. Thus we might expect a correlation between exchange current density and M–H bond strength which would distinguish between the two kinds of mechanism. The exchange current density for several metals is plotted as a function of the estimated M–H bond strength in Figure 5.10. The exchange current density increases with increasing bond strength in the series Tl – Co, presumably because step (i) of the mechanism above is getting faster, but is still rate limiting (*i.e.*, slower than steps iia or iib). In the series Pt – Ta, the exchange current density decreases with increasing bond strength, presumably because step (iia) or (iib) is rate limiting.

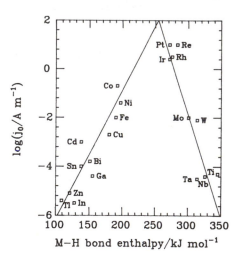

Figure 5.10 Exchange current density for hydrogen evolution from aqueous solutions for various electrode materials as a function of estimated metal–hydrogen bond strength. Data from Trasatti *(12)*.

5.3 *IRREVERSIBILITY IN VOLTAMMETRY*

In Section 5.1, we considered the limiting case where current through a cell is limited only by the kinetics of the electron–transfer process and assumed that transport to the electrode surface is fast. In Section 4.1, we considered the opposite case, where electron transfer is fast and the current is diffusion limited. The real world, of course, often contains intermediate cases and we now consider the situation where the rates of mass transport and electron transfer are of comparable magnitude.

The mathematical description of current–potential, current–time, or potential–time curves can become quite involved when the rates of transport and electron transfer are both important. In the case of the rotating disk electrode, the current is measured under steady–state conditions and the mathematics becomes relatively simple. We will treat the cases of quasi–reversible and irreversible r.d.e. voltammetric waves in some detail and, with this background, present (without derivation) the results of more complex calculations for potential step chronoamperometry, d.c. polarography, and cyclic voltammetry.

Steady–State Current at the Rotating Disk Electrode

When the rate of electron transfer is comparable to the rate of transport, equilibrium is not achieved at the electrode surface and the current must be calculated taking into account both the transport and electron–transfer steps. The problem can be understood most easily in terms of an analogy with homogeneous reaction kinetics. If, for a two–step reaction mechanism of the type

$$A \underset{k_{-1}}{\overset{k_1}{\rightleftarrows}} B \overset{k_2}{\rightarrow} C$$

B is a short–lived intermediate, the rate can be computed by applying the steady–state approximation to the rate of formation of B:

$$\frac{dC_B}{dt} = k_1 C_A - (k_{-1} + k_2)C_B = 0$$

Solving for C_B and substituting it in the expression for the net rate of the reaction

$$\text{rate} = \frac{dC_C}{dt} = k_2 C_B$$

we have

$$\text{rate} = \frac{k_1 k_2 C_A}{k_{-1} + k_2}$$

When the first step is rate limiting ($k_2 \gg k_{-1}$), the observed rate constant is $k_{obs} = k_1$ (independent of the rate of the second step); when the second step is slow, on the other hand, $k_{obs} = k_1 k_2/k_{-1}$.

In an electrode process, the transport of electroactive material to the electrode is reversible and analogous to step 1 of the mechanism above; electron transfer corresponds to step 2. When electron transfer is fast, we find that the rate (current) is limited by the rate of the transport process (the flux) and is given by eqs (4.7), (4.25), or (4.33) for a planar electrode, the d.m.e. or the r.d.e., respectively. The analog of the other limiting case of the homogeneous mechanism is slow electron transfer; in this case, the current is proportional to the electron–transfer rate constant but is also a function of the transport rate constants. Electron–transfer rate constants are functions of potential, so that when the electrode is made more negative, the electron–transfer rate increases (for a reduction process) and eventually becomes faster than the rate of transport. Thus at sufficiently negative potentials, a diffusion–limited current is usually observed, even for electron transfers which are intrinsically slow.

An exact solution to this problem would involve solving forced convection/diffusion equations, eq (4.29), with boundary conditions which include the finite rate of electron transfer. It is considerably easier, however, to follow the analogy with homogeneous kinetics and to solve the following reaction scheme using the steady–state approximation.

One–Electron Processes. Consider the one–electron reduction of O to R, represented by

$$O^* \underset{k_{DO}}{\overset{k_{DO}}{\rightleftarrows}} O_0 \underset{k_a}{\overset{k_c}{\rightleftarrows}} R_0 \underset{k_{DR}}{\overset{k_{DR}}{\rightleftarrows}} R^*$$

where $k_{DO} = D_O/x_{DO}$ and $k_{DR} = D_R/x_{DR}$ are the mass transport rate constants and k_c and k_a are the electron–transfer rate constants given by eqs (5.9). The steady–state approximation applied to the surface concentrations, $C_O(0)$ and $C_R(0)$, gives

$$\frac{dC_O(0)}{dt} = k_{DO}C_O{}^* + k_aC_R(0) - (k_{DO} + k_c)C_O(0) = 0$$

$$\frac{dC_R(0)}{dt} = k_{DR}C_R{}^* + k_cC_O(0) - (k_{DR} + k_a)C_R(0) = 0$$

Assuming that the bulk concentration of R is zero, we solve these equations for $C_O(0)$ and $C_R(0)$:

$$C_O(0) = \frac{C_O{}^*(1 + k_a/k_{DR})}{1 + k_a/k_{DR} + k_c/k_{DO}} \tag{5.29a}$$

$$C_R(0) = \frac{C_O{}^*(k_c/k_{DO})}{1 + k_a/k_{DR} + k_c/k_{DO}} \tag{5.29b}$$

The net current is

$$i = i_c + i_a = FA[k_cC_O(0) - k_aC_R(0)]$$

Substituting for the surface concentrations, we have (for $k_{DO} = k_{DR} = k_D$)

$$i = FA\left[\frac{k_DC_O{}^*}{1 + k_a/k_c + k_D/k_c}\right] \tag{5.30}$$

When the electrode potential is sufficiently negative, the cathodic rate constant k_c will be large compared with k_a or k_D and the current will be diffusion limited:

$$i_D = FAk_DC_O{}^*$$

which is identical to eq (4.32) since $k_D = D/x_D$. Dividing i_D by i and rearranging, we have

$$\frac{i_D - i}{i} = \frac{k_D + k_a}{k_c}$$

Substituting for k_c and k_a using eqs (5.6) and (5.9), we obtain

$$\frac{i_D - i}{i} = \frac{k_D}{k_0} \exp \frac{a F(E - E^\circ)}{RT} + \exp \frac{F(E - E^\circ)}{RT} \qquad (5.31)$$

When k_D/k_0 is small (fast electron transfer), the first term on the right–hand side of eq (5.31) can be neglected and we get the Heyrovský–Ilkovič equation, eq (4.10), with the half–wave potential given by eq (4.34). On the other hand, when electron transfer is slow, so that $k_D \gg k_0$, the first term dominates and we have a *completely irreversible wave*. Taking logs and rearranging, we get

$$E = E_{\frac{1}{2}} + \frac{RT}{a F} \ln \frac{i_D - i}{i} \qquad (5.32)$$

where the half–wave potential is given by

$$E_{\frac{1}{2}} = E^\circ + \frac{RT}{a F} \ln \frac{k_0}{k_D} \qquad (5.33)$$

In intermediate cases, where $k_D \approx k_0$, the electrode process is referred to as *quasi–reversible* and the wave shape must be calculated using the full equation, eq (5.31). Figure 5.11 shows the transition of the current–potential curve from reversible to quasi–reversible to irreversible as the electron–transfer rate decreases. Since $k_D = D/x_D$, we can use eq (4.31) to obtain

$$k_0/k_D = 1.61 \, k_0 D^{-\frac{2}{3}} \nu^{\frac{1}{6}} \omega^{-\frac{1}{2}}$$

This quantity can be varied experimentally over a factor of about 10, so that waves which appear reversible at slow rotation speeds may become quasi–reversible or irreversible at the fastest rotation rates. If the standard potential is known, the shift of half–wave potential may be used to compute k_0 and the transfer coefficient can be obtained from the slope of a plot of $E_{\frac{1}{2}}$ vs. $\ln \omega$.

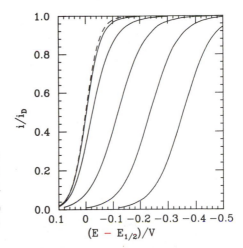

Figure 5.11 Plot of relative current at a rotating disk electrode *vs.* $E - E_{\frac{1}{2}}(\text{rev})$ for a reversible wave and for k_D/k_0 = 0.1, 1, 10, 100, and 1000, a = 0.5.

Two–Electron Processes. Consider now the two–electron reduction of O to R where we assume that, at least in principle, the process occurs in two one–electron steps (an EE process):

$$O + e^- \rightleftharpoons I$$

$$I + e^- \rightleftharpoons R$$

Following the foregoing procedure, we can compute the steady–state surface concentrations of O, I, and R as functions of the electron-transfer rate constants k_{1c}, k_{1a}, k_{2c}, and k_{2a}, and the mass transport rate constant k_D (assumed equal for all three species). Substitution into the expression for the net current

$$i = FA[k_{1c}C_O(0) + (k_{2c} - k_{1a})C_I(0) - k_{2a}C_R(0)]$$

gives

$$i = FAk_D C_O^* \left[\frac{2 + \theta_2 + k_D/k_{2c}}{(1 + \theta_1 + k_D/k_{1c})(1 + \theta_2 + k_D/k_{2c}) - \theta_1} \right] \quad (5.34)$$

where

$$\theta_1 = \exp \frac{F(E - E_1{}^\circ)}{RT} = \frac{k_{1a}}{k_{1c}}$$

$$\theta_2 = \exp \frac{F(E - E_2°)}{RT} = \frac{k_{2a}}{k_{2c}}$$

$$k_{1c} = k_{1,0} \exp \frac{-a_1 F(E - E_1°)}{RT}$$

$$k_{2c} = k_{2,0} \exp \frac{-a_2 F(E - E_2°)}{RT}$$

Equation (5.34) is a bit complicated; let us consider some limiting cases.

Case I. Assume that the first step is thermodynamically easier than the second, $E_2° < E_1°$. When $E \approx E_1°$, we should have ($\theta_2 + k_D/k_{2c}$) $\gg$ 1, so that eq (5.34) simplifies to

$$i = \frac{FAk_D C_O{}^*}{1 + \theta_1 + k_D/k_{1c}}$$

which is identical to eq (5.30) and represents a one–electron wave which may be reversible, quasi–reversible, or irreversible depending on the magnitude of k_D/k_{1c}.

When $E \approx E_2° < E_1°$, θ_1 will be very small and, if the first electron–transfer step is fast because of the large overpotential, k_D/k_{1c} will also be small. Equation (5.34) then becomes

$$i = FAk_D C_O{}^* \left[1 + \frac{1}{1 + \theta_2 + k_D/k_{2c}} \right]$$

which represents a one–electron wave in the vicinity of $E_2°$ which grows out of the one–electron limiting current of the first wave. Thus in Case I, we expect two essentially independent one–electron waves.

Case II. When $E_2° > E_1°$, the intermediate I is thermodynamically unstable and, if the electron–transfer rates are fast enough, we expect to see a two–electron wave. Since $E_2° > E_1°$, we have $\theta_2 \ll \theta_1$ whenever the current is nonnegligible. If k_D is small compared with k_{1c} and k_{2c} (the reversible case), eq (5.34) reduces to

$$i = \frac{2FAk_D C_O{}^*}{1 + \theta_1 \theta_2}$$

or, with $i_D = 2FAk_D C_O{}^*$, we have

$$\frac{i_D - i}{i} = \theta_1 \theta_2$$

which, on taking logs, becomes the Heyrovský–Ilkovič equation with $n = 2$.

Either of the two electron–transfer steps could be rate limiting. Consider first the case where step one is slow and step two fast. For $\theta_2 \ll 1$ and $k_D/k_{2c} \ll 1$, eq (5.34) becomes

$$i = \frac{2FAk_D C_O{}^*}{1 + \theta_1 \theta_2 + k_D/k_{1c}}$$

which is very similar to eq (5.30) except that the limiting current corresponds to two electrons. There is an important difference, however. In the reversible case ($k_D/k_{1c} \ll 1$), the Tomeš criterion gives

$$E_{\frac{1}{4}} - E_{\frac{3}{4}} = 56.5/n \text{ mV}$$

or 28.2 mV for a two–electron wave at 25°C. But when the process is completely irreversible, we have

$$E_{\frac{1}{4}} - E_{\frac{3}{4}} = 56.5/a \text{ mV}$$

or about 113 mV if $a = \frac{1}{2}$. Thus the increase in width with decreasing electron–transfer rate constant (or increasing mass transport rate) is rather more dramatic than in the one–electron case. Computed waves for $E_2{}^\circ - E_1{}^\circ = 0.2$ V and $a = \frac{1}{2}$ are shown in Figure 5.12 for several values of k_D/k_0.

When step 2 is rate limiting, the shape of the voltammetric wave is quite sensitive to the difference in potentials, $E_2{}^\circ - E_1{}^\circ$ and the electron–transfer rate. If the separation is large compared compared with the kinetic shift, a single symmetrical two–electron wave is expected with a width of

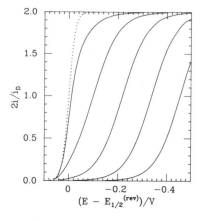

Figure 5.12 Plot of relative current at a r.d.e. *vs.* $E - E_{\frac{1}{2}}(\text{rev})$ for a reversible two–electron wave (dotted curve) and for a two–electron wave with step one rate limiting; (solid curves, left to right) k_D/k_0 = 0.1, 1, 10, 100, and 1000, a = 0.5, ΔE = 0.2 V.

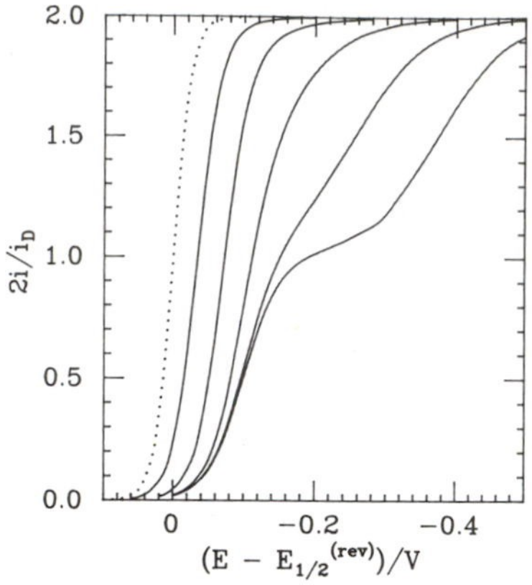

Figure 5.13 Plot of relative current at a r.d.e. *vs.* $E - E_{\frac{1}{2}}(\text{rev})$ for a reversible two–electron wave (dotted curve) and for a two–electron wave with step two rate limiting; (solid curves, left to right) k_D/k_0 = 1, 10, 100, 1000, and 10000, a = 0.5, ΔE = 0.2 V.

$$E_{\frac{1}{4}} - E_{\frac{3}{4}} = 56.5/(1 + a)\ \text{mV}$$

This is analogous to the two–electron process discussed in Section 5.2, where we found an apparent transfer coefficient of $\frac{3}{2}$.

When the kinetic shift is larger than the separation of standard potentials, a rather unsymmetrical wave results. If the second electron–transfer step is intrinsically very slow, the current from the first electron transfer may be nearly limiting before the overpotential is big enough to provide a significant rate for the second step. The transition from a reversible two–electron wave to this unsymmetrical situation is shown in Figure 5.13.

Criteria for Reversibility

Two methods are commonly used to ascertain the reversibility of an electrode process. The first, which is based on eq (4.10), uses a plot of E *vs.* $\ln[(i_D - i)/i]$, which should be linear with a slope of RT/nF if the process is reversible. Such a plot will not be linear if the process is quasi–reversible. An essentially equivalent check of reversibility is the Tomeš criterion (see Example 4.2): for a reversible process the difference in potentials at which $i = \frac{3}{4} i_D$ and $i = \frac{1}{4} i_D$ is

$$E_{\frac{1}{4}} - E_{\frac{3}{4}} = (RT/nF)\ln 9 = 56.5/n \text{ mV}$$

at 25°C. These methods can lead to misinterpretation, however. If the process is completely irreversible, a plot of E vs. $\ln[(i_D - i)/i]$ also gives a straight line but with slope RT/Fa_{app}. An irreversible two–electron process where the first step is rate limiting with $a = 0.5$ is thus indistinguishable from a irreversible one–electron process. This ambiguity can be avoided, of course, if n is known from the measured wave height or from a coulometric experiment (see Section 6.2).

A better method to check reversibility is to measure current–potential curves for different r.d.e. rotation speeds (or different d.m.e. drop times—see below). Since the diffusion layer thickness varies with rotation speed (or drop time), the ratio k_D/k_0 is affected. If the process is reversible, the shape of the curve is unaffected and the apparent half–wave potential is constant, but variations in $E_{\frac{1}{2}}$ will be found for quasi–reversible and irreversible waves.

Potential Step Chronoamperometry

In order to obtain the current–time response in a potential step experiment using a stationary planar electrode, the diffusion equations must be solved with the boundary condition

$$i = FA[k_c C_O(0,t) - k_a C_R(0,t)] \tag{5.35}$$

where k_c and k_a are functions of the potential and are given by eqs (5.9). The solution to the mathematical problem is fairly straightforward using Laplace transform techniques (see Appendix 4). For a one–electron process, the surface concentrations are found to be

$$C_O(0,t) = C_O^* - \frac{C_O^*}{1 + \theta}\left[1 - \frac{F(\lambda t^{\frac{1}{2}})}{\lambda (\pi t)^{\frac{1}{2}}}\right] \tag{5.36a}$$

$$C_R(0,t) = \frac{C_O^*}{1 + \theta}\left[1 - \frac{F(\lambda t^{\frac{1}{2}})}{\lambda (\pi t)^{\frac{1}{2}}}\right] \tag{5.36b}$$

where we have assumed equal diffusion coefficients. As usual, the parameter θ is

$$\theta = \exp \frac{F(E - E^\circ)}{RT} = \frac{k_a}{k_c}$$

and we have introduced the function

$$F(\lambda t^{\frac{1}{2}}) = \lambda (\pi t)^{\frac{1}{2}} \exp(\lambda^2 t)[1 - \text{erf}(\lambda t^{\frac{1}{2}})] \tag{5.37}$$

$$\lambda = (k_c + k_a)/D^{\frac{1}{2}} = (k_0/D^{\frac{1}{2}})(\theta^{-a} + \theta^\beta) \tag{5.38}$$

where we have assumed equal diffusion coefficients for O and R and zero bulk concentration of R. The function $F(\lambda t^{\frac{1}{2}})$, shown in Figure 5.14, varies from 0 to 1 with increasing $\lambda t^{\frac{1}{2}}$. In the limit of fast electron transfer and long times, $\lambda t^{\frac{1}{2}} \gg 1$, $F(\lambda t^{\frac{1}{2}})/\lambda (\pi t)^{\frac{1}{2}} \to 0$ and eqs (5.36) reduce to eqs (4.3), which were derived assuming a nernstian electrode process. The current is obtained by substituting the surface concentrations in eq (5.35), obtaining

$$i = FAk_c C_O^* \frac{F(\lambda t^{\frac{1}{2}})}{\lambda (\pi t)^{\frac{1}{2}}} \tag{5.39}$$

We can get some insight into the significance of eq (5.39) by considering two limiting cases. At short times, $F(\lambda t^{\frac{1}{2}}) \cong \lambda (\pi t^{\frac{1}{2}})$, and at $t = 0$, eq (5.39) becomes

$$i(0) = FAk_c C_O^*$$

so that the current is initially limited by the rate of electron transfer. When we make the overpotential sufficiently negative that $k_c \gg k_a$, $\lambda \cong k_c/D^{\frac{1}{2}}$ and eq (5.39) becomes

$$i = FAC_O^*(D/\pi t)^{\frac{1}{2}}F(\lambda t^{\frac{1}{2}})$$

For a sufficiently negative overpotential or at long times, $\lambda t^{\frac{1}{2}} \gg 1$, $F(\lambda t^{\frac{1}{2}}) \cong 1$, and eq (5.39) reduces to the Cottrell equation, eq (4.7); the current is then diffusion controlled.

In an experiment, the potential is stepped to some negative value and the current measured as a function of time. With increasing time,

$\lambda t^{\frac{1}{2}}$ increases, and $F(\lambda t^{\frac{1}{2}}) \to 1$. Thus for long enough time, current decay curves are expected to be independent of the electron–transfer rate constant; of course, the current may be very small indeed when that happens. Figure 5.15 shows the kinds of results expected for systems differing only in electron–transfer rates. Notice that for very slow electron–transfer rates, it is possible to extrapolate the current to zero time and thus to determine the electron–transfer rate constant k_c directly if the electrode area and the concentration of O are known.

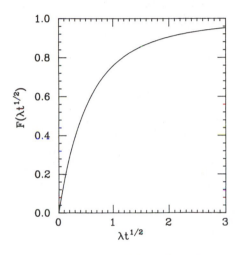

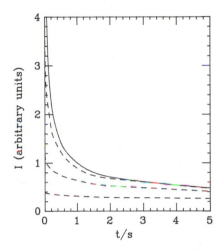

Figure 5.14 The function $F(\lambda t^{\frac{1}{2}})$ vs. $\lambda t^{\frac{1}{2}}$.

Figure 5.15 Current–time curves for potential step chronoamperometry with $a = 0.5$, $\eta = -0.1$ V, $D_O = 10^{-9}$ $m^2 s^{-1}$, and $k_0 = \infty$ (solid line) and $\log (k_0/m\ s^{-1}) = -5.0$, -5.5, and -6.0.

Irreversible Polarographic Waves

In Section 4.4, we corrected for the expanding mercury drop area by using an effective diffusion coefficient, $\frac{7}{3} D$, in the equations derived for a planar electrode. That approach is less accurate for an irreversible wave (although the right qualitative result is obtained). Meites and Israel *(13)* have shown that the shape of an irreversible polarographic wave is given approximately by

$$E = E_{\frac{1}{2}} + \frac{0.916\ RT}{a\,F}\ \ln \frac{i_D - i}{i} \tag{5.40}$$

where

$$E_{\frac{1}{2}} = E° + \frac{RT}{aF} \ln \frac{1.35\, k_0 t^{\frac{1}{2}}}{D_O^{\frac{1}{2}}}$$ (5.41)

where t is the drop time. Since x_D is proportional to $(D_O t)^{\frac{1}{2}}$, eqs (5.40) and (5.41) are functionally similar to eqs (5.32) and (5.33) for irreversible r.d.e. voltammetric waves. Irreversible polarographic waves are thus qualitatively similar to the curves shown in Figure 5.11. Again, if $E°$ is known, the shift in half-wave potential can be used to determine k_0.

Cyclic Voltammetry

It is easy to predict the qualitative effect of slow electron transfer on a cyclic voltammogram from our experience with chronoamperometry and r.d.e. voltammetry. We would expect a negative shift and broadening of the cathodic peak in a cyclic voltammogram (a positive shift of an anodic peak) with decreasing electron–transfer rate constant k_0. This expectation is fulfilled by calculations based on the mathematical model developed by Nicholson (14). The shape of a cyclic voltammogram is found to depend on the transfer coefficient a and on the dimensionless parameter

$$\lambda = k_0 \left[\frac{RT}{\pi DFv}\right]^{\frac{1}{2}}$$

where v is the potential scan rate. For $\lambda > 7$, the voltammogram appears essentially reversible, independent of λ and a, but for smaller values of λ, the expected peak shifts and broadening are predicted as shown in Figure 5.16a. When the transfer coefficient deviates from 0.5, cyclic voltammograms become asymmetric, as shown in Figure 5.16b. Thus when $a = 0.75$, for example, the cathodic peak is sharper than the anodic peak as expected from the Butler–Volmer equation and Figure 5.2.

Just as plots of E vs. $\ln[(i_D - i)/i]$ for voltammetric waves give straight lines for both reversible and completely irreversible processes, so plots of cyclic voltammetry peak currents vs. $\sqrt{v}$ are linear for both extremes. Such plots are not linear for quasi–reversible processes $(\lambda \approx 1)$ and so can be used (with care) for diagnostic purposes. The separation of the anodic and cathodic peaks, ΔE_p, increases with decreasing values of λ as shown in Figure 5.17, and this parameter can be used to obtain a rough measure of the electron–transfer rate constant; since ΔE_p also increases with increasing v because of solution iR drop, such a measurement must be interpreted with caution.

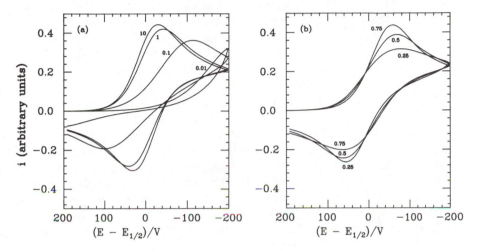

Figure 5.16 Computed cyclic voltammograms for (a) λ = 10, 1, 0.1, and 0.01, a = 0.5, and (b) λ = 0.3 and a = 0.25, 0.5, and 0.75.

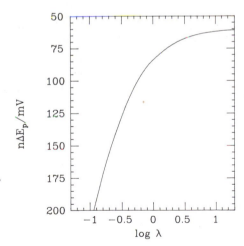

Figure 5.17 Separation of cyclic voltammogram peaks as a function of the parameter λ (a = 0.5) *(14)*.

5.4 FARADAIC IMPEDANCE

Some of the best methods for studying the rates of electron–transfer reactions at electrodes involve a.c. measurement of the faradaic impedance, the equivalent circuit parameter representing the barrier to current flow between electrode and solution. The faradaic impedance represents both the finite rate of electron transfer and the rate of transport of electroactive material to the electrode surface. Since (as we will see below) the current is expected to lag a bit behind the a.c. potential, the faradaic impedance is not purely resistive but has a capacitive component. The faradaic impedance acts in parallel with the double–layer capacitance C_d, and this combination is in series with the solution resistance R, as shown in Figure 3.1. In measurements of solution conductance, the cell is designed with large electrodes to maximize the double–layer capacitance and to minimize the faradaic impedance. Here we are concerned with the faradaic component and the opposite strategy is adopted. Low currents minimize the solution iR drop and small electrodes tend to amplify the faradaic contribution. In careful work, attention must be paid to the separation of the effects of solution resistance and double–layer capacitance from the measured impedance, but we will assume in this section that the faradaic impedance is measured directly.

Resistive and Capacitive Components

Suppose that the faradaic impedance is equivalent to a resistance R_s and capacitance C_s in series and that a time–dependent current flows through the circuit:

$$I(t) = I_0 \sin \omega t \qquad (5.42)$$

where ω is the angular frequency (2π times the frequency in hertz). The potential drop is

$$E = IR_s + Q/C_s$$

so that

$$\frac{dE}{dt} = R_s \frac{dI}{dt} + I/C_s$$

or

$$\frac{dE}{dt} = R_s I_0 \omega \cos \omega t + (I_0/C_s) \sin \omega t \qquad (5.43)$$

The solution to this differential equation can be written

$$E(t) = \Delta E \sin(\omega t + \phi) + E_{dc}$$

so that

$$\frac{dE}{dt} = \omega \Delta E \cos(\omega t + \phi) = \omega \Delta E (\cos \omega t \cos \phi + \sin \omega t \sin \phi)$$

Equating the coefficients of $\sin \omega t$ and $\cos \omega t$, we have

$$\omega \Delta E \cos \phi = \omega I_0 R_s$$

$$\omega \Delta E \sin \phi = I_0/C_s$$

so that the phase angle ϕ is given by

$$\cot \phi = \cos \phi / \sin \phi = \omega R_s C_s \qquad (5.44)$$

and the faradaic impedance Z_f is

$$| Z_f | = \Delta E/I_0 = [R_s^2 + 1/(\omega C_s)^2]^{\frac{1}{2}} \qquad (5.45)$$

Thus the magnitude of the faradaic impedance is given by the ratio of the amplitudes of the a.c. potential and a.c. current and the resistive and capacitive components can be separated if the phase angle is known.

Solution of the Diffusion Problem

For the electrode process

$$O + e^- \rightleftarrows R$$

the linear diffusion problem (see Section 4.1) can be solved with the periodic boundary condition

$$I_0 \sin \omega t = -FAJ_O(0,t) = FAJ_R(0,t)$$

to obtain the following expressions for the surface concentrations (see Appendix 4):

$$C_O(0,t) = C_O{}^* - \frac{I_0(\sin \omega t - \cos \omega t)}{FA(2\omega D_O)^{\frac{1}{2}}} \qquad (5.46a)$$

$$C_R(0,t) = C_R{}^* + \frac{I_0(\sin \omega t - \cos \omega t)}{FA(2\omega D_R)^{\frac{1}{2}}} \qquad (5.46b)$$

The surface concentrations undergo sinusoidal variations with amplitudes proportional to the current amplitude I_0 and inversely proportional to $(\omega D)^{\frac{1}{2}}$. Thus the amplitudes of the concentration variations decrease as the frequency or the rate of diffusion increases. Notice that the sinusoidal variation of $C_R(0,t)$ is 180° out of phase with $C_O(0,t)$; since

$$\sin \omega t - \cos \omega t = 2^{\frac{1}{2}} \sin(\omega t - \pi/4)$$

$C_R(0,t)$ lags 45° behind the current.

Calculation of the Faradaic Impedance

Driving these concentration variations leads to both dissipative (resistive) and reactive (capacitive) terms in the faradaic impedance. To see how these develop, we need to derive an expression for dE/dt for comparison with eq (5.43). Since the electrochemical system is assumed to be near equilibrium, η is very small and $C_O(0,t) \cong C_O{}^*$, $C_R(0,t) \cong C_R{}^*$. Thus eq (5.10) can be expanded, retaining only first-order terms, to obtain

$$I(t) \cong i_0 \left[\frac{C_O(0,t)}{C_O{}^*} - \frac{C_R(0,t)}{C_R{}^*} - \frac{F\eta(t)}{RT} \right]$$

or

$$\eta(t) \cong - \frac{RT}{F} \left[\frac{C_O(0,t)}{C_O{}^*} - \frac{C_R(0,t)}{C_R{}^*} + \frac{I(t)}{i_0} \right]$$

The time derivative of η can then be written:

$$\frac{d\eta}{dt} = R_{ct} \frac{dI}{dt} + \beta_O \frac{dC_O}{dt} + \beta_R \frac{dC_R}{dt} \tag{5.47}$$

where R_{ct} is the charge–transfer resistance

$$R_{ct} = \frac{\partial \eta}{\partial I} = -\frac{RT}{Fi_0} \tag{5.48}$$

and the other parameters are

$$\beta_O = \frac{\partial \eta}{\partial C_O} = -\frac{RT}{FC_O^*} \tag{5.49a}$$

$$\beta_R = \frac{\partial \eta}{\partial C_R} = +\frac{RT}{FC_R^*} \tag{5.49b}$$

The time derivatives of I, C_O, and C_R can be computed from eqs (5.42) and (5.46):

$$\frac{dI}{dt} = I_0\omega \cos \omega t \tag{5.50}$$

$$\frac{dC_O}{dt} = -\frac{I_0\omega(\cos \omega t + \sin \omega t)}{FA(2D_O\omega)^{\frac{1}{2}}} \tag{5.51a}$$

$$\frac{dC_R}{dt} = \frac{I_0\omega(\cos \omega t + \sin \omega t)}{FA(2D_R\omega)^{\frac{1}{2}}} \tag{5.51b}$$

Combining eqs (5.47), (5.50) and (5.51), we have

$$\frac{d\eta}{dt} = (R_{ct} + \sigma\omega^{-\frac{1}{2}}) I_0\omega \cos \omega t + \sigma\omega^{\frac{1}{2}}I_0 \sin \omega t \tag{5.52}$$

where

$$\sigma = \frac{1}{FA} \left[\frac{\beta_R}{(2D_R)^{\frac{1}{2}}} - \frac{\beta_O}{(2D_O)^{\frac{1}{2}}} \right]$$

or, using eqs (5.49),

$$\sigma = \frac{RT}{F^2 A} \left[\frac{1}{(2D_R)^{\frac{1}{2}}C_R{}^*} + \frac{1}{(2D_O)^{\frac{1}{2}}C_O{}^*} \right] \tag{5.53}$$

We can now identify the equivalent circuit parameters R_s and C_s by comparison of eqs (5.43) and (5.52):

$$R_s = R_{ct} + \sigma \omega^{-\frac{1}{2}} \tag{5.54}$$

$$C_s = 1/\sigma \omega^{\frac{1}{2}} \tag{5.55}$$

or

$$|Z_f| = (R_{ct}{}^2 + 2\sigma^2/\omega)^{\frac{1}{2}} \tag{5.56}$$

$$\cot \phi = 1 + (R_{ct}/\sigma)\, \omega^{\frac{1}{2}} \tag{5.57}$$

The faradaic impedance is seen to have two components, the charge-transfer resistance, which is inversely proportional to the exchange current i_0 and thus is a measure of the rate of electron transfer, and the parameter σ, which relates to mass transport and contributes equally to the resistive and reactive parts of the impedance. The impedance $\sigma(2/\omega)^{\frac{1}{2}}$ is sometimes called the *Warburg impedance* to distinguish it from the kinetic contribution to the faradaic impedance.

If the experiment gives the frequency dependence of the faradaic impedance and phase angle, eqs (5.56) and (5.57) can be used to separate the contributions from mass transport and charge transfer. Thus faradaic impedance measurements have been rather commonly used for determination of electron–transfer rate constants.

Direct measurement of the faradaic impedance for a system at equilibrium works well when both components of the system are chemically stable. There are many cases, however, where one of the components of the electrode couple is of limited stability so that a solution containing known amounts of O and R cannot be prepared. The general ideas of the faradaic impedance method can still be used; however, we start with a solution containing only O (or R) and apply a d.c. potential to adjust the O/R ratio at the electrode surface. This marriage of d.c. voltammetry and faradaic impedance methods can be applied using a variety of indicator electrodes, but the most common application has been with the dropping mercury electrode, a technique called a.c. polarography.

A.C. Polarography

In a.c. polarography, a small sinusoidal a.c. voltage is super-imposed on the d.c. ramp voltage and the a.c. component of the cell current is measured using a "lock–in detector" so that the in–phase and out–of–phase current components can be determined separately.[‡] In other words, the cell potential is modulated by an amount ΔE and the resulting current modulation Δi is measured. For small values of ΔE, the a.c. polarogram should approximate the first derivative of the d.c. polarogram (see Problems). The a.c. modulation voltage is typically in the audio–frequency range, ca. 10 Hz – 10 kHz, with an amplitude of 5–10 mV. In practice, the a.c. current oscillates with drop growth and dislodgement; the current is usually measured at maximum drop size, *i.e.*, just before drop fall.

We can derive the a.c. current – d.c. potential curve for a reversible process from the results of our discussion of the faradaic impedance. If the process is reversible, the charge transfer resistance is negligible and, according to eq (5.56), the faradaic impedance is

$$|Z_f| = \sigma (2/\omega)^{\frac{1}{2}}$$

In general the surface concentrations $C_O(0,t)$ and $C_R(0,t)$ will be very different from their bulk solution values. However, the sinusoidal variations are about mean values determined by the d.c. potential. Thus we can replace C_O^* and C_R^* in eq (5.53) by the mean values given by eqs (4.3), which we can rewrite in slightly more compact form in terms of the parameter

---

[‡] Alternatively, the magnitude of the a.c. current and the phase angle can be determined. Fourier transform methods can also be applied to a.c. polarography. For further details on the theory and practice of a.c. polarography, see Smith *(15)*.

$$\theta' = \exp \frac{F(E - E_{\frac{1}{2}})}{RT}$$

The mean surface concentrations (for $C_R{}^* = 0$) are

$$C_O(0,t) = C_O{}^* \frac{\theta'}{1 + \theta'},$$

$$C_R(0,t) = C_O{}^* \frac{(D_O/D_R)^{\frac{1}{2}}}{1 + \theta'}$$

Substituting these expressions in eq (5.53) in place of $C_O{}^*$ and $C_R{}^*$, we have

$$|Z_f| = \frac{RT}{F^2 A C_O{}^* (D_O\omega)^{\frac{1}{2}}} \frac{(1 + \theta')^2}{\theta'}$$

If the potential is

$$E(t) = E_{dc} + \Delta E \sin(\omega t)$$

the a.c. component of the current will be

$$I(t) = I_0 \sin(\omega t + \pi/4)$$

since $\cot \phi = 1$ for $R_{ct} = 0$. The current amplitude is

$$I_0 = \frac{\Delta E}{|Z_f|}$$

or

$$I_0 = \frac{F^2 A C_O{}^* (\omega D_O)^{\frac{1}{2}} \Delta E}{RT} \frac{\theta'}{(1 + \theta')^2} \tag{5.58}$$

Although the problem can be formulated to account for the expanding spherical drop of the d.m.e., it is usually more convenient to make the required corrections empirically by determining the ratio of the a.c. to d.c. currents, measured for the same drop time on the same solution. Using the Cottrell equation for i_D, we obtain

$$\frac{I_0}{i_D} = \frac{F(\pi\omega t)^{\frac{1}{2}}\Delta E\theta'}{RT(1 + \theta')^2} \tag{5.59}$$

The current at $E = E_{\frac{1}{2}}$ $(\theta' = 1)$ is

$$I_p = \frac{F^2 AC_O{}^*(\omega D_O)^{\frac{1}{2}}\Delta E}{4RT}$$

The shape of the current–potential curve can be seen more easily considering the ratio I/I_p:

$$\frac{I}{I_p} = \frac{\theta'}{(1 + \theta')^2}$$

The ratio can be inverted and written as

$$I_p/I = (1 + \theta')^2/4\theta'$$

or

$$(I_p - I)/I = (1 - \theta')^2/4\theta'$$

Taking the square root and subtracting these equations, we have

$$[I_p/I]^{\frac{1}{2}} - [(I_p - I)/I]^{\frac{1}{2}} = \theta'^{\frac{1}{2}}$$

Taking logs, we obtain

$$E = E_{\frac{1}{2}} \pm \frac{2RT}{F} \ln\{ [I_p/I]^{\frac{1}{2}} - [(I_p - I)/I]^{\frac{1}{2}} \} \tag{5.60}$$

This equation describes the shape of an a.c. polarogram for a reversible system, shown in Figure 5.18. Implicit in the derivation of eqs (5.58) and (5.60) has been the assumption that ΔE is small (5–10 mV in practice). At larger modulation amplitudes, the proportionality between peak current and ΔE breaks down and the peak potential is no longer equal to $E_{\frac{1}{2}}$.

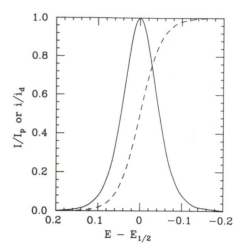

Figure 5.18 Comparison of the shapes of reversible a.c. and d.c. polarographic waves.

At low concentrations of electroactive material, the contribution of the double–layer capacitance is no longer negligible. However, because the resistive and capacitive components of the a.c. current are 90° out of phase, phase–sensitive detection of the a.c. signal can select the resistive component and almost wholly reject the contribution of the double–layer capacitive charging current. The sensitivity of a.c. polarography is therefore very good, comparable with differential pulse polarography. Even better selectivity for the resistive component, and correspondingly better analytical sensitivity, can be achieved with *second–harmonic a.c. polarography* where the a.c. current at twice the modulation frequency is detected. Since the faradaic impedance is nonlinear, the second harmonic signal is quite large; the nearly linear double–layer capacitive contribution gives very little second harmonic current. The second harmonic a.c. polarographic curve resembles the second derivative of the d.c. polarographic wave. Just as the resolution of an a.c. or differential pulse polarogram is better than that of a d.c. polarogram, a further improvement in resolution is achieved in second–harmonic a.c. polarography.

Example 5.5 Compute the ratio of the peak current amplitude for an a.c. polarogram to the d.c. diffusion–limited current. Assume a modulation amplitude of 5 mV, a modulation frequency of 1 kHz, and a drop time of 5 s.

At the peak potential, $\theta' = 1$, so that eq (5.59) becomes

$$\frac{I_p}{i_D} = \frac{F}{4RT} \ (\pi \omega t)^{\frac{1}{2}} \Delta E$$

Substitution of the numbers gives

$$I_0/i_D = 15.3$$

Since the a.c. signal can be amplified by a tuned amplifier, the signal–to–noise ratio for the a.c. polarographic peak current is expected to be very high.

So far we have assumed nernstian behavior in this discussion of a.c. polarography, but one of the most interesting applications of the technique is the measurement of electron–transfer rate constants. An a.c. polarogram will appear reversible when $(\omega D_O)^{\frac{1}{2}}/k_0 \ll 1$. When the rate of electron transfer is slower or the frequency higher, the height of the a.c. polarographic peak is decreased and the peak broadens and becomes somewhat unsymmetrical. The current can then be expressed by

$$I(\omega t) = I_0' \ \sin(\omega t + \phi)$$

where

$$I_0' = I_0 G(\omega) \left[\alpha (1 + 1/\theta') \frac{C_O(0)}{C_O^*} + \beta (1 + \theta) \frac{C_R(0)}{C_O^*} \right] \quad (5.61)$$

$$G(\omega) = \left[\frac{2}{1 + [1 + (2\omega)^{\frac{1}{2}}/\lambda]^2} \right]^{\frac{1}{2}} \quad (5.62)$$

α and β are the cathodic and anodic transfer coefficients, I_0 is given by eq (5.58), λ is given by eq (5.38),

$$\lambda = (k_0/D^{\frac{1}{2}})(\theta'^{-\alpha} + \theta'^{\beta})$$

and we have again assumed equal diffusion coefficients for O and R. If we use eqs (5.36) for the surface concentrations, the bracketed term in eq (5.61) becomes

$$1 + (a/\theta' - \beta) \; \frac{F(\lambda t^{\frac{1}{2}})}{\lambda (\pi t)^{\frac{1}{2}}}$$

where $F(\lambda t^{\frac{1}{2}})$ is given by eq (5.37) and t is the d.m.e. drop time. The phase angle ϕ is given by

$$\phi = \cot^{-1}[1 + (2\omega)^{\frac{1}{2}}\lambda] \tag{5.63}$$

These equations are sufficiently complex that pictures are more useful in gaining a feeling for the behavior of the system. Thus a.c. polarographic currents computed using eqs (5.61) and (5.62) for several values of k_0 and a are shown as functions of E_{dc} in Figure 5.19.

Figure 5.19 A.c. current *vs.* d.c. potential for $k_0 = \infty$ (dotted curve) and for (solid curves, top to bottom) $k_0 = 10^{-2}$, 10^{-3}, and 10^{-4} m s^{-1} with $a = 0.5$. The dashed curves with peaks displaced to positive and negative potentials correspond to $k_0 = 10^{-3}$ and 10^{-4} m s^{-1}, but with $a = 0.25$ or 0.75, respectively. Other parameters are $D_2 = 10^{-9}$ m^{2}s^{-1}, $A = 3.5$ mm^2, $C_O{}^* = 1$ mol m^{-3} (1 mM), $\omega = 2500$ s^{-1}, and $\Delta E = 5$ mV.

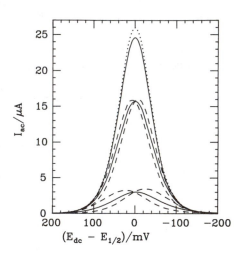

When the electron–transfer process is fast enough that the d.c. polarographic wave appears reversible, $F(\lambda t^{\frac{1}{2}})/\lambda (\pi t)^{\frac{1}{2}} \cong 0$ and the term in brackets in eq (5.61) reduces to 1. The function $G(\omega)$ varies from 1 when $\omega^{\frac{1}{2}}/\lambda \ll 1$ to $\lambda\omega^{-\frac{1}{2}}$ for high frequencies or slow rate constants. The predicted frequency dependence of the peak current for $a = 0.5$ is shown in Figure 5.20. Since I_0 is proportional to $\omega^{\frac{1}{2}}$, the a.c. current becomes frequency independent at high frequencies. Although this condition may not be experimentally attainable, it is often possible to extrapolate experimental data to infinite frequency where the a.c. current at $E = E_{\frac{1}{2}}$ (the peak current for $a = 0.5$), scaled by the d.c. diffusion current, is given by

$$\left[\frac{I_p}{i_D} \right]_{\omega=\infty} = \frac{F}{2RT}\ (\pi t/D)^{\frac{1}{2}}\Delta E\ k_0 \qquad (5.64)$$

The electron–transfer rate constant can also be extracted from the phase angle. According to eq (5.63), the cotangent of the phase angle is linear in $(2\omega)^{\frac{1}{2}}/\lambda$. If $a = 0.5$, the peak current corresponds to $\theta' = 1$ so that eq (5.38) reduces to $\lambda = 2k_0/D^{\frac{1}{2}}$. Thus a plot of cot ϕ vs. $\omega^{\frac{1}{2}}$ should have a zero–frequency intercept of 1 and a slope of $(D/2)^{\frac{1}{2}}/k_0$.

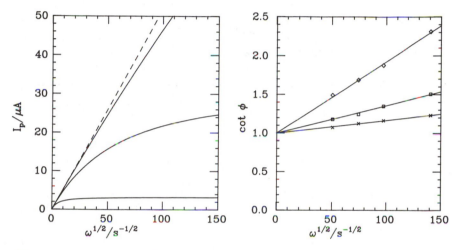

Figure 5.20 Frequency dependence of the a.c. polarographic peak current for $k_0 = \infty$ (dashed line) and $k_0 = 0.01$, 0.001, and 0.0001 m s^{-1}. Other parameters are as in Figure 5.19.

Figure 5.21 Plots of cot ϕ vs. $\omega^{\frac{1}{2}}$ for a.c. polarographic currents measured for *trans*-stilbene (crosses), trimethyl-*trans*-stilbene (squares), and hexamethyl-*trans*-stilbene (diamonds). Data from Dietz and Peover (16).

Example 5.6 Dietz and Peover (16) employed a.c. polarography to study the rate of reduction of *trans*-stilbene, 2,4,6–trimethyl-*trans*–stilbene, and 2,4,6,2',4',6'–hexamethyl-*trans*-stilbene in N,N–dimethylformamide solutions. The phase angle ϕ, measured using an impedance bridge, is plotted as cot ϕ vs. $\omega^{\frac{1}{2}}$ in Figure 5.21. Given diffusion coefficients, measured under the same conditions, the slopes of these plots were used with eq (5.63) to compute the electron–transfer rate constants, obtaining:

$$trans\text{–stilbene} \qquad k_0 = 1.2 \times 10^{-2} \text{ m s}^{-1}$$
$$\text{trimethyl–}trans\text{–stilbene} \qquad 5.2 \times 10^{-3}$$
$$\text{hexamethyl–}trans\text{–stilbene} \qquad 1.8 \times 10^{-3}$$

Since the activation energies obtained by comparison of the above data with values determined at $-20°C$ were about the same for trans–stilbene and the methyl derivatives (ca. 20 kJ mol^{-1}), Dietz and Peover suggested that the reduction in k_0 with increasing methyl substitution is due to steric requirements on the orientation of the molecule at the electrode prior to electron transfer, i.e., to changes in the entropy of activation. Electron transfer to trans–stilbene is apparently possible for more different orientations than for the more hindered methyl derivatives.

A.C. Cyclic Voltammetry

A relatively recent technical development is the combination of the methods of a.c. polarography and cyclic voltammetry. A low–frequency triangular–wave potential, modulated by a higher–frequency, small–amplitude, sinusoidal signal, is applied to the cell and the a.c. current at the frequency of the modulation signal is measured. The method has most of the advantages of a.c. polarography but is not restricted to the dropping mercury electrode. Since stationary planar electrodes can be used, the method is more general and the detailed theory somewhat easier. There are significant advantages over conventional cyclic voltammetry; interference of the double–layer capacitive charging current is much less important and quantitative measurements are considerably easier. Bond, O'Halloran, Ruzic, and Smith (17) have shown that the theory of a.c. polarography can be used without significant modification provided that the triangular–wave frequency is much less than the sinusoidal frequency,

$$\omega \geqslant v/\Delta E$$

Thus eqs (5.61)–(5.63) are valid for a.c. cyclic voltammetry. Unless the d.c. scan rate is very slow, however, eqs (5.36) cannot be used for the surface concentrations required in eq (5.61). Bond and co–workers used a digital simulation technique (see Appendix 5) to obtain these parameters, but an approach based on Nicholson's methods (14) would probably also work.

For electron–transfer processes which are reversible or quasi-reversible on the cyclic voltammetry time scale, a.c. cyclic voltammograms show current peaks on both the cathodic and anodic scans. The magnitudes and potentials of the peak currents are sensitive to the electron–transfer rate constant k_0 and to the transfer coefficient a. The latter parameter is particularly easy to measure by this technique. Since the current returns to zero following the peak, the zero of current is well defined and accurate peak current measurements are much easier than in conventional cyclic voltammetry.

5.5 MECHANISMS OF ELECTRODE PROCESSES

One of the most interesting applications of electrochemical techniques has been in the study of reactions initiated by oxidation or reduction at an electrode. In this and the following four sections, we will consider the mechanisms of electrode processes studied by voltammetric methods.[‡] Here we will discuss some general mechanistic principles and work through some examples which can be understood from qualitative observations. In these sections, we will assume that the electron–transfer processes, *per se*, are nernstian. Irreversibility can be introduced using the methods of Section 5.3, but the results would complicate our study unnecessarily.

Multielectron Processes

The mechanism of a chemical reaction is made up of a series of *elementary processes*, *i.e.*, uncomplicated single reaction steps in which the reactant and product species are exactly those specified. In electrode processes, we sometimes write a step such as

$$O + n\,e^- \rightleftharpoons R$$

and treat it as if it were an elementary process. In fact, electrode processes rarely, if ever, involve the concerted transfer of more than

---

[‡] See also reviews by de Montauzon, Poilblanc, Lemoine, and Gross *(18)* and Geiger *(19)* on organometallic electrochemistry, and Zuman *(20)* and Hawley *(21)* on organic electrochemistry.

one electron. Gas–phase electron attachment or ionization reactions always proceed in discrete one–electron steps. Even if the second electron goes into (or comes from) the same molecular orbital as the first, electron repulsion will cause the two steps to occur at well separated energies. For electron–transfer reactions at an electrode-solution interface, solvation effects may bring the two electron–transfer steps closer together in energy, but only in exceptional cases would we expect the two steps to coincide.

Electrode processes involving two or more electrons (and those involving two or more reactants) are analogous to rate laws with overall kinetic orders of three or more: they provide evidence for mechanisms of two or more elementary steps. However, a process involving two one–electron steps separated by a chemical step may be experimentally indistinguishable from a concerted two–electron transfer if the chemical step is very fast. In discussions of the mechanisms of electrode processes, we usually write the simplest mechanism consistent with the experiment at hand. If the process involves a chemical step which goes to completion in a time short compared with the characteristic time of the experiment, it is then reasonable to treat the process as if it involved a concerted transfer of more than one electron. It should be remembered, however, that an experiment with a shorter characteristic time may not be properly explained by the simplified mechanism.

Experimental Time Scales

Each of the experimental techniques of electrochemistry has a characteristic time scale. In assessing the effect of a coupled chemical reaction on the experimental response, we must ask how the characteristic time of the reaction, say the half–life $t_{\frac{1}{2}}$, compares with the characteristic time of the experiment. If the reaction time is very short compared with the experimental sampling time, the experiment will "see" the reaction at equilibrium. On the other extreme, if the reaction time is long on the experimental time scale, the reaction may have no discernible effect on the experimental response. When the reaction time lies within the experimental time scale, the experimental response will be sensitive to the rate of the chemical step and, at least in principle, the chemical rate constant will be determinable from the experiment.

For techniques such as chronoamperometry, chronopotentiometry, and d.c. polarography, where the current or potential is measured at a well–defined elapsed time after the start of the experiment, the time scale is straightforward. The accessible time scale for d.c. polarography is essentially the range of drop times, 1–10 s. Chronopotentiometric and chronoamperometric experiments may last as long as 100 s with potentials or currents measurable about 1 ms after beginning the experiment. Thus the time scale for these techniques is roughly 1 ms – 100 s. In practice, however, there are severe problems with vibration or other departures from pure diffusion for experiments that last more than 10 s or so.

Techniques such as r.d.e. voltammetry and a.c. polarography use a characteristic angular rotation or modulation frequency ω and the time scale of the experiment then is determined by $2\pi/\omega$, the period of the a.c. signal. For r.d.e. voltammetry, $10 < \omega < 1000$ rad s^{-1}, so that the accessible time scale is roughly 6–600 ms. For a.c. polarography, $10 < \omega < 10,000$ rad s^{-1}, so that the accessible time scale is 0.6–600 ms.

In cyclic voltammetry, the time scale is given by the range of the parameter RT/Fv, where v is the potential scan rate. If the range of v is roughly 0.02–200 V s^{-1}, the accessible time scale is on the order of 0.1–1000 ms.

Another way of looking at the effect of a chemical reaction on an electrochemical experiment is to compare the diffusion layer thickness, x_D, with the characteristic distance from the electrode through which a reactant or product diffuses before the chemical reaction comes to equilibrium. This so–called *reaction layer thickness* is typically on the order of magnitude

$$x_R \cong (D/k)^{\frac{1}{2}}$$

where D is the diffusion coefficient and k is the first–order (or pseudo–first–order) rate constant for the chemical step. If the reaction layer is much thicker than the diffusion layer, $x_R \gg x_D$, the reaction is very slow and the effects of the reaction rate on the experimental response may be negligible. When $x_R \ll x_D$, the reaction comes to equilibrium very close to the electrode surface and the experimental response to the equilibrium mixture may again be independent of the rate of the chemical step. When x_R and x_D are comparable in magnitude, the experimental response is expected to depend critically on the rate of the chemical step.

Chemical Reactions Initiated by Electron Transfer

Consider an electron–paired molecule in which main group atoms obey the octet rule and transition metals (if any) obey the 18–electron rule. The electron added in a one–electron reduction usually occupies an antibonding molecular orbital.[‡] The resulting radical anion may be stable on the time scale of the experiment, in which case the electrode process is chemically reversible. If the electron–transfer step is fast, the process will also be electrochemically reversible. Often, however, the initially formed radical anion undergoes a rapid chemical transformation. There are three possibilities for the primary chemical step:

(1) Cleavage of a chemical bond. The antibonding electron may weaken a bond to the point that the activation barrier to bond cleavage can be surmounted by thermally accessible energies. Bond cleavage will then be rapid. For example, one–electron reduction of an aryl halide is followed by rapid loss of halide ion:

$$ArX + e^- \rightarrow ArX^{\cdot-}$$

$$ArX^{\cdot-} \rightarrow Ar\cdot + X^-$$

Similarly, organometallic molecules and coordination complexes often lose a ligand on reduction:

$$ML_n + e^- \rightarrow ML_n^{\cdot-}$$

$$ML_n^{\cdot-} \rightarrow ML_{n-1}\cdot + L^-$$

In either case, an electron–deficient radical is formed which almost always undergoes a secondary reaction. Common follow–up reactions include further reduction, *e.g.*,

$$Ar\cdot + e^- \rightarrow Ar^-$$

Generally speaking, further reduction occurs if the bond cleavage reaction is fast enough that the radical species is formed at the

---

[‡] Transition metal complexes often have empty essentially nonbonding orbitals which can accommodate the added electron; somewhat different reactions then may occur.

electrode surface. If the neutral radicals are formed less rapidly, but still fast enough that radical–radical encounter is likely, we expect dimerization:

$$2 \text{ Ar} \cdot \;\rightarrow\; \text{Ar–Ar}$$

If the radical is formed more slowly and farther from the electrode, it may abstract a hydrogen atom from the solvent or supporting electrolyte,

$$\text{Ar} \cdot + \text{SH} \;\rightarrow\; \text{ArH} + \text{S} \cdot$$

or react with a nucleophile,

$$\text{Ar} \cdot + \text{Nu}^- \;\rightarrow\; \text{ArNu} \cdot^-$$

Further reactions are expected from most of these secondary products. The general scheme accommodates many other apparently disparate electrode processes. Thus, for example, the two–electron reduction of dicobalt octacarbonyl can be understood in terms of the mechanism

$$\text{Co}_2(\text{CO})_8 + e^- \;\rightarrow\; \text{Co}_2(\text{CO})_8 \cdot^-$$

$$\text{Co}_2(\text{CO})_8 \cdot^- \;\rightarrow\; \text{Co}(\text{CO})_4^- + \text{Co}(\text{CO})_4 \cdot$$

$$\text{Co}(\text{CO})_4 \cdot + e^- \;\rightarrow\; \text{Co}(\text{CO})_4^-$$

Here the metal–metal bond is broken in the primary step and the resulting radical is reduced in the secondary step.

(2) Electrophilic attack. A radical anion generally is a strong nucleophile, highly susceptible to electrophilic attack. In protic solvents, the most readily available electrophile is usually the hydrogen ion. Thus the first reaction step of unsaturated radical anions often is protonation. Anthracene, for example, undergoes an overall two–electron reduction to 9,10–dihydroanthracene. The sequence of steps is

$$\text{C}_{14}\text{H}_{10} + e^- \;\rightarrow\; \text{C}_{14}\text{H}_{10} \cdot^-$$

$$\text{C}_{14}\text{H}_{10} \cdot^- + \text{H}^+ \;\rightarrow\; \text{C}_{14}\text{H}_{11} \cdot$$

$$\text{C}_{14}\text{H}_{11} \cdot + e^- \;\rightarrow\; \text{C}_{14}\text{H}_{11}^-$$

$$\text{C}_{14}\text{H}_{11}^- + \text{H}^+ \;\rightarrow\; \text{C}_{14}\text{H}_{12}$$

The neutral radical produced by protonation of the anthracene anion radical is more easily reduced than anthracene itself, so that the second electron–transfer step is fast at the potential of the first step. Similarly, nitro compounds undergo a six–electron, six–proton reduction to amines:

$$Ar\text{-}NO_2 + 6\,H^+ + 6\,e^- \rightarrow Ar\text{-}NH_2 + 2\,H_2O$$

Depending on the relative rates of the protonation steps, the nitroso (Ar–NO) and hydroxylamino (Ar–NHOH) intermediates may be isolable.

   (3) Rearrangement. A radical anion can sometimes undergo a change in structure which reduces the energy of the molecule by better distributing the excess charge. If the rearranged radical is stable on the time scale of the experiment, reoxidation would produce the starting compound in a different conformation and would occur at a different potential than the reduction. Conformational rearrangement is probably a very common primary process; however, if the rearranged radical then undergoes bond cleavage or electrophilic attack, there may be no direct evidence of the rearrangement step in the experimental results.

   A similar analysis can be applied to electrode processes initiated by an oxidation step. The resulting electron–deficient cation radical may be stable, but more often will react by dimerization, by bond cleavage (this time the bond is weakened by removal of a bonding electron), by nucleophilic attack, or by rearrangement. See Alder *(22)* for an interesting discussion of electrophilic and nucleophilic substitution reactions initiated by electron transfer.

   While voltammetric techniques provide the first line of attack in mechanistic studies, the successful investigator rarely confines his work to electrochemical methods. Spectroscopy has been particularly important in sorting out electrochemical mechanisms. Electron spin resonance spectroscopy has played a key role in the identification of odd–electron reduction or oxidation products *(23)*. Ultraviolet–visible and/or infrared spectrophotometry using optically transparent electrodes have given similarly important insights into the nature of intermediates in electrode processes *(24,B12,B13)*. Victory in the mechanistic _game goes to the scientist who is best able to piece together many bits of evidence from spectroscopic, synthetic, structural and/or kinetic studies, as well as from electrochemical work, to construct a mechanism capable of explaining all the results. The kinds of thought required are very similar to those involved in establishing mechanisms for ordinary homogeneous reactions.

Example 5.7 Pickett and Pletcher *(25)* studied the reduction of $Cr(CO)_6$ in acetonitrile solution using cyclic voltammetry with a platinum indicator electrode and tetrabutylammonium tetrafluoroborate supporting electrolyte. Controlled potential electrolysis of $Cr(CO)_6$ gave a nearly quantitative yield of $Cr_2(CO)_{10}^{2-}$, so that the overall electrode process is

$$2\ Cr(CO)_6 + 2\ e^- \rightarrow Cr_2(CO)_{10}^{2-} + 2\ CO$$

The detailed mechanism is somewhat more complex.

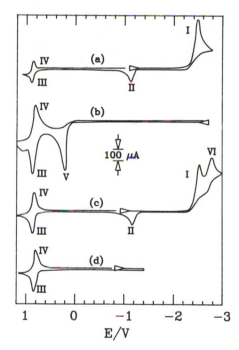

Figure 5.22 Cyclic volt-ammograms of acetonitrile solutions of (a) 5 mM $Cr(CO)_6$, (b) 2.5 mM $Cr_2(CO)_{10}^{2-}$, (c) and (d) 5 mM $Cr(CO)_6$ after UV irradiation.

The cyclic voltammogram of $Cr(CO)_6$ in CH_3CN solution at room temperature is shown in Figure 5.22a. The primary reduction peak (I, -2.66 V *vs.* aqueous s.c.e.) has a peak current proportional to $C_O^* v^{\frac{1}{2}}$, consistent with an electrochemically reversible process, eq (4.22). Since the peak current is comparable to that given by known one–electron reduction processes at the same scan rate and concentration, the process must be the reversible one–electron reduction of $Cr(CO)_6$ to form the radical anion, $Cr(CO)_6^{-}$. If the radical anion were stable on the experimental time scale, a reoxidation peak would be expected on the reverse scan, but no such peak is observed in Figure 5.22a ($v = 300$ mV s^{-1}) or in voltammograms at higher scan rates or at temperatures down to $-45°C$. Thus the radical anion must undergo a fast irreversible reaction. Two oxidation peaks (II and III) were observed on the reverse (positive–going) scan, and another reduction peak (IV) was seen on the second negative–going scan. These features were not observed when the scan was reversed before the primary reduction and thus must correspond to oxidation of $Cr(CO)_6$ reduction products.

The cyclic voltammogram of the ultimate product, $Cr_2(CO)_{10}^{2-}$, is shown in Figure 5.22b. Since the primary oxidation peak (V) is not seen in Figure 5.22a, we conclude that $Cr_2(CO)_{10}^{2-}$ is not produced in significant amounts on the cyclic voltammetry time scale. Thus there must be one or more relatively long–lived intermediates in the reduction scheme.

Photolysis of a solution of $Cr(CO)_6$ in CH_3CN is known to produce $Cr(CO)_5NCCH_3$. Partial conversion of $Cr(CO)_6$ to the CH_3CN derivative by photolysis at 356 nm gave a solution which exhibited the cyclic voltammograms of Figure 5.22c and d. Reduction peak I corresponds to unconverted $Cr(CO)_6$ while the new peak (VI) corresponds to the chemically irreversible reduction of the CH_3CN derivative. As we see in Figure 5.22d, peak III corresponds to the oxidation of the CH_3CN derivative and peak IV to reduction of the resulting cation radical. $Cr(CO)_6$ itself is reversibly oxidized to a relatively stable cation radical at $+1.5$ V (not shown in the figure) so that oxidation of $Cr(CO)_5NCCH_3$ at $+0.8$ V is reasonable (see Example 4.9). Apparently, $Cr(CO)_5NCCH_3$ is formed in the oxidation of $Cr_2(CO)_{10}^{2-}$, but since peak VI is not seen in Figure 5.22a, $Cr(CO)_5NCCH_3$ is not formed directly in the reduction of $Cr(CO)_6$ and hence must have been the product of the oxidation represented by anodic peak II.

All these observations are accommodated by the scheme shown in Figure 5.23. Note that only electrode process III/IV is chemically reversible. The others have fast following chemical reactions (EC electrode processes). The two–electron oxidation of $Cr_2(CO)_{10}^{2-}$ is probably not a concerted process but occurs by the ECE mechanism:

$$Cr_2(CO)_{10}^{2-} \rightarrow Cr_2(CO)_{10}^{-\cdot} + e^-$$

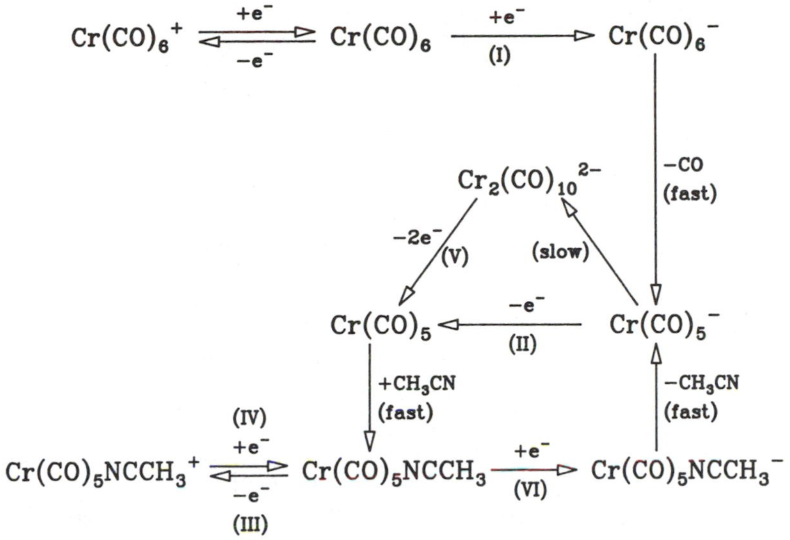

Figure 5.23 Proposed reaction scheme to explain the results shown in Figure 5.22. The roman numerals correspond to the cyclic voltammogram peaks shown in Figure 5.22.

$$Cr_2(CO)_{10}^{\cdot -} \rightarrow Cr(CO)_5 + Cr(CO)_5^{\cdot -}$$

$$Cr(CO)_5^{\cdot -} \rightarrow Cr(CO)_5 + e^-$$

In this case, we know the potential of the second electron–transfer step; it is peak II in Figure 5.22a. The potential at which the apparent two–electron transfer occurs (peak V in the figure) is more positive than that of peak II; thus the second electron–transfer step should be fast and essentially irreversible at the potential of peak V.

Example 5.8 Bond, Colton, and McCormick *(26)* studied the electrochemical oxidation of $Mn(CO)_3dpmCl$ (dpm = $Ph_2PCH_2PPh_2$). The cyclic voltammogram of this species at a Pt electrode in acetonitrile solution (0.1 M Et_4NClO_4, 500 mV s^{-1}) is shown in Figure 5.24. At room temperature, the one–electron oxidation is nearly reversible, but a small reduction peak, matched by an oxidation on the second anodic scan, suggests some decomposition of the Mn(II) product. The cyclic voltammogram at $-35°C$ shows complete chemical reversibility of the oxidation with no trace of the second electrode process. Half–wave potentials—*i.e.*, $E_{\frac{1}{2}} = \frac{1}{2}(E_{pa} + E_{pc})$—for the primary and secondary processes were

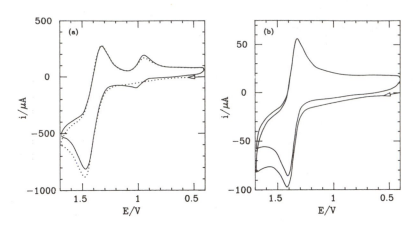

Figure 5.24 Cyclic voltammograms of $Mn(CO)_3dpmCl$ in acetonitrile solution at (a) 22°C and (b) -35°C. In (a), the dotted curve represents the first cycle and the solid curve the second and subsequent cycles.

found to be $+1.44$ and $+0.98$ V, respectively (*vs.* Ag/AgCl in CH_3CN). Very similar results were obtained in acetone solutions.

Controlled potential electrolytic oxidation gave confusing results (see below) but chemical oxidation by $NOPF_6$ in CH_2Cl_2 gave a red compound which analyzed as $[Mn(CO)_3dpmCl]PF_6$. When this compound was dissolved in acetonitrile, the red color slowly faded to yellow and the solution showed a chemically reversible oxidation at $+0.98$ V with no trace of the $+1.44$ V oxidation. In another experiment, $Mn(CO)_3dpmCl$ was oxidized by $NOPF_6$ in CH_3CN at $-35°C$ to give a dark green solution. When this was warmed to room temperature, the solution rapidly changed to red and then more slowly to yellow. Both the red and green solutions were paramagnetic (the yellow solution was diamagnetic) and gave e.s.r. spectra with the six lines expected for hyperfine coupling to ^{55}Mn (nuclear spin of $\frac{5}{2}$).

Apparently, the red and green species are isomeric forms of $Mn(CO)_3dpmCl^+$. Only two isomers are possible for these complexes, the facial (*fac*) and meridional (*mer*) isomers shown in Figure 5.25a. Since the starting material was known to be *fac*, the green form of the Mn(II) complex, which is related to the Mn(I) starting material by reversible electron transfer, is presumably also *fac*. The red form must then be the *mer* isomer. Infrared spectra of

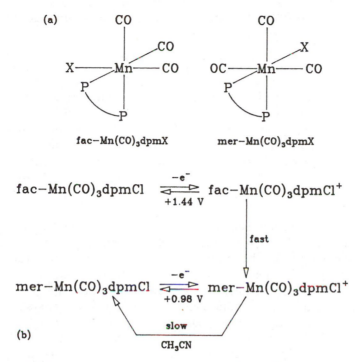

Figure 5.25   (a) Schematic representation of the facial and meridional isomers of $Mn(CO)_3dpm.Cl$. (b) Mechanistic scheme summarizing the experimental results.

the red Mn(II) complex and its reduced form were consistent with *mer* stereochemistry.

The results described above indicate that *fac*–$Mn(CO)_3dpmCl^+$ is rapidly converted to the *mer* isomer at room temperature but only slowly at –35°C. The *mer*–$Mn(CO)_3dpmCl^+$, however, slowly oxidized acetonitrile (but not CH_2Cl_2) at room temperature to produce *mer*–$Mn(CO)_3dpmCl$. (This reaction caused the controlled-potential electrolysis experiments to lead only to confusion.) The results can then be summarized by the scheme shown in Figure 5.25b.

Several additional points of interest should be noted. Whereas the $fac^+ \rightarrow mer^+$ isomerization is fast at room temperature, the interconversion of the isomeric forms of the d^6 Mn(I) species is very slow. Given the well-known kinetic stability of the isoelectronic Co(III) complexes, this suggests that the Mn(I) species are low spin. If we take the $E_{1/2}$ values to be approximations to the standard reduction potentials, the standard free energy change for the process

$$mer + fac^+ \rightleftharpoons mer^+ + fac$$

is $\Delta G° = -0.46\ F$ or $-44\ kJ\ mol^{-1}$, corresponding to an equilibrium constant of 6×10^7. The *mer* isomer is clearly the thermodynamically favored form of the Mn(II) species, and since it is normally obtained in preparative work, the *fac* isomer is probably the favored form of the Mn(I) complex. (Rationalization of this result in terms of molecular orbital theory arguments is left to the reader as an exercise.)

The spectroscopic results yield one further insight. High–spin d^5 Mn(II) complexes are usually nearly colorless, whereas the Mn(II) complexes in this case were highly colored (the red *mer*$^+$ and green *fac*$^+$ isomers). Furthermore, well–resolved e.s.r. spectra of high–spin Mn(II) complexes are seen only for highly symmetrical octahedral complexes. Reduction of symmetry splits the spin levels in such a way as to produce a complex (and usually very broad line) spectrum. The e.s.r. spectra in this case were sharp. Both these results provide further evidence that the Mn(II) species are low spin.

Standard Mechanistic Schemes

In the examples just discussed, the mechanisms could be deduced from essentially qualitative observations—the lack of chemical reversibility in a cyclic voltammogram or the identification of secondary electrode processes with certain intermediates. Mechanistic work often proceeds on the basis of just such qualitative arguments. Electrochemistry is capable of providing more quantitative information about rates if the measured currents or potentials are affected by the rates of the chemical steps. Although the details of the chemistry may vary enormously from one system to another, electrochemists have found that most electrode processes fall into one of a rather small number of mechanistic schemes. In general an electrode process consists of a series of electron–transfer steps, designated by the symbol E, and chemical steps, designated by C. In Sections 5.5–5.8, we will discuss the effects of the following mechanistic schemes on voltammetric experiments:

(1) Preceding Reaction (CE). The electroactive species O is in equilibrium with an electroinactive precursor A:

$$A \rightleftarrows O$$

$$O + e^- \rightleftarrows R$$

(2) Following Reaction (EC). The electroactive product is converted to an electroinactive species Z:

$$O + e^- \rightleftarrows R$$

$$R \rightleftarrows Z$$

(3) Catalytic Reaction (EC'). A special case of the EC mechanism where the following reaction is the oxidation of R to O by an electroinactive species A:

$$O + e^- \rightleftarrows R$$

$$R + A \rightarrow O + B$$

In this scheme, the reduction of A must be thermodynamically more favorable than that of O, $E^\circ_{A/B} > E^\circ_{O/R}$, but the electrochemical reduction of A is too slow to be observable.

(4) Following Reaction with Electroactive Product (ECE). The product of an electron–transfer step is converted to a species which is electroactive. There are two possibilities:

(i) The product of the chemical step is reducible (the $\overrightarrow{E}\overrightarrow{C}E$ mechanism):

$$O_1 + e^- \rightleftarrows R_1 \qquad (E_1{}^\circ)$$

$$R_1 \rightleftarrows O_2$$

$$O_2 + e^- \rightleftarrows R_2 \qquad (E_2{}^\circ)$$

If E is less than either $E_1{}^\circ$ or $E_2{}^\circ$ and the $R_1 \rightarrow O_2$ conversion is fast enough, the overall process may appear to involve two electrons. If $E_1{}^\circ > E > E_2{}^\circ$, the electrode process corresponds to the simple EC mechanism.

(ii) The product of the chemical step is oxidizable (the $\overrightarrow{\text{E}}\overleftarrow{\text{C}}\text{E}$ mechanism):

$$O_1 + e^- \rightleftarrows R_1 \qquad\qquad (E_1°)$$

$$R_1 \rightleftarrows R_2$$

$$O_2 + e^- \rightleftarrows R_2 \qquad\qquad (E_2°)$$

When E is less than either $E_1°$ or $E_2°$ this mechanism corresponds to a simple EC process, but with $E_1° > E > E_2°$, R_2 is oxidized to O_2. Thus if the $R_1 \rightarrow R_2$ conversion is fast enough, the current may approach zero.

In either of the ECE variations, homogeneous disproportionation may be important:

$$R_2 + O_1 \rightleftarrows O_2 + R_1$$

These four mechanisms, or combinations thereof, cover most known electrode processes.

5.6 REACTION PRECEDING ELECTRON TRANSFER

Consider an electrode process in which a reversible homogeneous reaction precedes the electron–transfer step (the CE mechanism):

$$A \underset{k_{-1}}{\overset{k_1}{\rightleftarrows}} O$$

$$O + n\,e^- \rightleftarrows R$$

We assume that species A is not electroactive and that the reduction of O is reversible.

Rotating Disk Electrode Voltammetry

Since a rotating disk voltammetric experiment is performed under steady–state conditions, it is relatively easy to derive equations for the electrode current as a function of potential. In Appendix 6, it is shown that the r.d.e. current for the CE mechanism is given by

$$i = \frac{nFADKC^*}{x_R' + Kx_D + \theta(1 + K)x_D} \tag{5.65}$$

where $K = k_1/k_{-1}$, x_D is the diffusion layer thickness given by eq (4.31), and

$$\theta = \frac{C_O(0,t)}{C_R(0,t)} = \exp \frac{nF(E - E^\circ)}{RT} \tag{4.1}$$

x_R' is the effective reaction layer thickness given by eq (A.26). If the reaction is fast $(x_R' \ll x_D)$, then

$$x_R^2 = \frac{D}{k_1 + k_{-1}} \tag{5.66}$$

When the electrode is sufficiently negative that every O molecule arriving at the surface is reduced, $\theta \cong 0$, and the limiting current is

$$i_L = \frac{nFA(D/x_D)KC^*}{K + x_R'/x_D} \tag{5.67}$$

Dividing i by i_L and rearranging, we obtain the familiar Heyrovský–Ilkovič equation, eq (4.10), where $E_{\frac{1}{2}}$ now is

$$E_{\frac{1}{2}} = E^\circ - \frac{RT}{nF} \ln \frac{1 + K}{K + x_R'/x_D} \tag{5.68}$$

The behavior of the CE reaction scheme can be characterized in terms so–called *kinetic zones* corresponding to limiting cases of eqs (5.67) and (5.68). When $x_R'/x_D \ll K$, either because the reaction is very fast or because K is large, the current is limited by the rate of transport and is proportional to D/x_D. When $K \gg 1$, the half–wave potential is also unaffected. In this case A is really unimportant to the electrode process; the kinetic zone of pure diffusion control (DP) thus corresponds to the limit:

$$\text{DP:} \quad K > 10$$

When $K < 10$ but $x_R'/x_D \ll K$, the reaction is fast and equilibrium is achieved close to the electrode surface. The current is still diffusion limited, but the half-wave potential then depends on K and is more negative than $E°$:

$$E_{\frac{1}{2}} = E° - \frac{RT}{nF} \ln \frac{1 + K}{K} \tag{5.69}$$

This modified diffusion–controlled zone (DM) is then characterized by the limits:

$$\text{DM:} \quad 10x_R'/x_D < K < 10$$

In the DM zone, half-wave potential data can be used to determine formation constants and formulas of complex ions as discussed in Section 4.7; eq (5.69) is a simplified form of eq (4.39).

When K is small and the reaction is slow such that $x_R'/x_D \gg K$, the current is controlled by the rate of the A → O conversion,

$$i_L = nFAK(D/x_R')C^*$$

and the half-wave potential is shifted,

$$E_{\frac{1}{2}} = E° + \frac{RT}{nF} \ln \frac{x_R'}{x_D} \tag{5.70}$$

This is the pure kinetic zone (KP) with limits:

$$\text{KP:} \quad 10\,K < x_R'/x_D < 1$$

Since i_L is independent of x_D in the KP zone, the limiting current will be independent of the rotation speed ω.

In the kinetic zone intermediate between pure diffusion and pure kinetic control, the intermediate kinetic zone (KI), the current and half–

wave potential are influenced both by the rate of transport and by the rate of the reaction. The kinetic zones are summarized in Figure 5.26.

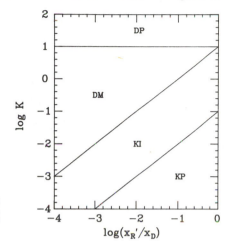

Figure 5.26 Kinetic zone diagram for the CE mechanism.

Example 5.9 Albery and Bell (27) employed r.d.e. voltammetry to measure the rate of proton transfer from acetic acid to water, taking advantage of the fact that the free acid is electroinactive and that the reduction of $H^+(aq)$ is quasi–reversible at a platinum electrode. The electrode process is

$$HOAc \underset{k_{-1}}{\overset{k_1}{\rightleftarrows}} H^+(aq) + OAc^-(aq)$$

$$H^+(aq) + e^- \rightarrow \tfrac{1}{2} H_2(g)$$

In order to partially compensate for reduction of $H^+(aq)$ from the dissociation of water, the experiment used two cells, each equipped with a rotating platinum disk electrode arranged in a bridge circuit such that the applied potentials and currents were identical when the bridge was balanced. One cell contained 0.005 M HCl in 0.1 M KCl, the other 0.02 M acetic acid in 0.1 M potassium acetate. The rotation speed of the r.d.e. in the strong acid cell, ω_H, was adjusted to balance the bridge, holding ω_{HA} constant. Since the electron–transfer process is only quasi–reversible, the applied potential included the necessary overpotential to bring the cell current onto the limiting plateau. Since the currents and electrode areas were identical, eq (5.67) gives

$$\frac{i_L}{FA} = (D_H/x_H)C_H = \frac{(D_{HA}/x_{HA})KC_{HA}}{K + x_R'/x_{HA}}$$

Using eq (4.31) for the diffusion layer thicknesses, x_H and x_{HA}, and eq (5.66) for x_R', this expression can be rearranged to

$$\frac{\omega_{HA}^{\frac{1}{2}}}{\omega_H^{\frac{1}{2}}} = \frac{D_H^{\frac{2}{3}}C_H}{D_{HA}^{\frac{2}{3}}C_{HA}}\left[1 + \frac{\omega_{HA}^{\frac{1}{2}}(D_{HA}/\nu)^{\frac{1}{6}}}{1.61(Kk_1)^{\frac{1}{2}}}\right]$$

Thus a plot of $(\omega_{HA}/\omega_H)^{\frac{1}{2}}$ vs. $\omega_{HA}^{\frac{1}{2}}$, such as shown in Figure 5.27, gives a straight line. The ratio of the slope to the intercept is

$$\frac{\text{slope}}{\text{intercept}} = \frac{(D_{HA}/\nu)^{\frac{1}{6}}}{1.61\,(Kk_1)^{\frac{1}{2}}}$$

With separately determined values of D_{HA} and ν, Albery and Bell found $Kk_1 = 17.0\ \text{s}^{-1}$ at 25°C. Thus with $K = 1.75 \times 10^{-5}$,

$$k_1 = 9.7 \times 10^5\ \text{s}^{-1},\ k_{-1} = 5.5 \times 10^{10}\ \text{M}^{-1}\text{s}^{-1}$$

These rate constants are in good agreement with similar parameters determined by the temperature–jump method.

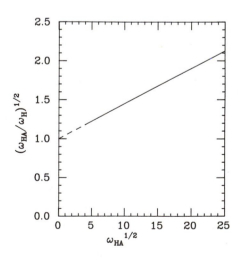

Figure 5.27 Plot of frequency ratio vs. frequency used in the determination of the rate of proton transfer from acetic acid to water.

Potential Step Chronoamperometry

Solution of the linear diffusion problem, using the homogeneous kinetics of the $A \rightleftharpoons O$ equilibrium as a boundary condition, gives a familiar result (B12). For $K \ll 1$ and the electrode potential sufficiently negative that $C_O(0) = 0$,

$$\frac{i_L}{i_D} = F(\lambda t^{\frac{1}{2}}) \qquad (5.71)$$

where i_D is the diffusion–limited current given by the Cottrell equation, eq (4.7), $F(\lambda t^{\frac{1}{2}})$ is given by eq (5.37) and

$$\lambda = (k_1 K)^{\frac{1}{2}} \qquad (5.72)$$

Thus the experimental behavior of an electrode process with a prior equilibrium step is identical to that of a slow electron–transfer process. For short times such that $\lambda t^{\frac{1}{2}} \ll 1$,

$$F(\lambda t^{\frac{1}{2}}) \cong \lambda (\pi t)^{\frac{1}{2}}$$

so that the initial current is

$$i(0) = nFAD^{\frac{1}{2}}C^*(k_1 K)^{\frac{1}{2}} \qquad (5.73)$$

For small λ, this current will decay relatively slowly and extrapolation to zero time may be possible, allowing determination of the rate parameter $(k_1 K)^{\frac{1}{2}}$.

Polarography

The effect of a prior equilibrium on a d.c. polarographic wave is qualitatively similar to that found above for r.d.e. voltammetry. The limiting current is given, at least approximately, by eq (5.71), where t is now the drop time. When $k_1 Kt \ll 1$, the system is in the pure kinetic zone (KP) and the current is expected to be independent of drop time. Since the time scale for d.c. polarography is so long, however, only relatively slow reactions will meet this requirement.

Cyclic Voltammetry

Based on the response of r.d.e. voltammetry and potential step chronoamperometry to the CE mechanism, we can predict the qualitative appearance of a stationary electrode voltammogram or a cyclic voltammogram. The relevant kinetic parameter is essentially the ratio of the chemical reaction rate to the scan rate v:

$$\lambda = \frac{RT}{nF} \frac{k_1 + k_{-1}}{v} \tag{5.74}$$

For $\lambda < 0.1$ and/or $K > 10$ (the DP zone), we expect a reversible cyclic voltammogram with current peaks proportional to the equilibrium concentration of O. In the DM zone ($K < 10$, $K\lambda^{\frac{1}{2}} > 10$), the current peaks will be proportional to the total concentration of A and O and the peak potential shift will be given by eq (5.67). In the KI or KP zones, we expect an increase in peak current and a negative shift of the peak potential with increasing λ (decreasing scan rate). This case has been analyzed in detail by Nicholson and Shain (28), who showed that the shape of cyclic voltammograms depends on $K\lambda^{\frac{1}{2}}$. Curves of similar shape but different K will be shifted by amounts given by eq (5.68). Nicholson and Shain provide working curves which relate peak current ratios and peak potentials to the parameter $K\lambda^{\frac{1}{2}}$. Cyclic voltammograms for $K = 0.2$ and $\lambda = 25$, 2.5, and 0.25 are shown in Figure 5.28.

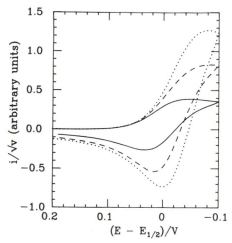

Figure 5.28 Cyclic voltammograms at constant scan rate for the CE reaction scheme for $K = 0.2$ and $\lambda = 0.25$ (solid line), 2.5 (dashes), and 25 (dots).

5.7 *FOLLOWING CHEMICAL REACTION*

When R, the product of the electron–transfer step, is consumed by a chemical reaction, the resulting EC mechanism can be written as

$$O + n\,e^- \rightleftarrows R$$

$$R \quad \overset{k_1}{\underset{k_{-1}}{\rightleftarrows}} \quad Z$$

When $E \ll E^\circ$, the current is limited by transport of O to the electrode surface and should be unaffected by the following reaction. Thus we expect that the limiting current will be diffusion controlled and independent of the rate of the chemical step. However, if R is consumed fast enough that its concentration at the electrode surface is significantly reduced, we expect a positive shift in the half–wave potential. The EC mechanism is probably the most commonly encountered of all the mechanistic schemes, and we will look at examples of EC processes studied by several techniques.

R.D.E. Voltammetry and Polarography

Using the general method of Appendix 6, it is easy to show (see Problems) that a following chemical reaction has no effect on the diffusion current or wave shape of an r.d.e. voltammogram and that these are given by the Levich equation, eq (4.33), and the Heyrovský–Ilkovič equation, eq (4.10), respectively. The half–wave potential, however, is given by

$$E_{\frac{1}{2}} = E^\circ - \frac{RT}{nF} \ln \frac{1 + K x_R'/x_D}{1 + K} \tag{5.75}$$

where $K = k_1/k_{-1}$ Since $x_R'/x_D \leq 1$, $E_{\frac{1}{2}}$ is shifted to positive potentials by an amount which depends on the homogeneous rate constant and on the rotation speed.

If the $R \rightarrow Z$ reaction is completely irreversible, $1/K = 0$ and eq (5.75) becomes

$$E_{\frac{1}{2}} = E^\circ - \frac{RT}{nF} \ln \frac{x_R{'}}{x_D} \tag{5.76}$$

Referring to Figure A.1b, we see that $E_{\frac{1}{2}}$ will depend on the rotation speed when k_1 is on the order of $1\ s_1^{-1}$ or greater. When $x_R/x_D < 0.5$ ($k/\omega > 0.15$), $x_R{'}/x_D$ is linear in $\omega^{\frac{1}{2}}$ with a slope proportional to the chemical rate constant. The half–wave potential then will be linear in $\ln \omega$. If E° is known, or if k/ω can be made small enough that the nonlinear region of Figure A.1b is well defined, the rate constant k can be determined.

The same qualitative result is obtained for the d.m.e. The polarographic diffusion current and wave shape are not affected by a following reaction, but the half–wave potential is shifted by an amount depending on the rate constant and the drop time. According to Kern (29) the half–wave potential (for kt > 2) is

$$E_{\frac{1}{2}} = E^\circ + \frac{RT}{2nF} \ln(5.36kt) \tag{5.77}$$

Double Potential Step Chronoamperometry

Reversal techniques such as double potential step chronoamperometry, cyclic voltammetry, and current reversal chronopotentiometry are particularly appropriate to the study of chemical reactions which follow electron transfer. Since these techniques sample the concentration of the product of the electrode process a short time after it is formed, they lead to a straightforward measurement of rate.

In the case of double potential step chronoamperometry, the potential is stepped to a point where every O molecule is reduced on arrival at the electrode. At time τ, the potential is stepped to a point where every R is oxidized on arrival (see Figure 4.16). If the cathodic current is measured at time t after the first potential step and the anodic current is measured at time t after the second step (total time t + τ), eqs (4.20) can be combined to give the ratio i_a/i_c:

$$\frac{i_a}{i_c} = -\phi(k\tau, t/\tau) + [t/(t + \tau)]^{\frac{1}{2}} \tag{5.78}$$

where $\phi(k\tau, t/\tau) = 1$ when $k = 0$. When a following chemical step consumes R, the function $\phi(k\tau, t/\tau)$ becomes smaller, approaching $[t/(t + \tau)]^{\frac{1}{2}}$ as k increases. Determination of the function requires solution of the linear diffusion problem under the appropriate boundary conditions, a task accomplished by Schwartz and Shain (30). The function turns out to be an infinite series of hyperbolic Bessel functions which must be evaluated numerically. Schwartz and Shain provide a set of working curves, reproduced in Figure 5.29, from which a value of $k\tau$ can be read given the anodic–to–cathodic current ratio measured for certain values of t/τ. The experiment works best for $k\tau \cong 1$ and for t/τ less than 0.5.

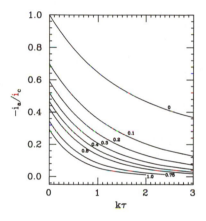 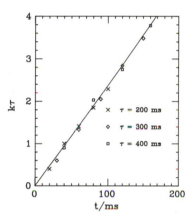

Figure 5.29 Theoretical working curves for double potential step chronoamperometry applied to the EC mechanism. The curves correspond to various values of t/τ, where τ is the time of the second potential step and t is the current measurement time after each potential step.

Figure 5.30 Kinetic plot for the rearrangement of hydrazobenzene followed by double potential step chronoamperometry: 1.0 mM azobenzene in 1.98 M $HClO_4$ in 50% aqueous ethanol; $\tau = 200$ ms (crosses), 300 ms (diamonds), and 400 ms (squares).

Example 5.10 Schwartz and Shain (30) employed double potential step chronoamperometry to study the rate of the benzidine rearrangement of hydrazobenzene. Azobenzene undergoes a fast ECEC reduction to hydrazobenzene,

⟨◯⟩—N=N—⟨◯⟩ $\xrightarrow[\text{2 H}^+]{\text{2 e}^-}$ ⟨◯⟩—NH—NH—⟨◯⟩

which undergoes acid–catalyzed rearrangement to benzidine

⟨◯⟩—NH—NH—⟨◯⟩ $\xrightarrow{\text{k}}$ H$_2$N—⟨◯⟩—⟨◯⟩—NH$_2$

Hydrazobenzene is oxidized at about the same potential where azobenzene is reduced, but benzidine is electroinactive in this potential range. Thus the double potential step method could be used to monitor the amount of hydrazobenzene remaining at various times after formation. Anodic–to–cathodic current ratios were measured for t/τ = 0.1, 0.2, 0.3, 0.4, and 0.5 for various values of τ, the switching time. Values of $k\tau$ could then be read off the working curves, multiplied by t/τ to obtain kt, and these numbers plotted $vs.$ t as shown in Figure 5.30 for data obtained for τ = 200, 300 and 400 ms, [azobenzene] = 1 mM in 1.98 M HClO$_4$ in 50% aqueous ethanol solutions. The slope of this plot is the rate constant k = 23 s^{-1}. Similar plots were obtained for other acid concentrations with rate constants ranging from 0.6 to 90 s^{-1}.

Cyclic Voltammetry

Cyclic voltammograms of a system perturbed by a following chemical reaction are expected to show smaller anodic current peaks when the lifetime of R is comparable to the scan time. For an irreversible chemical step, the relevant kinetic parameter is

$$\lambda = \frac{kRT}{\upsilon n F} \tag{5.79}$$

Some computer–simulated cyclic voltammograms are shown in Figure 5.31. As expected, the anodic peak disappears for λ > 0.1. In favorable cases, the anodic–to–cathodic peak current ratio can be used to estimate the rate constant. A working curve, obtained from the work of Nicholson and Shain (28), is shown in Figure 5.32 (the parameter τ is the time required to scan the potential from the cathodic peak and the switching point).

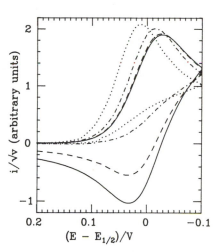

Figure 5.31 Simulated cyclic voltammograms at constant scan rate for a chemical reaction following a reversible one–electron process with λ = 0.01 (solid curve), 0.1 (dashed curve), 1.0 (dot–dash curve), and 10 (dotted curve).

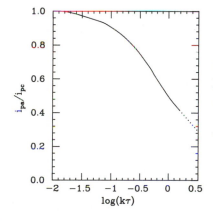

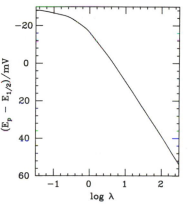

Figure 5.32 Ratio of anodic to cathodic peak currents for an EC process as a function of $k\tau$.

Figure 5.33 Shift of the cathodic peak potential as a function of λ for an EC process.

As we might have expected from the results for r.d.e. voltammetry, the cathodic peak shifts to more positive potentials for very fast following chemical steps. If the unperturbed peak potential is known, the peak potential shift can be used to estimate the rate constant. A working curve, obtained from the work of Nicholson and Shain (28), is shown in Figure 5.33.

Current–Reversal Chronopotentiometry

In chronopotentiometry, a constant current is applied to the electrochemical cell and, when the electroactive material available near the electrode is no longer able to supply the required current, the potential swings to that of a new couple (e.g., the solvent or supporting electrolyte). This transition time τ is proportional to the square of the concentration of the electroactive species, eq (4.21a). In the current reversal technique, the current is reversed before the cathodic transition time is reached and an anodic transition is observed which corresponds to the reoxidation of R. If R is consumed in a chemical step, the anodic transition time will be correspondingly shorter.

Theoretical analysis of the EC reaction scheme for chronopotentiometry is straightforward using Laplace transform techniques, and this case is discussed in some detail by Bard and Faulkner (B12). It can be shown that the reverse current transition time τ and the time of current reversal t are related by

$$\mathrm{erf}[k(t + \tau)]^{\frac{1}{2}} = 2\,\mathrm{erf}(k\tau)^{\frac{1}{2}} \tag{5.80}$$

Solutions of this equation are plotted as τ/t vs. kt in Figure 5.34. This plot serves as a working curve whereby measured values of τ/t can be used to find kt, and knowing t, the rate constant k.

Example 5.11 Testa and Reinmuth (31) used chronopotentiometry to measure the rate of hydrolysis of p–benzoquinone imine:

$$O=\!\!\left\langle \underset{}{\bigcirc} \right\rangle\!\!=NH \quad \xrightarrow[-NH_3]{\substack{k\\ +H_2O}} \quad O=\!\!\left\langle \underset{}{\bigcirc} \right\rangle\!\!=O$$

Since this reaction is relatively rapid, the imine cannot be kept in solution without rapid hydrolysis. To overcome this problem, the imine was produced in situ by the electrolytic oxidation of p–aminophenol:

$$HO\!\!-\!\!\left\langle \underset{}{\bigcirc} \right\rangle\!\!-NH_2 \quad \xrightarrow[-2\ H^+]{-2\ e^-} \quad O=\!\!\left\langle \underset{}{\bigcirc} \right\rangle\!\!=NH$$

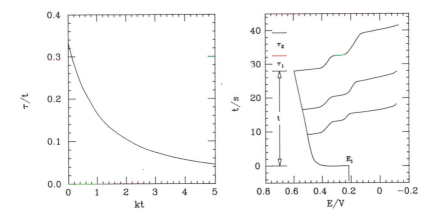

Figure 5.34 Working curves for determination of rate constants from current–reversal chronopotentiometric measure- ments for the EC reaction scheme. τ is the transition time after current reversal at time t.

Figure 5.35 Current–reversal chronopotentiometric curves for the oxidation of p–aminophenol (1 mM in 0.1 M H_2SO_4 solution) and the hydrolysis of p–benzoquinoneimine; t is the time of anodic electrolysis, τ_1 and τ_2 are the transition times corresponding to reduction of the imine and quinone.

A constant anodic current of 1.29 mA was passed through a cell with a platinum electrode ($A = 12.6$ cm^2) for times ranging from 3 to 30 s; the current was then reversed and the amounts of imine and quinone in solution near the electrode measured chronopotentiometrically. Some typical experimental curves are shown in Figure 5.35. Notice that the ratio of the quinone transition time, τ_2, to the imine transition time, τ_1, increases with increasing time of electrolysis. The ratio τ_1/t measured from a chronopotentiogram was used with the working curve of Figure 5.34 to determine a value of kt. Data for various electrolysis times gave a rate constant $k = 0.103 \pm 0.003$ s^{-1} for the hydrolysis reaction in 0.1 M H_2SO_4 at 30°C. The transition time corresponding to the quinone, τ_2, provides an internal check on the experiment since the sum $\tau_1 + \tau_2$, should be exactly t/3, independent of the rate of the chemical step.

A.C. Polarography

In principle it is possible to use a.c. polarography to measure the rates of many types of chemical reactions coupled to electron–transfer processes. The technique has not been as widely used as the reversal methods we have just discussed primarily because a.c. polarography is not readily used in a qualitative way to characterize a system; quantitative measurements often grow out of qualitative studies using the less precise methods. Nonetheless, for accurate rate constant measurements on a well–characterized system, particularly for very fast rates, a.c. polarography is the method of choice.

In general, theoretical expressions for the a.c. polarographic current are extremely complex, involving integrals which must be evaluated numerically. Although this can be done, in practice it is usually much easier to measure the phase angle, which can be expressed as an analytical function of rate constants, potential and frequency. Furthermore, the phase angle is dimensionless and independent of electrode area and geometry. For an irreversible chemical step following electron transfer, it can be shown that

$$\cot \phi = V/U \tag{5.81}$$

where

$$V = \frac{(2\omega)^{\frac{1}{2}}}{\lambda} + \frac{1 + G_+(k/\omega)\,\theta'}{1 + \theta'} \tag{5.82a}$$

$$U = \frac{1 + G_-(k/\omega)\,\theta'}{1 + \theta'} \tag{5.82b}$$

wtih

$$G_\pm(k/\omega) = \left[\frac{[1 + (k/\omega)^2]^{\frac{1}{2}} \pm k/\omega}{1 + (k/\omega)^2} \right]^{\frac{1}{2}} \tag{5.83}$$

and λ is given by eq (5.38). The effect of the added terms in $G_+(k/\omega)$ is to produce a hump in the $\cot \phi$ vs. $\sqrt{\omega}$ plot at $\omega/k \cong 0.39$. If the electron–transfer rate is fast enough, this hump may be resolved and the rate constant of the chemical step can be determined quite simply. When the electron–transfer rate is slow enough to contribute a significant slope in the absence of the chemical step, however, the hump may be manifested only by a change in slope.

A.c. cyclic voltammetry is particularly well suited to the study of EC processes. Bond, O'Halloran, Ruzic, and Smith *(32)* have shown that the theory of a.c. polarography applies provided that the a.c. modulation frequency is much greater than the triangular–wave frequency. The ratio of the a.c. current peaks on the forward and reverse scans can be accurately measured and this ratio is very sensitive to the rate constant of the following chemical reaction.

Example 5.12 Moraczewski and Geiger *(33)* studied the reduction of $(\eta^4$–cyclooctatetraene)–$(\eta^5$–cyclopentadienyl)cobalt, (COT)CoCp, using the techniques of d.c. and a.c. polarography and cyclic voltammetry.

Cyclic voltammograms of (COT)CoCp in DMF solution, shown in Figure 5.36, have two cathodic peaks ($E_{\frac{1}{2}} = -1.82$ and -2.05 V *vs.* s.c.e.). These reductions are assigned to the isomeric forms in which the cyclooctatetraene ligand is bound through the 1,3 and 1,5 double bonds, respectively. Since the peak currents are proportional to concentration, the peak current ratio gives the equilibrium constant

$$K = [1,5]/[1,3]$$

Thus $K = 3.3$ at 298 K and 1.46 at 343 K ($\Delta H° = -13$ kJ mol^{-1}). On the reverse scans of the cyclic voltammograms, a single anodic peak is observed, apparently corresponding to the oxidation of the radical anion of the 1,3–isomer. Furthermore, on repetitive cycles, the cathodic peak due to the 1,5–isomer decreases at the expense of the peak due to the 1,3–isomer until the cyclic voltammogram shows an apparently reversible pattern, centered at –1.82 V, with only a trace of the more negative peak remaining. Isomerization of neutral (COT)CoCp is quite slow but (COT)CoCp$\cdot^-$ seems not to exist at all as the 1,5–isomer. These results could be accommodated by either of the two schemes shown in Figure 5.37. In scheme I, electron transfer to (1,5–COT)CoCp is presumed to be reversible but perturbed by very rapid isomerization of the radical anion to the more stable 1,3–isomer. In scheme II, isomerization is presumed to be concerted with electron transfer. If scheme II is correct, the electron–transfer rate constant might be expected to be slower than normal. Indeed, a.c. polarograms show that the peak due to

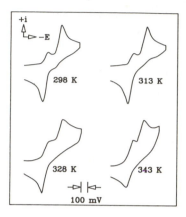

Figure 5.36 Cyclic voltammograms at a hanging mercury drop electrode ($v =$ 19 V s^{-1}) of (COT)CoCp in DMF solution at various temperatures.

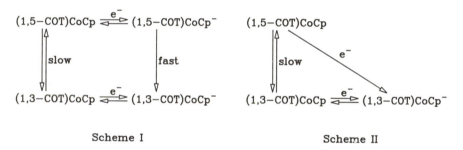

Scheme I   Scheme II

Figure 5.37 Mechanistic schemes for the reduction of (COT)CoCp.

reduction of the 1,5–isomer is much smaller than that due to the 1,3–isomer. Plots of cot ϕ vs. $\sqrt{\omega}$ were linear for both peaks for frequencies up to about 200 Hz ($\sqrt{\omega} < 35$ s$^{-\frac{1}{2}}$). Poor signal-to-noise ratios precluded work at higher frequencies. The slopes of the cot ϕ vs. $\sqrt{\omega}$ plots suggested that electron transfer to the 1,5–isomer is about 10 times slower than to the 1,3–isomer.

It was later realized that, because of the limited frequency range, the a.c. polarographic result might be consistent with Scheme I. Accordingly, the experiment was repeated using a fast Fourier-transform technique which permitted measurements at considerably higher frequencies *(34)*. This work gave the cot ϕ vs. $\sqrt{\omega}$ plots shown in Figure 5.38. The data for the 1,3–isomer give a good straight line with slope corresponding to an electron-transfer rate constant $k_0 = 2.8 \times 10^{-3}$ m s^{-1}. The 1,5–isomer data, on the

Figure 5.38 Plots of cot ϕ vs. $\sqrt{\omega}$ for a.c. polarographic currents measured for (COT)CoCp in DMF solution. The squares correspond to the 1,5–isomer, the diamonds to the 1,3–isomer. The dashed line corresponds to the electron–transfer rate constant $k_0 = 6 \times 10^{-4}$ m s^{-1}.

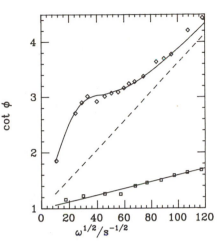

other hand, show considerable kinetic control at low frequencies; cot ϕ asymptotically approaches the linear dependence on $\sqrt{\omega}$ at high frequencies. The data of Figure 5.38 can be interpreted in terms of an electron–transfer rate constant $k_0 = 6 \times 10^{-4}$ m s^{-1}, somewhat slower than the value for the 1,3–isomer, but with a following chemical step with rate constant $k = 2000$ s^{-1}.

These results suggest that electron–transfer rate constants in general must be viewed with caution. There is always the chance that a rate constant appears to be slow not because the electron–transfer step is slow, but because a very fast chemical transformation is coupled to electron transfer.

5.8 CATALYTIC REACTIONS

When the product of the electron–transfer reaction R reacts with another solution species A to regenerate the substrate O, we have the so–called catalytic mechanism, EC':

$$O + n e^- \rightleftarrows R$$

$$R + A \xrightarrow{k} O + B$$

If the reduction of A by R is spontaneous, then according to thermodynamics, A is more easily reducible than O. Indeed, the kinetic scheme requires that the reduction of A at the electrode be much slower than the homogeneous redox reaction of R and A. This might seem an unlikely situation, but it is actually rather common. For example, the electrolytic reduction of hydrogen peroxide is slow, but the homogeneous reaction of H_2O_2 with Fe^{2+} is fast. Thus the electrode process when Fe^{3+} is reduced in the presence of H_2O_2 is

$$Fe^{3+} + e^- \rightleftarrows Fe^{2+}$$

$$Fe^{2+} + \tfrac{1}{2} H_2O_2 + H^+ \rightarrow Fe^{3+} + H_2O$$

In effect, the Fe^{3+}/Fe^{2+} couple catalyzes the electrolytic reduction of H_2O_2.

Rotating Disk Electrode Voltammetry

Using the steady–state equations appropriate to r.d.e. voltammetry, it is easy to show (see Appendix 6) that the limiting current for the EC' mechanism is

$$i_L = nFAJ_O(0) = nFAC_O{}^*(D/x_R') \tag{5.84}$$

For a slow reaction, $x_R' \cong x_D$ and eq (5.84) reduces to eq (4.32), i.e., the voltammetric response is unaffected by the coupled reaction. When the reaction is fast, $x_R'^2 \cong D/k'$, where $k' = kC_A$ is the pseudo–first-order rate constant. The limiting current then is independent of x_D and thus independent of the r.d.e. rotation speed. The limiting current gives an absolute measure of the rate constant independent of the details of r.d.e. theory, provided that the electrode area and the diffusion coefficent are known. A more practical method is to measure the limiting current in the presence and absence of A; the ratio of these limiting currents is x_D/x_R' and the rate constant can then be extracted from eqs (A.24) and (A.26).

Chronoamperometry

When the potential is stepped to a sufficiently negative value that $C_O(0) = 0$, the limiting current can be shown to be *(35)*

$$i_L = nFAD_O{}^{\frac{1}{2}}C_O{}^* [(\pi t)^{-\frac{1}{2}} \exp(-k't) + k'^{\frac{1}{2}} \operatorname{erf}(k't)^{\frac{1}{2}}] \tag{5.85}$$

or

$$i_L/i_D = \exp(-k't) + (\pi k't)^{\frac{1}{2}}\mathrm{erf}(k't)^{\frac{1}{2}} \qquad (5.86)$$

where i_D is the unperturbed diffusion–limited current, eq (4.7). For short times, the error function is near zero, the exponential close to one, and $i_L \cong i_D$. For long times, k't ≫ 1, the exponential approaches 0, the error function goes to 1, and eq (5.85) becomes

$$i_L(t \to \infty) = nFAC_O{}^*(k'D)^{\frac{1}{2}} \qquad (5.87)$$

independent of time. The predicted behavior is shown in Figure 5.39 for several values of k'. The current should decay at very long times since A is consumed and pseudo–first–order conditions no longer prevail; however, if [A] ≫ [O] and semi–infinite linear diffusion applies, the current may be constant for quite a long time. Thus if the rate constant is to be extracted, it is best to measure the current in the presence and absence of A and to compute i_L/i_D. In this way, k' = kC_A can be determined without having to know the diffusion coefficient or electrode area. A plot of i_L/i_D vs. $(k't)^{\frac{1}{2}}$, shown in Figure 5.40, serves as a working curve for such an experiment. Figure 5.40 also serves as a reaction zone diagram with the pure diffusion (DP) zone defined by k't < 0.05, the pure kinetic (KP) zone by k't > 1.5, and the intermediate kinetic (KI) zone by 0.05 < k't < 1.5.

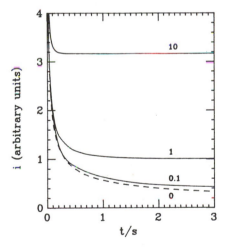

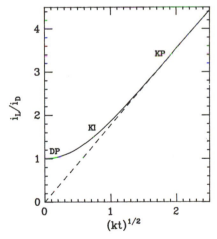

Figure 5.39 Chronoampero- metric curves for EC' process with k' = 0, 0.1, 1 and 10 s^{-1}.

Figure 5.40 Chronoampero- metric response curve for catalytic electrode processes.

Polarography

The polarographic limiting current is also increased by a catalytic process. The current enhancement is given approximately by eq (5.86), where t is now the drop time. When k't > 1.5, the limiting current is proportional to $\sqrt{k'}$ and thus to $\sqrt{C_A}$. Polarography thus can be used for the analysis of electroinactive species which give catalytic current enhancements with electroactive catalysts such as Fe^{3+} or for the determination of trace amounts of a catalyst by adding a known amount of an electroinactive oxidant such as H_2O_2.

For k't $\gg$ 1, eq (5.86) gives $i_L/i_D = (\pi k't)^{\frac{1}{2}}$. An analysis specific to the d.m.e. by Birke and Marzluff *(36)* gives a slightly different result:

$$i_L/i_D = 1.160\,(k't)^{\frac{1}{2}} \qquad (5.88)$$

Birke and Marzluff also give equations to correct for departure from pseudo–first–order conditions.

Example 5.13 Birke and Marzluff *(36)* measured polarographic limiting currents for $Cr(NH_3)_6^{3+}$ in the presence of chlorite ion in NH_4^+/NH_3 aqueous buffer solution,

$$Cr(III) + e^- \rightleftharpoons Cr(II)$$

The catalytic reaction,

$$4\,Cr(II) + ClO_2^- + 4\,H^+ \rightarrow 4\,Cr(III) + Cl^- + 2\,H_2O$$

produces four Cr(III) for each ClO_2^- consumed. Thus $k[ClO_2^-] = 4k'$. The experimental data and the second–order rate constants derived therefrom are presented below (the drop time was 4.85 s).

[Cr(III)]/mM	[ClO_2^-]/mM	i_L/i_D	$k/10^4\ M^{-1}s^{-1}$
0.050	1.00	19.4	1.44
0.075	1.00	20.4	1.59
0.100	1.00	16.7	1.07
0.050	0.50	14.6	1.63
0.050	0.25	10.5	1.69

Cyclic Voltammetry

Linear scan voltammograms at a stationary electrode are closely analogous to potential step chronoamperograms for catalytic processes. The kinetic zone parameter is again

$$\lambda = RTk'/nFv$$

The pure kinetic zone, defined by $\lambda > 1$, is characterized by a curve resembling a polarographic wave with limiting current given by eq (5.86), independent of scan rate, $E_{p/2} = E_{\frac{1}{2}}$, and the wave shape given by the Heyrovský–Ilkovič equation. The reverse scan in cyclic experiments virtually retraces the current–potential curve of the forward scan. When the homogeneous rate becomes slower (or the scan faster) cathodic and anodic peaks develop and the half-peak potentials shift toward their unperturbed values ($E_{\frac{1}{2}} \pm$ 28.5/n mV at 25°C). Some representative curves are shown in Figure 5.41.

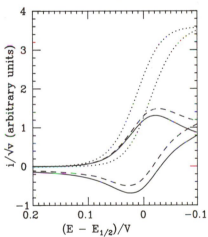

Figure 5.41 Simulated cyclic voltammograms at constant scan rate for a EC' process with λ = 0.01 (solid curve), 0.1 (dashed curve), and 1.0 (dotted curve).

5.9 THE ECE MECHANISMS

When the product of a chemical step following electron transfer is electroactive in the potential range of interest, the response of a voltammetric experiment can be rather interesting. The nature of the ECE mechanism depends on whether the product of the chemical step is reducible or oxidizable and on whether the half–wave potential of the second step is greater or less than that of the first step. The mechanisms can be written (for one–electron–transfer steps):

$\overrightarrow{ECE}$ Mechanism: $O_1 + e^- \rightleftarrows R_1$ $E_1°$

$R_1 \overset{k}{\rightarrow} O_2$

$O_2 + e^- \rightleftarrows R_2$ $E_2°$

$\overrightarrow{E}\overleftarrow{C}\overrightarrow{E}$ Mechanism: $O_1 + e^- \rightleftarrows R_1$ $E_1°$

$R_1 \overset{k}{\rightarrow} R_2$

$O_2 + e^- \rightleftarrows R_2$ $E_2°$

For an $\overrightarrow{E}\overrightarrow{C}\overrightarrow{E}$ process with $E_1° < E_2°$ (case IA), O_2 is reduced at potentials where R_1 is formed and, if the chemical step is fast enough, the overall process will appear to involve two electrons. When $E_1° > E_2°$ (case IB), O_2 is stable at potentials near $E_1°$, and the process is indistinguishable from an EC mechanism. However, O_2 will be reduced at more negative potentials, so that for $E < E_2°$, case IB is identical to case IA.

For an $\overrightarrow{E}\overleftarrow{C}\overrightarrow{E}$ process with $E_1° < E_2°$ (case IIA), R_2 is stable at potentials where R_1 is formed and the process is again indistinguishable from the EC mechanism. Case IIA can be distinguished from an EC process by reversal techniques since the oxidation of R_2 would then contribute to the reversal current. When $E_1° > E_2°$ (case IIB), O_1 is converted to O_2 via R_1 and R_2. The overall process then is neither an oxidation nor a reduction and, if k is large enough, the current may be nearly zero.

In all of these cases, the homogeneous electron–transfer (disproportionation) reaction

$$O_1 + R_2 \rightleftarrows O_2 + R_1$$

may be important. The effect of this reaction depends on the case and on the technique involved.

Rotating–Disk Voltammetry

The steady–state equations for the $\overrightarrow{E}\overrightarrow{C}\overrightarrow{E}$ mechanism are easily solved (Appendix 6) to obtain

$$i = FA \left[\frac{DC_{O1}^*}{x_D + \theta_1 x_R'} \right] \left[1 + \frac{1 - x_R'/x_D}{1 + \theta_2} \right] \tag{5.89}$$

The limiting current, obtained at sufficiently negative potentials ($\theta_1 \cong \theta_2 \cong 0$), is

$$i_L = FAC_{O1}^*(D/x_D)(2 - x_R'/x_D) \tag{5.90}$$

When the chemical step is fast, $x_R'/x_D \lll 1$, the process corresponds to a two–electron wave. When the reaction is slow, $x_R'/x_D \cong 1$ and we expect a one–electron wave.

For the $\overrightarrow{E}\overleftarrow{C}E$ mechanism, the current is found to be

$$i = FA \left[\frac{DC_{O1}^*}{x_D + \theta_1 x_R'} \right] \left[1 - \frac{\theta_2(1 - x_R'/x_D)}{1 + \theta_2} \right] \tag{5.91}$$

When the electrode potential is sufficiently negative that θ_1 and θ_2 are both near zero, the current is diffusion limited and eq (5.91) reduces to eq (4.32). The current then corresponds to one electron, independent of the rate of the chemical step. We now consider each of the four cases and assess the effect of the homogeneous disproportionation process.

Case IA. When $E_1° < E_2°$ for the $\overrightarrow{E}\overrightarrow{C}E$ mechanism, θ_2 is very small whenever θ_1 is small enough that significant current flows. Thus eq (5.89) reduces to

$$i = \frac{FADC_{O1}^*(2 - x_R'/x_D)}{x_D + \theta_1 x_R'}$$

which corresponds to a wave of shape given by the Heyrovský–Ilkovič equation with i_L given by eq (5.90) and $E_{\frac{1}{2}}$ by eq (5.76). Thus the apparent number of electrons is $(2 - x_R'/x_D)$ and the wave is shifted to a more positive potential, as expected from the EC part of the mechanism. The disproportionation reaction

$$O_1 + R_2 \rightleftarrows O_2 + R_1$$

has an equilibrium constant

$$K = \exp \frac{F(E_1{}^\circ - E_2{}^\circ)}{RT}$$

and, for case IA, $K \ll 0$. Consider the situation when $E \cong E_1{}^\circ$. The effect of the $R_1 \to O_2$ reaction is to reduce the R_1 concentration at the electrode surface, resulting in a positive shift of the half–wave potential and an increase in current (O_2 is irreversibly reduced to R_2). When disproportionation is allowed, however, O_2 is scavenged by R_1 to produce O_1 and R_2. Since the reduction of O_1 is less complete than that of O_2, the current drops somewhat and $E_{\frac{1}{2}}$ shifts back toward $E_1{}^\circ$. Voltammograms computed from eq (5.90) and by a digital simulation method (assuming that the disproportionation reaction is infinitely fast) are shown in Figure 5.42a for $x_R/x_D = 5$, 0.5, and 0.05.

Case IB. When $E_1{}^\circ > E_2{}^\circ$ for the $\vec{E}\vec{C}E$ mechanism, two waves are expected from eq (5.90), the first of which is a one–electron wave indistinguishable from that given by an EC mechanism; the second wave has $E_{\frac{1}{2}} = E_2{}^\circ$ and (ignoring disproportionation) an apparent number of electrons equal to $(1 - x_R'/x_D)$. In this case, the disproportionation equilibrium constant is very large. However, no R_2 is produced until the potential approaches $E_2{}^\circ$. Thus the first wave is expected to be unperturbed by disproportionation. When $E \cong E_2{}^\circ$, the O_2 produced from R_1 is reduced and we expect a current increase. The resulting R_2, however, reacts with O_1 to regenerate O_2 and R_1. Thus a further current enhancement is expected. Computed voltammograms for infinitely slow and infinitely fast disproportionation are shown in Figure 5.42b. The effects are more noticeable in this case.[‡]

Case IIA. When $E_1{}^\circ < E_2{}^\circ$ for the $\vec{E}\overset{\leftarrow}{C}E$ mechanism, $\theta_1 \gg \theta_2$, and the second bracketed term of eq (5.91) will be close to 1 whenever the first bracketed term is significantly greater than zero. Thus eq (5.91) reduces to eq (5.75) and case IIA is really just an EC process as far as r.d.e. voltammetry is concerned. The effects of disproportionation are completely negligible. A one–electron wave is

---

[‡] Curves showing n_{app} as a function of the rate constant of the chemical step have been computed by Marcoux, Adams, and Feldberg *(37)* for cases IA and IB for various values of the equilibrium constant K.

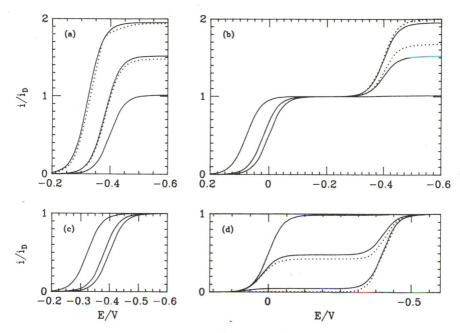

Figure 5.42 Computed r.d.e. voltammograms for ECE mechanisms for x_R/x_D = 5, 0.5 and 0.05 (left to right). The solid and dotted lines correspond, respectively, to infinitely slow and infinitely fast disproportionation equilibria. (a) Case IA; (b) case IB; (c) case IIA; and (d) case IIB. For cases IA and IIA, $E_1° = -0.5$ V, $E_2° = 0.0$ V, and for Cases IB and IIB, $E_1° = 0.0$ V, $E_2° = -0.5$ V.

then expected, shifted to a more positive potential as shown in Figure 5.42c.

Case IIB. When $E_1° > E_2°$ for the $\overset{\rightarrow\;\leftarrow}{\text{ECE}}$ mechanism and the potential is in the vicinity of $E_1°$, we have $\theta_1 \ll \theta_2$. A wave with apparent number of electrons equal to x_R'/x_D is then expected with half-wave potential shifted as expected for an EC process. A second wave is expected with $E_{\frac{1}{2}} = E_2°$ and total limiting current corresponding to one electron. The effects of disproportionation are greatest in this case when $E_1° > E > E_2°$. Here O_1 is reduced irreversibly to R_1, which reacts to form R_2. At the electrode, R_2 is oxidized and the net current is diminished. Disproportionation increases the efficiency of this process since any R_2 which escapes from the electrode is then oxidized by O_1, leading to a further decrease in the current. Simulated r.d.e. voltammograms for this case are shown in Figure 5.42d.

In practice, the disproportionation reaction is not instantaneous (although homogeneous electron–transfer processes are often diffusion controlled) so that experimental results for a given value of x_R/x_D would be expected to lie between the limits suggested by the curves of Figure 5.42. Kinetic information, obtained from half–wave potential shifts or limiting current ratios, thus is at best semiquantitative. However, the range of rate constants which can be estimated is quite wide. If we assume that the limit of accurate measurement of wave height ratios corresponds to

$$0.1 < x_R'/x_D < 0.9 \quad (0.1 < x_R/x_D < 1.7)$$

then eq (A.29) gives

$$0.26 < \omega/k < 75$$

and since ω has a practical range of 10–1000 rad s^{-1}, the range of rate constant estimable by r.d.e. voltammetry is roughly

$$0.1 < k < 4 \times 10^3 \text{ s}^{-1}$$

Chronoamperometry

The solution of the linear diffusion problem for the $\overrightarrow{E}\overrightarrow{C}E$ mechanism is relatively straightforward for a planar electrode if the homogeneous disproportionation step is ignored. Alberts and Shain (38) showed that the current–time response follows the Cottrell equation when the potential is stepped to $E < E_1°, E_2°$:

$$i = n_{app}FAC^*(D/\pi t)^{\frac{1}{2}}$$

where

$$n_{app}/n = 2 - \exp(-kt) \tag{5.92}$$

and n is the number of electrons involved in the individual electron–transfer steps. Feldberg and co–workers (39,40) have examined the effect of homogeneous disproportionation on the current response in chronoamperometry using digital simulation techniques. They found that eq (5.92) is accurate for kt < 0.2. In case IA ($E_1° < E_2°$, $K \ll 1$), n_{app}/n may be somewhat less that the value predicted by eq (5.92), and in case IB, it may be somewhat greater. In the later case, n_{app}/n can be somewhat greater than 2 when kt $\cong$ 3.

For case IIA (the $\overrightarrow{E}\overleftarrow{C}E$ mechanism), a potential step to $E < E_1°$, $E_2°$ is expected to involve only n electrons regardless of the rate of the chemical step and so should be independent of the existence of the homogeneous electron–transfer process. In case IIB, on the other hand, the Alberts–Shain theory predicts

$$n_{app}/n = \exp(-kt) \tag{5.93}$$

but Feldberg and Jeftič *(40)* find that for kt > 1, n_{app}/n is significantly smaller than predicted by eq (5.93) when disproportionation is considered and even becomes negatively for kt > 2.5. Early in the experiment, the current is due mostly to the reduction of O_1 to R_1. Later when O_1 is polarized at the electrode, the R_2 oxidation dominates and the current becomes negative if the conversion of R_1 to R_2 is fast enough. Some computed working curves from Feldberg's work are shown in Figure 5.43.

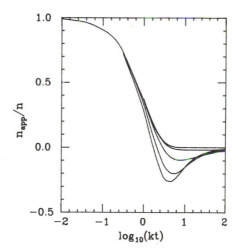

Figure 5.43 $\overrightarrow{E}\overleftarrow{C}E$ chrono–amperometric working curve showing n_{app}/n as a function of log kt for (top to bottom) $k_{disp}C^*/k$ = 0, 0.1, 1, 10, and ∞.

Example 5.14 Alberts and Shain *(38)* used chronoamperometric measurements to study the reduction of p–nitrosophenol in 20% aqueous ethanol solutions buffered to pH 4.8. The electrode process consists of the two–electron reduction of p–nitrosophenol to p–hydroxylaminophenol, followed by dehydration to p–benzoquinoneimine and the reduction of the imine to p–aminophenol:

$$HO-\!\!\langle\bigcirc\rangle\!\!-NO \longrightarrow HO-\!\!\langle\bigcirc\rangle\!\!-NHOH$$

$$HO-\!\!\langle\bigcirc\rangle\!\!-NHOH \xrightarrow[-H_2O]{k} O=\!\!\langle\bigcirc\rangle\!\!=NH$$

$$O=\!\!\langle\bigcirc\rangle\!\!=NH \xrightarrow[2\ H^+]{2\ e^-} HO-\!\!\langle\bigcirc\rangle\!\!-NH_2$$

The corrected data[‡] are plotted in Figure 5.44 as $i/FAC^*D^{\frac{1}{2}}$ vs. $t^{-\frac{1}{2}}$, along with curves, computed using eq (5.92) for k = 0.4, 0.6, and 0.8 s^{-1}. The data fit the curve for k = 0.6 s^{-1} reasonably well; Alberts and Shain report k = 0.59 ± 0.07 s^{-1}. The small discrepancy between theory and experiment is probably due to neglect of the homogeneous disproportionation reaction.

Cyclic Voltammetry

Cyclic voltammetry is a particularly powerful technique for the investigation of systems where an ECE process is operative, since the O_1/R_1 and O_2/R_2 couples are observed separately. The relative size of the O_1 and O_2 reduction peaks thus can usually give some indication of the magnitude of the rate constant for the chemical step. Examples of cyclic voltammograms for several values of the rate parameter λ, given by eq (5.79), are shown in Figures 5.45 (cases IA and IB) and 5.46 (cases IIA and IIB).

[‡] The experiment employed a hanging mercury drop electrode; the theory for a spherical electrode is considerably more complex than that for a planar electrode, although the qualitative effect is similar to that described by eq (5.92). The data plotted in Figure 5.44 have been corrected to correspond to a planar electrode.

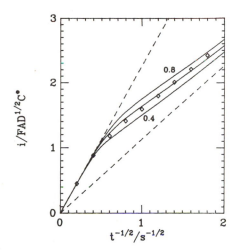

Figure 5.44 Chronoamperometric data for the reduction of p–nitrosophenol in 20% aqueous ethanol solution, pH 4.8.

   In case IA (the $\overrightarrow{ECE}$ mechanism with $E_1° < E_2°$, Figure 5.45a–d), the first negative–going scan shows only the $O_1 \to R_1$ reduction. On the reverse scan, however, there are several features worth noting. The R_1 oxidation peak decreases in size and then disappears, as the rate of the chemical step increases; an oxidation peak due to R_2 grows with increasing chemical rate. The current between these peaks is particularly interesting. When the rate is slow (Figure 5.45b), R_1 survives long enough to contribute a negative current in the region between $E_1°$ and $E_2°$. When the chemical step is faster, so that no R_1 survives when $E > E_1°$, but slow enough that most of the O_2 is formed some distance from the electrode, diffusion of O_2 to the electrode gives a net reduction current in the interpeak region (Figure 5.45c). In this case, the current on the positive–going scan is actually greater than that on the negative–going scan and the curves cross. When the chemical step is so fast that most O_2 is formed near the electrode, it is reduced quickly, leading to a much larger cathodic peak and decreasing the interpeak current to near zero (Figure 5.45d).

   The disproportionation reaction

$$O_1 + R_2 \rightleftarrows O_2 + R_1$$

has $K \ll 1$ in case IA, so that O_2 and R_1 tend to scavenge one another. The principal effect of disproportionation on case IA cyclic voltammograms is to suppress the curve crossing seen for intermediate rates (40,41). Amatore, Pinson, Savéant, and Thiebault (42) have pointed out that disproportionation can have the opposite effect when the chemical step is exceedingly fast. If K is not too small and R_1 is

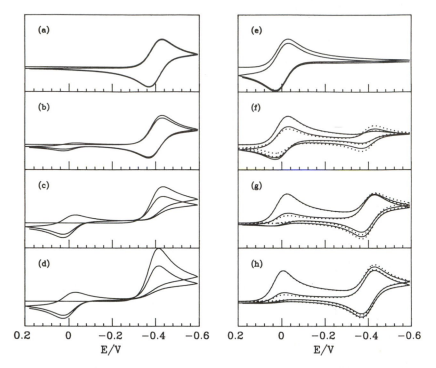

Figure 5.45 Cyclic voltammograms computed by digital simulation for cases IA (a–d) and IB (e–h) of the ECE mechanism. The rate parameter λ = 0 (a and e), 0.025 (b and f), 0.25 (c and g), and 2.5 (d and h). For case IA, E_1° = -0.4 V, E_2° = 0.0 V, and for case IB, the potentials are reversed. The solid lines were computed neglecting disproportionation. The dotted lines in (f) – (h) show the effect of infinitely fast disproportionation.

very short–lived, the disproportionation reaction may be pulled to the right (uphill energetically) to supply more O_2 and R_1 (which is quickly converted to O_2). The effect is again to give a cathodic current in the interpeak region.

In case IB (the $\overrightarrow{E}\overrightarrow{C}E$ mechanism with $E_1^\circ > E_2^\circ$, Figure 5.45e–h), the first negative–going scan shows cathodic peaks at both E_1° and E_2°. The height of the second peak approaches that of the first peak as the reaction rate increases. The reverse scan shows an R_2 oxidation peak comparable in size to the O_2 reduction; if the reaction is fast, little R_1 remains to be oxidized, so that this peak may be missing. Disproportionation has little effect on the first negative–going scan, but the R_1 oxidation peak is somewhat smaller on the reverse scan. When the $R_1 \rightarrow O_2$ conversion is not very fast (Figure 5.45f), the disproportionation reaction enhances the peaks due to the O_2/R_2 couple.

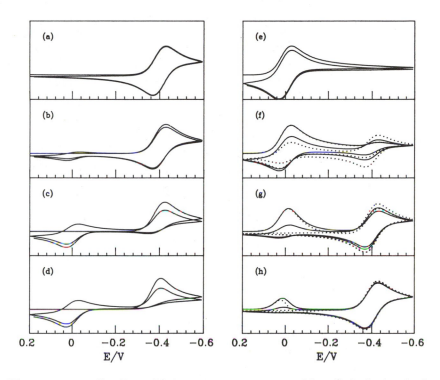

Figure 5.46 Cyclic voltammograms computed by digital simulation for cases IIA (a–d) and IIB (e–f) of the ECE mechanism. The rate parameters and standard potentials are as described for Figure 5.45.

In case IIA (the $\vec{E}\overset{\leftarrow}{C}E$ mechanism with $E_1° < E_2°$, Figure 5.46a–d), the O_1 reduction peak shifts toward positive potentials with increasing chemical rate and the R_1 oxidation peak disappears as expected, since we have essentially an uncomplicated EC process for $E < E_2°$. The R_2 oxidation peak and (on the second scan) the O_2 reduction peak grow with increasing chemical rate. As we found for r.d.e. voltammetry and chronoamperometry, disproportionation has a negligible effect for case IIA.

In case IIB (the $\overrightarrow{E}\overleftarrow{C}E$ mechanism with $E_1° > E_2°$, Figure 5.46e–h), the first negative–going scan is significantly perturbed by the chemical step and by disproportionation. The chemical step converts R_1 to R_2, but R_2 is oxidized when $E > E_2°$. Thus the current drops more rapidly than normal after the O_1 reduction peak when the rate is fast. This effect is enhanced by disproportionation as shown in Figure 5.46g and h *(40)*. If the O_1 reduction peak is smaller than normal on the first scan, it will be much smaller (perhaps not even detectable) on the second and subsequent scans. The O_2/R_2 reduction/oxidation peaks grow with increasing rate (enhanced by the disproportionation reaction) and, for a fast reaction, appear as an unperturbed reversible couple.

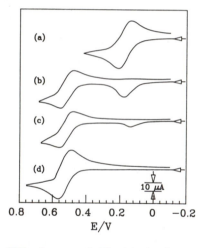

Figure 5.47 Cyclic voltammograms of MnL in acetonitrile solution. (a–c) L = CH_3CN with (a) no added PPh_3, (b) 1 equivalent, and (c) 9 equivalents of PPh_3 added. (d) L = PPh_3.

Example 5.15 Hershberger, Klingler, and Kochi *(43)* used cyclic voltammetry to study the oxidation of $(\eta^5\text{-}CH_3C_5H_4)Mn(CO)_2L$ (MnL). Cyclic voltammograms of the derivatives with L = $NCCH_3$ and PPh_3 (traces a and d, respectively, in Figure 5.47) are uncomplicated and apparently reversible with peak current ratios of 1.0 and normal peak potential separations. However, addition of PPh_3 to a solution of $Mn(NCCH_3)$ results in a dramatic change (traces b and c of Figure 5.47). On the first positive–going scan, the oxidation peak due to the substrate decreases in size and an oxidation peak due to the product, $MnPPh_3$, appears. On the reverse scan, the product cation is reduced but the cathodic peak expected for the substrate cation is completely absent. Comparison of traces b and c of Figure 5.47 with trace d of Figure 5.46 suggests that these results can be understood as an ECE process—the analog of case IIB, making appropriate allowance for the fact that the first step is an oxidation rather than a reduction. The mechanism can be written as follows:

$$Mn(NCCH_3) - e^- \quad \rightleftarrows \quad Mn(NCCH_3)^+ \qquad E_1° = 0.19 \text{ V}$$

$$Mn(NCCH_3)^+ + PPh_3 \quad \overset{k_1}{\rightarrow} \quad Mn(PPh_3)^+ + CH_3CN$$

$$Mn(PPh_3)^+ + e^- \quad \rightleftarrows \quad Mn(PPh_3) \qquad E_2° = 0.52 \text{ V}$$

$$Mn(PPh_3)^+ + Mn(NCCH_3) \quad \overset{k_2}{\rightarrow} \quad Mn(PPh_3) + Mn(NCCH_3)^+$$

As we have seen in Figure 5.46, the shape of a cyclic voltammogram for such a system is very sensitive to the rate constant k_1 and somewhat dependent on k_2 as well. Hershberger *et al.* carried out a systematic investigation using digital simulation techniques and were able to establish that $k_1 = (1.3 \pm 0.2) \times 10^4$ $M^{-1}s^{-1}$, $k_2 > 10^4 M^{-1}s^{-1}$.

Since the disproportionation step is exoergic ($\Delta G = -32$ kJ mol^{-1}) and fast, it represents the propagation step of a homogeneous chain reaction initiated by oxidation of a few substrate molecules. Indeed, Hershberger *et al.* found that controlled current oxidation of $Mn(NCCH_3)$ in the presence of PPh_3 (continued until the potential approached 0.5 V) resulted in virtually quantitative yields of substitution product with the passage of less than 0.001 Faraday of charge per mole of substrate. Thus chain lengths in excess of 1000 are apparently attained.

Similar electrocatalytic processes, related in general to case IIB of the ECE mechanism, have been found for a number of other organic *(44)* and organometallic *(45)* systems.

REFERENCES

(Reference numbers preceded by a letter, *e.g. (E5)*, refer to a book listed in the Bibliography.)

1. R. A. Marcus, *J. Chem. Phys.* **1965**, *43*, 679; *Ann. Rev. Phys.*

Chem. **1964,** *15,* 155.

2. J. A. V. Butler, *Trans. Faraday Soc.* **1924,** *19,* 734.

3. T. Erdey–Gruz and M. Volmer, *Z. phys. Chem.* **1930,** *150A,* 203.

4. R. D. Allendoerfer and P. H. Rieger, *J. Am. Chem. Soc.* **1965,** *87,* 2336.

5. B. J. Heubert and D. E. Smith, *J. Electroanal. Chem.* **1971,** *31,* 333.

6. A. N. Frumkin, *Z. phys. Chem.* **1933,** *164A,* 121.

7. R. Parsons, *Adv. Electrochem. Electrochem. Engin.* **1961,** *1,* 1.

8. J. Tafel, *Z. phys. Chem.* **1905,** *50,* 641.

9. K. J. Vetter and G. Manecke, *Z. phys. Chem.* **1950,** *195,* 270.

10. K. J. Vetter, *Z. phys. Chem.* **1952,** *199,* 285.

11. J. O'M. Bockris, D. Drazic, and A. R. Despic, *Electrochim. Acta* **1961,** *4,* 325.

12. S. Trasatti, *J. Electroanal. Chem.* **1972,** *39,* 163.

13. L. Meites and Y. Israel, *J. Electroanal. Chem.* **1964,** *8,* 99.

14. R. S. Nicholson, *Anal. Chem.* **1965,** *37,* 1351.

15. D. E. Smith, *Electroanalytical Chemistry* **1966,** *1,* 1; *Crit. Rev. Anal. Chem.* **1971,** *2,* 247.

16. R. Dietz and M. E. Peover, *Disc. Faraday Soc.* **1968,** *45,* 154.

17. A. M. Bond, R. J. O'Halloran, I. Ruzic, and D. E. Smith, *Anal. Chem.* **1976,** *48,* 872.

18. D. de Montauzon, R. Poilblanc, P. Lemoine, and M. Gross, *Electrochem. Acta* **1978,** *23,* 1247.

19. W. E. Geiger in *Laboratory Techniques in Electroanalytical Chemistry,* P. T. Kissinger and W. R. Heineman, eds, New York: Marcel Dekker, 1984, p 483.

20. P. Zuman, *The Elucidation of Organic Electrode Processes,* New York: Academic Press, 1969.

21. M. D. Hawley in *Laboratory Techniques in Electroanalytical Chemistry,* P. T. Kissinger and W. R. Heineman, eds, New York: Marcel Dekker, 1984, p 463.

22. R. Alder, *J. Chem. Soc., Chem. Commun.* **1980,** 1184.

23. I. B. Goldberg and T. M. McKinney in *Laboratory Techniques in Electroanalytical Chemistry,* P. T. Kissinger and W. R. Heineman, eds, New York: Marcel Dekker, 1984, p 675.

24. H. Tachikawa and L. R. Faulkner in *Laboratory Techniques in Electroanalytical Chemistry,* P. T. Kissinger and W. R. Heineman, eds, New York: Marcel Dekker, 1984, p 637.

25. C. J. Pickett and D. Pletcher, *J. Chem. Soc., Dalton Trans.* **1976,** 749.

26. A. M. Bond, R. Colton, and M. J. McCormick, *Inorg. Chem.* **1977,** *16,* 155,

27. W. J. Albery and R. P. Bell, *Proc. Chem. Soc.* **1963,** 169.

28. R. S. Nicholson and I. Shain, *Anal. Chem.* **1964**, *36*, 706.
29. D. M. H. Kern, *J. Am. Chem. Soc.* **1954**, *76*, 1011.
30. W. M. Schwartz and I. Shain, *J. Phys. Chem.* **1965**, *69*, 30.
31. A. C. Testa and W. H. Reinmuth, *Anal. Chem.* **1960**, *32*, 1512.
32. A. M. Bond, R. J. O'Halloran, I. Ruzic, and D. E. Smith, *J. Electroanal. Chem.* **1982**, *132*, 39.
33. J. Moraczewski and W. E. Geiger, *J. Am. Chem. Soc.* **1981**, *103*, 4779.
34. M. Grzeszczuk, D. E. Smith, and W. E. Geiger, *J. Am. Chem. Soc.* **1983**, *105*, 1772.
35. P. Delahay and G. L. Steihl, *J. Am. Chem. Soc.* **1952**, *74*, 3500.
36. R. L. Birke and W. F. Marzluff, Jr., *J. Electroanal. Chem.* **1968**, *17*, 1.
37. L. S. Marcoux, R. N. Adams, and S. W. Feldberg, *J. Phys. Chem.* **1969**, *73*, 2611.
38. G. S. Alberts and I. Shain, *Anal. Chem.* **1963**, *35*, 1859.
39. M. D. Hawley and S. W. Feldberg, *J. Phys. Chem.* **1966**, *70*, 3459.
40. S. W. Feldberg and L. Jeftič, *ibid.* **1972**, *76*, 2439.
41. S. W. Feldberg, *J. Phys. Chem.* **1971**, *75*, 2377.
42. C. Amatore, J. Pinson, J. M. Savéant, and A. Thiebault, *J. Electroanal. Chem.* **1980**, *107*, 59.
43. J. W. Hershberger, R. J. Klingler, and J. K. Kochi, *J. Am. Chem. Soc.* **1983**, *105*, 61.
44. J. M. Savéant, *Accts. Chem. Res.* **1980**, *13*, 323.
45. G. J. Bezems, P. H. Rieger, and S. J. Visco, *J. Chem. Soc., Chem. Commun.* **1981**, 265.

PROBLEMS

5.1 Show that either eq (5.11a) or (5.11b) can be rearranged (with the help of the Nernst equation) to

$$i_0 = FAk_0 C_O{}^\beta C_R{}^\alpha$$

5.2 Derive an equation for the current, analogous to eq (5.30), for the case that $C_O{}^*$ and $C_R{}^*$ are both nonzero.

5.3 Derive eqs (5.23) from eqs (5.4) and (5.11).

5.4 Show that the data discussed in Example 5.2 are consistent with the simple mechanism

$$Mn^{3+} + e^- \rightleftarrows Mn^{2+}$$

when subjected to the more complete analysis of Example 5.3.

5.5 Figure 5.48 shows a Tafel plot for the reduction of H^+(aq) in a 0.1 M solution of HCl at a nickel electrode in the presence of 1 bar H_2(g) at 20°C.
(a) Determine the exchange current density j_0 and the apparent transfer coefficient.
(b) Compute the electron–transfer rate constant k_0.

5.6 Derive eqs (5.26) by an alternative route in which the derivatives

$$\left[\frac{\partial \ln i_0}{\partial \ln C_O} \right]_{C_R} \quad \text{and} \quad \left[\frac{\partial \ln i_0}{\partial \ln C_R} \right]_{C_O}$$

are first computed.

5.7 Bockris, Drazic, and Despic (see Example 5.4) found that the exchange current density for the oxidation of iron depends on the concentration of Fe^{2+} according to

$$\frac{\partial \ln |j_0|}{\partial \ln[Fe^{2+}]} = 0.8$$

Show that the mechanism discussed in Example 5.4 leads to

$$\frac{\partial \ln |j_0|}{\partial \ln C_O} = 1 - a/2$$

and thus is consistent with the experimental result.

5.8 The exchange current for an EE reduction with $E_2° \gg E_1°$ and the first electron–transfer step rate–limiting is given by eqs (5.27) and the net current is given by the Butler–Volmer equation with $\beta_{app} = 1 + \beta$ where β is the anodic transfer coefficient for the rate–limiting step. Derive analogous expressions for the case where the second electron–transfer step is rate–limiting.

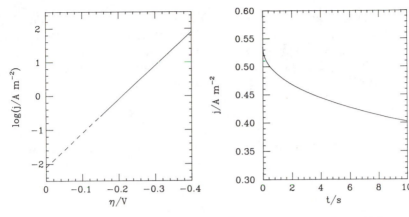

Figure 5.48. Tafel plot for the reduction of H^+ at a nickel electrode.

Figure 5.49 Potential step chronoamperogram for a CE process.

5.9   Equation (33) gives the half–wave potential of an irreversible wave for $k_0/k_D \ll 1$. For quasi–reversible waves, $E_{\frac{1}{2}}$ must be computed from eq (5.31).
(a) Compute $E_{\frac{1}{2}} - E^\circ$ for $k_D = k_0$, $a = 0.5$.
(b) What is the maximum value of k_0/k_D for which eq (5.33) gives $E_{\frac{1}{2}} - E^\circ$ correct to within 1 mV?

5.10 The Tomeš criterion for a reversible one–electron wave is $\Delta E = E_{\frac{1}{4}} - E_{\frac{3}{4}} = 56.5/n$ mV at 25°C.
(a) Compute ΔE for several values of k_D/k_0 in the range 0.1 to 2 using eq (5.31) ($a = 0.5$) and plot ΔE vs. $\log(k_0/k_D)$.
(b) A r.d.e. voltammogram gave a value of $\Delta E = 75$ mV after correction for solution iR drop. If $\nu = 10^{-6}$ m^{2}s^{-1}, $D = 10^{-9}$ m^{2}s^{-1}, and $\omega = 500$ s^{-1}, what is the approximate value of k_0?

5.11 (a) Show that the chronocoulometric charge for a potential step experiment (see Problem 4.11) involving an electron–transfer step of finite rate is

$$Q = FAC_O^* k_c \lambda^{-2} \{ \exp(\lambda^2 t)[1 - \mathrm{erf}(\lambda t^{\frac{1}{2}})] + 2\lambda (t/\pi)^{\frac{1}{2}} - 1 \}$$

where λ is given by eq (5.38). Hint: Integrate eq (5.39) by parts.
(b) For sufficiently long times, the first term in the bracketted

expression goes to zero and Q becomes linear in $\sqrt{t}$. The linear region of a plot of Q $vs.$ $\sqrt{t}$ can be extrapolated to t = 0 or to Q = 0. Obtain expressions for Q extrapolated to t = 0 and for $\sqrt{t}$ extrapolated to Q = 0. Describe how such data could be used to obtain the electron–transfer rate constant k_0.

5.12 The following data were obtained for a potential step chronocoulometry experiment. The solution contained 1 mM O which is reduced by one electron, and the potential step was to E = $E_{\frac{1}{2}}$ – 50 mV. Assume that D_O = 10^{-9} m^2s^{-1}, a = 0.5, A = 3.1×10^{-6} m^2.

t/s	$Q/\mu C$
1	1.13
3	2.10
5	2.78
7	3.33
9	3.81

Plot Q $vs.$ $\sqrt{t}$, extrapolate to zero charge and estimate the electron–transfer rate constant k_0 using the equations derived in Problem 5.11.

5.13 Derive expressions for the resistive and capacitive components of the faradaic impedance in terms of $|Z_f|$ and ϕ.

5.14 (a) Show that eq (5.58) can be derived (within a multiplicative factor) by differentiation of the Heyrovský–Ilkovič equation to obtain di/dE.
(b) What is the width at half height of a reversible one–electron a.c. polarographic wave?

5.15 (a) Find the shape of a second–harmonic a.c. polarogram by computing d^2i/dE^2 and prepare a plot of this function $vs.$ the d.c. potential.
(b) What is the width (between current extrema) of a reversible one–electron second–harmonic a.c. wave?

5.16 Given the time scales accessible to the various experimental techniques (Section 5.5), what is the fastest electron–transfer rate constant k_0 which could be measured for a simple one–electron reduction using:
(a) Potential step chronoamperometry, $(E - E^\circ)$ = –0.2 V?

(b) R.d.e. voltammetry?
(c) A.c. polarography?
Assume that $D = 10^{-9}$ m^2s^{-1}, $\nu = 10^{-6}$ m^2s^{-1}, $a = 0.5$.

5.17 The cyclic voltammogram of 0.001 M ferrocene, 0.5 M tetra–n–butylammonium hexafluorophosphate, in dichloromethane at a platinum microelectrode (10 μm diameter) shows a peak separation of 147 mV when the scan rate $v = 200$ V s^{-1} [N. J. Stone, private communication]. Since the peak current under these circumstances was about 24 nA, ohmic potential drop in solution should be negligible. Assuming that the peak separation can be attributed to the ferrocene electron–transfer rate, what is k_0? Assume that T = 293 K, $D \cong 2 \times 10^{-10}$ m^2s^{-1}.

5.18 A CE electrode process studied by potential step chronoamperometry gave the current density – time curve shown in Figure 5.49. If the concentration of the precursor species is 1 mM (1 mol m^{-3}), the diffusion coefficient is 10^{-9} m^2s^{-1}, and the reduction is by two electrons, estimate $k_1 K$ for the reaction preceding electron transfer.

5.19 Potential step chronoamperometry is to be used to measure the concentration of an electroinactive analyte A through the catalytic intermediacy of a one–electron couple, O/R.
(a) Because of vibration and convective mixing, the current cannot be trusted to remain diffusion–controlled for more than 10 s. If the rate constant for the reaction of R with A is 2 x 10^5 M^{-1}s^{-1}, what is the minimum A concentration accessible to the analysis? Assume a diffusion coefficient D = 5 x 10^{-10} m^2s^{-1}.
(b) What limiting current density would be expected at the limiting A concentration? If the available current measuring device can read down to 0.1 μA with the accuracy needed for the analysis, what is the minimum electrode area needed?
(c) In order to maintain pseudo–first–order conditions, it is necessary that C_O^* be at least 10 times smaller than C_A. What concentration of O should be used? What problems might be expected from this result?

5.20 What requirements must be met in order that eq (5.68) reduces to the Lingane equation, eq (4.40)?

5.21 Assuming that the reduction of Pb(II) in basic aqueous solution actually proceeds via the mechanism

$$HPO_2^- + H_2O \rightarrow Pb^{2+} + 3\,OH^-$$

$$Pb^{2+} + 2\,e^- \rightleftarrows Pb(s)$$

as assumed in Example 4.10 and Problem 4.23, use the results of Problem 5.20 to estimate a lower bound to the rate of the homogeneous step. What conclusions can be drawn from the result of this calculation?

5.22 Derive eq (5.75) for the half–wave potential of an r.d.e. voltammogram for an EC process using the methods of Appendix 6.

5.23 Show that the r.d.e. current for an EE process where both steps are assumed to be nernstian, *i.e.*,

$$O + e^- \rightarrow R_1 \qquad\qquad E_1^\circ$$

$$R_1 + e^- \rightarrow R_2 \qquad\qquad E_2^\circ$$

is given by

$$i = FAC_O^*(D/x_D) \left[\frac{2 + \theta_2}{1 + \theta_2 + \theta_1\theta_2} \right]$$

where θ_1 and θ_2 are the nernstian concentration concentration ratios for the first and second couples.

5.24 (a) Given the result of Problem 5.23, derive an expression for $(i_L - i)/i$.
(b) If $E_2^\circ \geq E_1^\circ$, a single wave is expected. Compute the half–wave potential.
(c) Under what circumstances does the answer to part (a) reduce to the Heyrovský–Ilkovič equation for a two–electron wave?
(c) Compute $E_{\frac{1}{4}}^1 - E_{\frac{3}{4}}^3$ for $E_2^\circ - E_1^\circ = 0, 50, 100$, and 200 mV. What value would be expected for a two–electron wave according to the Tomeš criterion for reversibility?

5.25 Derive eq (5.34) for the current in an irreversible EE process. Either follow a procedure analogous to that used in the derivation of eq (5.30) or use the flux equations introduced in Appendix 6.

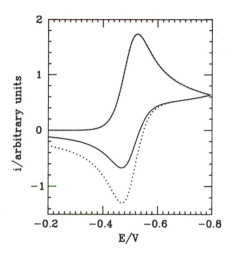

Figure 5.50 Cyclic voltammogram of an EC electrode process with scan rate $v = 400$ mV s^{-1}. The dotted curve shows a cyclic voltammogram of the same system uncomplicated by the following chemical reaction.

5.26 The cyclic voltammogram shown in Figure 5.50 corresponds to an EC electrode process. Estimate the rate constant k. You need not correct for iR drop or for capacitive current.

5.27 Digital simulation of cyclic voltammograms for case IA of the ECE mechanism leads to the working curves for $E_p - E°$ and $i_p/i_p(0)$ shown in Figure 5.51, where $i_p(0)$ is the expected cathodic peak current for k = 0. Both electrode processes involve one electron.
(a) Why does the peak current increase by more than a factor of two? Give a qualitative explanation. Hint: What is the ratio $C_O(0)/C_R(0)$ at the electrode surface at the potential of the peak current if the homogeneous rate is zero?
(b) Why is there a small negative shift in the peak potential before the expected positive shift? (See Figure 5.33.)

5.28 Cyclic voltammograms for an ECE process are shown in Figure 5.52 for three different scan rates.
(a) Identify each peak in the cyclic voltammograms with the reduction or oxidation of a species in the mechanism.
(b) Estimate the rate constant of the chemical step. You can assume that the electron–transfer steps are nernstian and that capacitive current and ohmic potential drop are negligible.

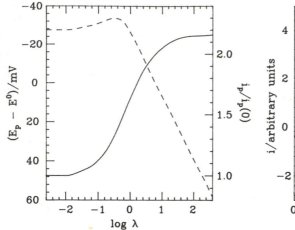

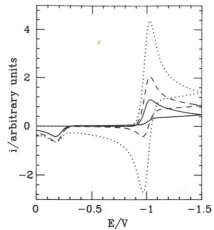

Figure 5.51 Theoretical working curves for cyclic voltammograms for case IA of the ECE mechanism: peak potential shift (dashed curve) and peak current (solid curve) vs. the kinetic parameter $\lambda = RTk/Fv$.

Figure 5.52 Cyclic voltammograms for an ECE process. The potential scan rate v is 100 mV s^{-1} (solid curves), 500 mV s^{-1} (dashed curves), and 2.5 V s^{-1} (dotted curves).

6

ELECTROLYSIS

In this chapter we will be concerned with electrolysis experiments in which a significant fraction of the electroactive material is oxidized or reduced. We begin in Section 6.1 with a discussion of some general aspects of electrolysis, and we proceed to some analytical applications of electrolysis in Section 6.2. In Sections 6.3 and 6.4, we explore some applications of electrolysis to chemical synthesis, including examples from industrial chemistry. A short–circuited galvanic cell is a self-powered electrolysis cell. This notion will give us a useful model for a discussion of corrosion in Section 6.5.

6.1 BULK ELECTROLYSIS

Volta's invention of the first practical source of electrical power in 1800 stimulated a flurry of activity among scientists. Indeed, the ideas for many bulk electrolysis experiments can be traced to experiments done in the first decade of the nineteenth century. We have already mentioned the electrolysis of water by Nicholson and Carlisle. This work led to Erman's discovery that the conductivity of salt solutions depends on concentration. In another series of experiments following the Nicholson–Carlisle discovery, Cruickshank[‡] found that when a solution of a salt such as copper sulfate is electrolyzed, metallic copper is plated onto the negative electrode and no hydrogen is evolved *(1)*. Many other important discoveries were made in the following years; in particular, between 1806 and 1809, Davy conducted electrolysis experiments on many salt solutions and discovered six new elements: Na, K, Mg, Ca, Ba, and Sr. Davy's work on electrolysis was extended and quantified by Faraday, who, in 1833 *(2)*, summarized his results in two statements which have come to be known as *Faraday's laws of electrolysis*:

> The weight of metal plated on the cathode during passage of current through a solution of the metal salt is proportional to: (1) the charge passed through the solution; and (2) the equivalent weight (atomic weight/oxidation number) of the metal.

In modern terms, we would say that one mole of electrons (one Faraday) will reduce one mole of Na^+, one–half mole of Ca^{2+}, or one-third mole of Al^{3+}. Faraday's work not only put electrolysis on a quantitative basis but laid the groundwork for the more complete understanding of the role of electrons in chemistry which began to emerge 50 years later.

Exhaustive Electrolysis

When an electrolyte solution is subjected to electrolysis, either to separate a component of the solution or to oxidize or reduce a substrate, rather large amounts of electrical charge must be passed through the cell. As we saw in Chapter 4 (Example 4.7), the net electrolysis at a small electrode (*e.g.*, a d.m.e.) in an unstirred solution is negligible, even after several hours. Thus our first concern is to see how the current can be increased to obtain significant net electrolysis in a reasonable time.

[‡] William Cruikshank was Lecturer in Chemistry at the Royal Military Academy, Woolwich, from 1788 until 1804. Little is known of Cruickshank's life, but he made important contributions to early electrochemistry and to the chemistry of simple gases (he is credited with the discovery of carbon monoxide).

One obvious way to increase the current through an electrolysis cell is to stir the solution, thus increasing the rate of mass transport. In a stirred solution, the diffusion layer thickness can be reduced to as little as tens of microns. Suppose that the electrode process is

$$O + n\,e^- \rightarrow R$$

and that a linear concentration gradient in O is set up across a diffusion layer of constant thickness. If the potential of the working electrode is sufficiently negative that the concentration of O is zero at the electrode surface, then the current is given by eq (4.14b),

$$i = nFAk_DC_O^* \qquad (4.14b)$$

where $k_D = D/x_D$ is the mass transport rate constant. The number of moles of O reduced is related to the charge transferred by

$$dn_O = -\frac{dQ}{nF}$$

Since $dQ = i\,dt$ and $dn_O = VdC_O^*$ (V is the solution volume), the change in bulk concentration on passage of current for a time interval dt is

$$dC_O^* = -\frac{i}{nFV}\,dt$$

Substituting eq (4.14b) for i, we have

$$dC_O^* = -(Ak_DC_O^*/V)dt$$

Integration of this expression gives

$$C_O^*(t) = C_O^*(0)\exp(-kt) \qquad (6.1)$$

where

$$k = k_DA/V \qquad (6.2)$$

Thus we expect the bulk concentration C^* to fall exponentially with a rate proportional to the rate of mass transport and to the ratio of the electrode area to the solution volume. Thus if the process is to be completed in minimum time, the general strategy for exhaustive electrolysis should include efficient stirring and a high area–to–volume ratio. Since the stirring must be particularly efficient at the working electrode surface, careful attention to electrode geometry is required for optimum performance.

Example 6.1 Suppose that 250 mL of 0.1 M $CuSO_4$ is electrolyzed using a cathode of area 250 cm^2 and an initial current of 5 A. How long will it take to remove 99% of the copper from solution? 99.99%? Assume that the cathode process is rate limiting.

From eq (4.14b), we obtain the diffusional rate constant

$$k_D = i/2FAC_O{}^*$$

Substituting the electrode area and the initial current and concentration, we have

$$k_D = 1.04 \times 10^{-5} \text{ m s}^{-1}$$

With the area–to–volume ratio of 100 m^{-1}, eq (6.2) gives

$$k = 1.04 \times 10^{-3} \text{ s}^{-1}$$

To reduce the copper concentration from 0.1 M to 0.001 M thus requires a time

$$t = -[\ln(0.01)]/k = 4440 \text{ s} \quad (74 \text{ minutes})$$

To reduce the concentration to 10^{-5} M amounts to another factor of 100 decrease in concentration and thus will require another 74 minutes or 148 minutes in all.

Electroseparation

Electrolysis is often used to separate the components of a mixture. For example, a desired metallic component might be plated out at the cathode or a nonmetal could be oxidized to a gas or deposited as an insoluble oxide at the anode. Alternatively, we might try to selectively remove an impurity from solution by oxidation or reduction.

The factors which govern the efficiency of electroseparations are important in both the analytical and synthetic applications of electrolysis. If another electrode process competes with the one of interest, then nonproductive current flows. Electrode side reactions lead to departures from a simple application of Faraday's laws and thus invalidate analytical procedures based on the proportionality of analyte consumed to charge passed. Side reactions also introduce unwanted contaminations in electrosynthetic methods and constitute an important waste of electrical energy in industrial applications. A useful measure of the significance of competing electrode processes is the *current efficiency*, defined by

$$\phi \;=\; \frac{n \;\times\; \text{moles of product formed}}{\text{Faradays of charge passed}} \tag{6.3}$$

where n is the numbers of Faradays consumed per mole of product formed.

As a simple model for separation efficiency, we consider two couples,

$$O_1 + n_1\,e^- \;\rightarrow\; R_1$$

$$O_2 + n_2\,e^- \;\rightarrow\; R_2$$

both involving solution species only, and assume reversible behavior. We then ask what separation between standard (formal) potentials is required in order for one species, say O_1, to be 99.9% reduced to R_1 without reducing O_2 by more than 0.1%. The Nernst equation can be written for the first couple as

$$E = E_1^{\,\circ} + \frac{RT}{n_1 F}\,\ln\frac{C_{O1}}{C_{R1}}$$

Defining X_1 as the fraction in the reduced form

$$X_1 = \frac{C_{R1}}{C_{R1} + C_{O1}}$$

the Nernst equation can be reformulated:

$$E = E_1^\circ + \frac{RT}{n_1 F} \ln \frac{1 - X_1}{X_1} \tag{6.4a}$$

and similarly for species 2,

$$E = E_2^\circ + \frac{RT}{n_2 F} \ln \frac{1 - X_2}{X_2} \tag{6.4b}$$

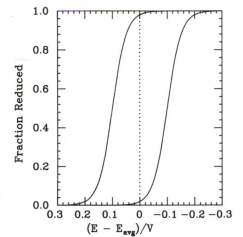

Figure **6.1** Optimum separation is achieved in electroseparations when the potential is set midway between half-wave potentials.

The fractions X_1 and X_2 are plotted *vs.* potential in Figure 6.1 for a difference in standard potential, $E_1^\circ - E_2^\circ = 0.20$ V and $n_1 = n_2 = 1$. It is fairly easy to prove (see Problems) that, when $n_1 = n_2$, the optimum separation is achieved, (*i.e.*, $X_1 - X_2$ is maximized) when

$$E = \tfrac{1}{2}(E_1^\circ + E_2^\circ) \tag{6.5}$$

When this condition is met, $X_2 = 1 - X_1$ and

$$E_1^\circ - E_2^\circ = \frac{RT}{2nF} \ln \frac{X_1}{X_2} \tag{6.6}$$

or, at 25°C,

$$E_1^\circ - E_2^\circ = \frac{0.1183}{n} \log \frac{X_1}{X_2}$$

Thus for $X_1/X_2 = 999$ (*i.e.*, $X_1 = 0.999$, $X_2 = 0.001$), the difference in standard potentials must be $0.355/n$ V.

In practice the situation is usually more complicated. Processes are rarely strictly nernstian when large currents are flowing, and if the interfering couple has very slow electrode kinetics, the potential separation can be considerably smaller than predicted from eq (6.6). If one or both of the couples involves deposition of a metal on the cathode, the activity of the plated metal initially may be proportional to the fraction reduced, but as a monolayer of atoms is formed on the cathode, the activity approaches unity, *i.e.*, the electrode surface approachs the property of the pure plated metal. In the limit of $a_R = 1$, the electrode potential is

$$E = E° + \frac{RT}{nF} \ln C_O$$

so that the electrode potential for small C_O may be rather more negative than expected from eq (6.4a).

Despite these complications, the simple model is useful in predicting electroseparation efficiencies when both couples have similar properties. Thus if both couples involve the deposition of a metal and they have similar rates, eq (6.6) gives a semiquantitative guide to the required difference in standard potentials.

6.2 ANALYTICAL APPLICATIONS OF ELECTROLYSIS

The electroanalytical techniques discussed in Chapters 1, 3, and 4 may be divided into two general classes: (1) titration methods where the endpoint is detected electrochemically, *e.g.*, potentiometric, conductometric, and amperometric titrations; and (2) methods in which little or no chemical transformation occurs and the electrochemical measurement gives a direct measure of concentration, *e.g.*, direct potentiometry, polarography, chronopotentiometry, etc. Most analytical methods based on electrolysis are similar to titration methods in that the analyte undergoes a stoichiometric chemical transformation, but here the role of electrochemistry is to affect the transformation. The equivalence point of the transformation may or may not be determined electrochemically. In some methods, the reaction is judged to be complete when the electrolysis current falls to zero, while other methods use an independent endpoint detection method (visual, spectrophotometric, potentiometric, etc.).

We will discuss four general methods. *Electrogravimetry* involves the controlled potential reduction of a solution of a metal salt which is terminated when the current falls to near zero; the analysis is based on the increase in weight of the metal–coated cathode. Thus electrolysis is used solely as a means of chemical transformation. *Controlled potential coulometry* also detects the endpoint by the decay of current to zero, but here the analysis is based on the net charge transferred. *Constant current coulometry* detects the endpoint by some method independent of the electrolysis process and bases the analysis on the net charge transferred. Finally, a technique known as *stripping voltammetry* uses electrolysis to deposit the analyte on an electrode surface (or in a mercury drop) and then determines the surface (or amalgam) composition by a voltammetric technique.

Electrogravimetry

Although the first electroplating experiments were reported by Cruikshank in 1801, it was not until 1864 that Wolcott Gibbs[‡] adapted the techniques of electroplating to quantitative chemical analysis *(3)*. Nonetheless, electrogravimetry is the oldest of electroanalytical techniques, having become well established by the end of the nineteenth century. In electrogravimetric analysis, a solution of a metal salt is exhaustively electrolyzed, plating the metal onto a previously weighed cathode. The increase in weight of the cathode affords a direct determination of the metal content *(4,D2)*.

As an example of the technique, consider the electrogravimetric determination of copper. An acidic solution of a copper salt is electrolyzed using platinum electrodes. Metallic copper plates out on the cathode and we suppose that oxygen is produced at the anode. Thus the cell reaction is

$$Cu^{2+}(aq) + H_2O \rightarrow Cu(s) + \tfrac{1}{2}O_2(g) + 2H^+(aq)$$

[‡] Oliver Wolcott Gibbs (1822–1908) taught at City College of New York from 1849 until his appointment as Rumford Professor at Harvard in 1863. The following decade saw many contributions to the analytical chemistry of the platinum metals as well as the development of electrogravimetry. A curricular change at Harvard in 1871 effectively ended Gibbs' career as a research scientist.

The reversible potential of the cell is given by the Nernst equation

$$E = E° - \frac{RT}{2F} \ln \frac{[H^+]^2 P(O_2)^{\frac{1}{2}}}{[Cu^{2+}]}$$

where $E° = -0.889$ V. When $[H^+] = 1$ M, $P(O_2) = 0.2$ bar and $[Cu^{2+}] = 0.1$ M, $E = -0.91$ V. For every tenfold reduction in the copper concentration, the equilibrium cell potential becomes more negative by about 30 mV. Thus when the Cu^{2+} concentration is reduced to 10^{-4} M, $E = -1.00$ V. Because the evolution of O_2 is inherently slow, an activation overpotential of about 0.4 V is required to give reasonable current; iR drop in the electrolyte solution is typically about 0.1 V so that an applied potential of $(-1.0 - 0.4 - 0.1) = -1.5$ V is required to deposit the copper. With this applied potential, we might expect hydrogen evolution since the standard potential for the decomposition of water is -1.23 V. Some hydrogen evolution may occur early in the experiment, but the exchange current density for reduction of H^+ is about 200 times slower on a copper surface than on platinum. Thus hydrogen evolution is expected to be negligible once the cathode surface is coated with copper.

If the electrodeposition is carried out at constant cell potential (say -1.5 V), the cathode potential will become more negative late in the experiment when the current density decreases. This effect results from the decrease in magnitude of the solution iR drop and anode overpotential with decreasing current. Thus some H^+ may then be reduced. If copper is being plated out from a solution containing other reducible metal ions, the more negative cathode potential may lead to plating of other metals, such as lead. A positive gravimetric error would result. For this reason, electrogravimetry is normally carried out in a three–electrode cell (see Section 4.3) with the cathode potential controlled such that only the metal of interest is plated out.

The nature of the metal deposit on the cathode in electrodeposition turns out to be critically dependent on the identity of the metal species in solution. It has been found that addition of complexing agents such as tartrate ion (for Cu^{2+}) or cyanide ion (for Ag^+) results in dense, lustrous deposits, relatively free of inclusions, which adhere well to the platinum cathode.

Hydrogen evolution at the cathode is generally undesirable since hydrogen atoms tend to be included in the deposit, leading directly to a gravimetric error, but more important, the deposit is often flaky or spongy and thus more prone to solvent or electrolyte inclusions or to loss of metal particles during washing. Hydrogen reduction can be suppressed either by buffering the solution to pH 5–6 or by controlling the cathode potential.

If no more easily oxidizable species is present, water will be oxidized at the anode. In solutions containing halide ions, halogens will be produced at the anode. Complexing agents, added to improve the characteristics of the cathode deposit, are often oxidizable and may also contribute to the anode process. To one degree or another, all these processes are undesirable, either because of the production of large amounts of corrosive gases or the consumption of complexing agents or because the anode potential must be inconveniently high. Furthermore, if the anode products are reduced at the cathode, the current efficiency of the electrolysis is reduced. Thus it is common practice in electrogravimetry to add a species (called an anodic depolarizer) which is easily oxidized to an innocuous product. A common choice is hydrazine, added as the hydrochloride salt, which is oxidized to nitrogen gas:

$$N_2H_5^+ \rightarrow N_2(g) + 5\,H^+ + 4\,e^-$$

Hydroxylamine hydrochloride is also used, but the anode products then are a mixture of nitrous oxide, nitrous acid, and nitrate ion, some or all of which may be reducible at the cathode.

Since electrogravimetry depends only on the completeness of electrolysis and on the increase in cathode weight for its success, current efficiencies of less than 100% can be tolerated. Cathode side reactions have no effect on the results as long as they do not result in inclusions in the deposit or cause loss of metal on washing. Thus as long as the anode process does not produce a reducible metal, the two electrodes need not be isolated; indeed, the resistance of a membrane separating the anolyte from the catholyte would decrease the current and increase the analysis time. Cells for electrogravimetry then can be designed for maximum current with large electrodes and efficient stirring. A typical arrangement is shown in Figure 6.2.

Electrogravimetry has been used most frequently in the analysis of copper–based alloys. By careful control of the cathode potential, the pH and concentration of complexing agents, Cu, Bi, Pb, and Sn can be successively plated from the same solution. For example, electrolysis at –0.30 V (vs. s.c.e.) of a solution, buffered to pH 5 and containing 0.25 M tartrate, deposits copper. Further reduction of the same

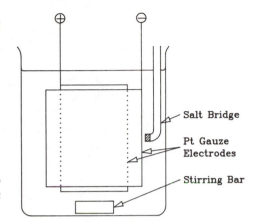

Salt Bridge

Pt Gauze
Electrodes

Stirring Bar

Figure 6.2 Electrolysis cell for electrogravimetric determination of metals.

solution at –0.40 V deposits bismuth, and at –0.60 V, lead. If the solution is then acidified with HCl and electrolysis continued at –0.65 V, tin is deposited. If the cathode is removed, washed, dried, and weighed at each step, the amount of each of the four metals in the original sample is determined.

Like any gravimetric method, electrogravimetry works well only when accurately weighable amounts of the analyte are present in the sample. The techniques lacks the sensitivity required for trace metal analysis, but it may still be useful as the first step in the analysis of an alloy. After removal of the bulk constituents by electrolysis, the trace constituents can be analyzed by more sensitive electroanalytical methods.

Constant Potential Coulometric Analysis

Electrogravimetric analysis uses the gain in weight of the cathode as the analytical sensor in determining the amount of metal ion in a solution. While the method works well for copper and certain other metals, it is limited to the analysis of metals which can be plated out and weighed. An alternative approach is to base the analysis on Faraday's laws of electrolysis, *i.e.*, to measure the charge passed through the electrolysis cell. Thus exhaustive electrolysis of n_O moles of species O,

$$O + n\,e^- \rightarrow R$$

should consume

$$Q = n_O n F \text{ coulombs}$$

If Q is measured and n is known, n_O can be determined. The major limitation on this strategy is that the current efficiency must be very near 100%. That is, all the current passed through the cell must go to the reduction of O; no electrode side reactions can be tolerated. Although the basis for coulometric analysis has been around since 1833, it was not until 1938 that Szebelledy and Somogyi (5) suggested coulometry as a general analytical technique. Extensive work on perfection of analytical methods really began only after World War II.

The requirement of 100% current efficiency generally means that the anode and cathode must be separated. If the product of the cathode reaction is a solution species, it must not be reoxidized at the anode and anode products must not be reduced at the cathode. Thus, with a few exceptions, coulometric methods employ cells in which the anode and cathode are separated by a salt bridge, membrane, diaphragm or some such barrier to prevent mixing of anolyte and catholyte. A typical cell design is shown in Figure 6.3.

Until relatively recently, an electrochemical coulometer was used for the measurement of total charge passed in an electrolysis experiment. A coulometer is simply an electrolysis cell designed to operate at 100% current efficiency to give a weighable cathode deposit, a titratable solution product, or a gas, the volume of which can be measured using a gas buret. The ultimate device was (and for very precise work, still is) the silver coulometer. In this device, silver is deposited on the cathode which can be removed, washed, dried, and weighed. Other practical coulometers employed the electrolysis of water,

$$2 \, H_2O \rightarrow 2 \, H_2(g) + O_2(g)$$

to produce a mixture of hydrogen and oxygen gases, or the electrolysis of an aqueous acidic solution of hydrazine,

$$N_2H_5^+ \rightarrow H^+ + 2 \, H_2(g) + N_2(g)$$

to produce a mixture of hydrogen and nitrogen. In both the H_2/O_2 and the H_2/N_2 coulometers, 0.75 mole of gas is produced per mole of electrons, so that, at 298 K, 1 bar pressure, 0.193 mL of gas is obtained per coulomb of charge passed through the coulometer. If gas

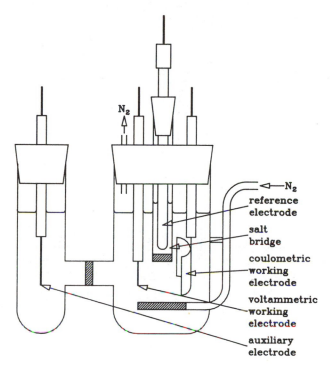

Figure 6.3 Cell for controlled potential coulometric analysis.

volumes can be measured with a precision of ± 0.01 mL, a measurement of charge with $\pm 0.1\%$ precision would require passage of at least 50 coulombs.

Nowadays, electronic instrumentation allows the measurement of much smaller charges with comparable precision and considerably less bother. There are two common approaches to the problem. The most straightforward arrangement is to record a current–time curve using a strip–chart recorder. Graphical integration of the current–time curve then gives the charge passed through the cell. This procedure works well but is tedious in routine applications. If such measurements are to be made frequently, it is more convenient to use an electronic coulometer built around a voltage integrator circuit similar to that used to generate a voltage ramp (see Figure 4.6). In the coulometer circuit shown in Figure 6.4, the cell current is divided so that

$$i_{cell} = i_1 + i_2$$

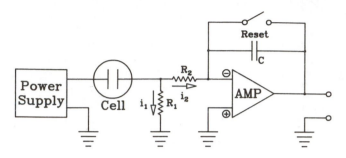

Figure 6.4 Electronic coulometer based on an operational amplifier integrator circuit.

Because of the capacitive feedback loop, the potential at the inverting input of the operational amplifier is a virtual ground, 0 V. Thus the potentials across resistors R_1 and R_2 must be equal:

$$i_1 R_1 = i_2 R_2$$

With this relationship, we obtain

$$i_{cell} = i_2 \frac{R_1 + R_2}{R_1}$$

Since the input impedance of the operational amplifier is very large, the current i_2 must produce a charge: Q_2 on the capacitor C:

$$i_2 = \frac{dQ_2}{dt}$$

But the potential at the output is proportional to this charge

$$Q_2 = C\Phi_{out}$$

Thus

$$i_2 = -C \frac{d\Phi_{out}}{dt} = \frac{R_1 i_{cell}}{R_1 + R_2}$$

Rearranging, we have

$$i_{cell}dt = - \frac{R_1 + R_2}{R_1} C \, d\Phi_{out}$$

and integration gives

$$Q_{cell} = \int_0^t i_{cell}dt = - \frac{R_1 + R_2}{R_1} C \int_0^{\Phi_{out}} d\Phi_{out}$$

or

$$Q_{cell} = - \frac{R_1 + R_2}{R_1} C \, \Phi_{out}$$

By selecting appropriate values of R_1, R_2, and C, charges ranging from less than $1 \, \mu C$ up to about 1 C are easily measured. The method requires accurately known resistances R_1 and R_2 and capacitance C as well as an accurate potential measurement.

Controlled potential coulometric analysis (6,7) can be used in much the same way as electrogravimetry to determine reducible metal ions. Controlled potential coulometry at a mercury cathode is a particularly attractive variation. The advantages are that competition from hydrogen reduction is negligible (because of the large overpotential) and that most metals dissolve in the mercury to form amalgams.

The coulometric method has considerably greater scope than electrogravimetry. In particular, it is not limited to reduction of metal ions but can be used for any electrode process which can be operated at near 100% current efficiency. For example, controlled potential coulometry has been used to determine trichloroacetic acid. The reduction

$$Cl_3CCOOH + H^+ + 2 \, e^- \rightarrow Cl_2CHCOOH + Cl^-$$

is sufficiently well separated from the reduction of di- and monochloroacetic acid that a mixture of these acids can be accurately analyzed, a feat impossible to contemplate with a pH titration.

Coulometry is not limited to reductions—the analytical reaction can just as well occur at the anode. Thus controlled potential coulometry at a silver anode is a convenient method for the determination of halide ions or other species forming insoluble silver salts. In effect, the Ag^+ generated coulometrically acts as a titrant for halides just as in the classical Volhard titration and its variants.

Controlled potential coulometry is commonly used in mechanistic electrochemical studies to determine the number of electrons involved in an oxidation or reduction. Determination of n from a measured voltammetric diffusion current depends on a reasonably accurate value for the diffusion coefficient. When n might be either 1 or 2, D need be known only to $\pm 10\%$ or so to get a satisfactorily unambiguous result. When n is larger or when slow following chemical steps lead to other electroactive species, the results from voltammetry may be less satisfactory. A coulometric measurement gives a value for n which includes the effects of follow–up chemical reactions but it does not depend on the diffusion coefficient, electrode area, or other unknowns (other than the current efficiency).

Example 6.2 An extreme case of voltammetric uncertainty arises in the polarographic reduction of picric acid. The diffusion current constant is

$$I_D = 27 \, \mu A \, mM^{-1} (mg \, s^{-1})^{-\frac{2}{3}} s^{-\frac{1}{6}}$$

in 0.1 M aqueous HCl. If we assure that D is about the same as for the 2+ metal ions of Table 4.1, comparison of the values of I_D suggests that n is about 18 (but obviously with a large uncertainty). Since the overall reduction is expected to yield 2,4,6–triaminophenol,

n = 18 is a reasonable result. Lingane *(8)* determined n coulometrically in 1945 and found n = 17.07. This result was confirmed by others but Meites *(9)* later showed that the value of n depends on the initial concentration of picric acid and on the HCl concentration. When the initial concentration was less than 0.1 mM and/or the HCl concentration is 3 M or more, coulometry gave the expected value, n = 18. For higher picric acid concentrations and/or lower HCl concentrations, smaller values of n were obtained. Apparently, side reactions compete with protonation of radical intermediates. The dependence on total picric acid concentration suggests that a radical coupling reaction is important and Lingane suggested that the eventual product is a substituted hydrazobenzene,

but other reactions must also be important since Meites found values of n less than 17 under some conditions.

Constant Current Coulometry

There are a number of coulometric methods which employ an electrolysis cell operated at constant current rather than constant working electrode potential. One might think that such an arrangement could not work. Surely, the analyte would be quickly exhausted near the electrode and the potential would swing to begin consuming the solvent or supporting electrolyte. The result would be analogous to chronopotentiometry (Section 4.3), where a constant current is passed through an unstirred solution. Perhaps (we might reason) the method could be made to work if the current were so small that mass transport could keep up with consumption at the electrode, even when the analyte is nearly exhausted in the bulk solution. However, to make this approach work, the current would have to be small and the time required for electrolysis would then be impossibly long.

There is a trick involved. Constant current coulometry invariably involves another electrode couple, the product of which reacts rapidly and stoichiometrically with the analyte. For example, suppose that we want to determine Fe(II) coulometrically. A large excess of Ce(III) is added to the cell and a constant current is passed. Initially, the Fe(II) concentration is high enough that the oxidation

$$Fe^{2+} \rightarrow Fe^{3+} + e^-$$

is the major contributor to the current. However, as the Fe(II) concentration begins to drop, the potential swings positive and Ce(III) begins to be oxidized:

$$Ce^{3+} \rightarrow Ce^{4+} + e^-$$

However, Ce(IV) reacts rapidly with Fe(II),

$$Ce^{4+} + Fe^{2+} \rightarrow Ce^{3+} + Fe^{3+}$$

so the effect is as if Fe(II) were oxidized directly. As the Fe(II) concentration drops, more and more of the current is carried by the Ce(III) oxidation process. The net effect, nonetheless, is the oxidation of Fe(II), at least up to the point that the Fe(II) is entirely consumed. After the equivalence point, current continues to flow, of course, and the coulometer continues to run, clocking up the charge passed. Clearly, some means is required to note the time at which the last Fe(II) was consumed.

A moment's reflection will show that the situation is exactly analogous to a titration in homogeneous solution of Fe(II) with Ce(IV) solution added from a buret. What is needed is a means of detecting the endpoint of the titration. In this example, the obvious solution is to add another pair of electrodes—an indicator electrode and a reference electrode—to the anode compartment of the electrolysis cell. If the potential of this pair is monitored, it should show a sharp positive swing when the last of the Fe(II) is consumed and the Ce(IV) concentration starts to increase.

A constant current coulometry experiment is often referred to as a *coulometric titration* to emphasize the similarity to ordinary volumetric analysis. There is another analogy which should be pointed out. The overall process can be written (for a net reduction)

$$O + n\,e^- \rightarrow R$$

$$R + A \rightarrow O + B$$

where A is the analyte and R is the internally generated titrant. This electrode process is identical to the EC' (catalytic) process discussed in Section 5.8. In the voltammetric context, we supposed that A was not reduced at the electrode because of slow electron transfer kinetics. Here the rate–limiting step is mass transport, but in either case, coupling of the O/R and A/B redox couples results in a catalytic current enhancement.

Instrumentation for constant current coulometry can be quite simple. A satisfactorily constant current can be produced by

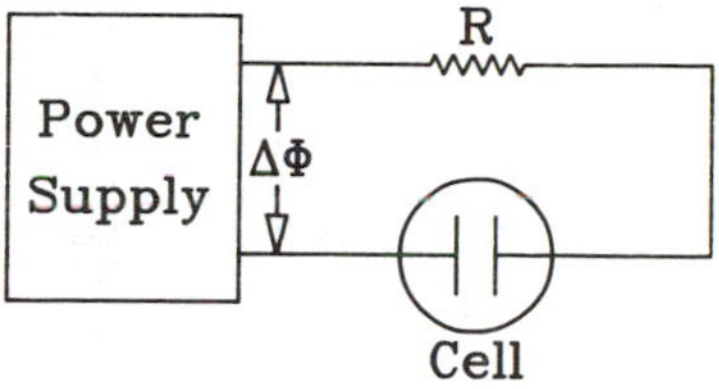

Figure 6.5 A simple circuit for constant current electrolysis.

connecting a high–voltage d.c. power supply across the cell in series with a resistance R as shown in Figure 6.5. The current through the cell is

$$i = \frac{\Delta\Phi}{R + R_{cell}}$$

If R is much larger than R_{cell}, the current will be nearly independent of the cell resistance and thus nearly constant. For example, if $\Delta\Phi = 500$ V and $R = 500$ kΩ, then $i = 1$ mA when the cell is out of the circuit. When $E_{cell} = -1$ V and $R_{cell} = 1$ kΩ, $i = 0.996$ mA, only 0.4% different. Variations of E_{cell} and R_{cell} during electrolysis will cause similarly small changes in i.

The charge passed through the cell generally is determined by measuring the time between the application of current and the appearence of the titration endpoint, $Q = it$. If 1% analytical accuracy is acceptable, a current variation on the order of a few tenths of a percent is tolerable. If more constant current is required, an operational amplified–based constant current source (a galvanostat) is the next step up in sophistication. A galvanostat circuit is shown in Figure 4.6, but this circuit is not well adapted to passage of large currents since the reference source supplies the current. A better high–current circuit is shown in Figure 6.6. Here the current is supplied entirely by the operational amplifier. As usual with operational amplifier feedback circuits, the potential at the inverting input is at virtual ground (0 V). Thus the potential across the resistor R must be equal to the battery potential $\Delta\Phi$, and the current through R must be $i = \Delta\Phi/R$. But since the input impedance of the operational amplifier is high, virtually all the current through R flows through the cell, independent of the cell's internal resistance and potential. Thus provided that R_{cell} is much less than the input impedance of the operational amplifier and $i(R_{cell} + R)$ is within the range of the operational amplifier output voltage, the circuit will supply a constant

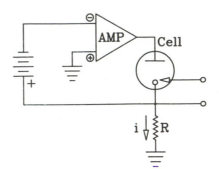

Figure   6.6   Constant
current source based on
an operational amplifier.

current.   With the addition of a reference electrode, the working
electrode potential can be monitored if desired.

Once we realize that constant current coulometry is equivalent to
a titration, the possibilities open up in all directions.   We have
mentioned the electrochemical generation of Ce(IV), but many other
oxidants can be produced.   Thus Mn(III) or Ag(II) can be obtained by
oxidizing Mn(II) or Ag(I) solutions.   Chlorine, bromine, or iodine can be
generated from Cl$^-$, Br$^-$, or I$^-$ solutions.   The halogens are excellent
oxidizing agents, but, because of their volatility, are relatively difficult
to use in volumetric analysis.   Similarly, reductants such as Cr(II) or
Ti(III) can be produced (from Cr(III) or Ti(IV), respectively).
Coulometry is not even limited to generation of oxidants or reductants.
Hydroxide or hydrogen ions can be produced by reducing or oxidizing
water at a Pt electrode, so that acid–base titrations are possible.
Precipitation titrations using Ag(I) or Hg(I) can be done coulometrically
as well.

Coulometric titrations have some significant advantages over
volumetric methods.   We have mentioned the use of titrants which
would be difficult to use under ordinary circumstances, but there are a
number of other considerations.   Since the titrant is generated
internally and only in the required amounts, there is some savings in
expensive reagents; furthermore, no standardization of titrant is
necessary.   Coulometric methods are much more easily automated
than volumetric procedures.   Since charges on the order of
microcoulombs are easily measured with precision, much smaller
samples can be used in a coulometric titration than in the analogous
volumetric method.   See a review by Curran *(10)* for further details.

Example 6.3 The standard analytical method for the determination of water in organic solvents employs a titration using the Karl Fischer reagent—a pyridine/methanol solution of iodine and sulfur dioxide. The reagent contains the monomethyl sulfite ion *(11)*, formed by the reaction of methanol with SO_2, pyridine acting as a proton acceptor:

$$CH_3OH + SO_2 + C_5H_5N \rightleftharpoons CH_3OSO_2^- + C_5H_5NH^+$$

This species is oxidized by iodine in the titration reaction, which requires one mole of water per mole of iodine:

$$CH_3OSO_2^- + I_2 + H_2O + 2\,C_5H_5N \rightarrow$$

$$CH_3OSO_3^- + 2\,I^- + 2\,C_5H_5NH^+$$

Needless to say, Karl Fischer reagent is moisture sensitive and is difficult to use in normal volumetric analysis. Coulometric generation of iodine (or I_3^-) from KI affords a substantial improvement in the method. The endpoint is normally detected potentiometrically; the platinum indicator electrode responds to the I_3^-/I^- couple.

While potentiometric endpoint detection is probably most common in coulometric titrations, other electrochemical methods are often employed. One particularly simple method is called *biamperometric* endpoint detection. In this method, a small potential (10–100 mV) is imposed between two identical platinum electrodes, and the current through this circuit is measured. If both members of a reversible couple, O and R, are present in solution, a small current (typically less than 100 μA) will flow as O is reduced at the negative electrode and R is oxidized at the positive electrode. When either O or R is absent, no current flows. In the example of the Karl Fischer titration, the I_3^-/I^- couple is reversible, so that current flows when both I_3^- and I^- are present. The $CH_3OSO_3^-/CH_3OSO_2^-$ couple, on the other hand, is irreversible, so that the current is negligibly small up to the endpoint and increases more or less linearly with time beyond the endpoint. This approach is applicable when the titrant is added from a buret; the technique is then called a *biamperometric titration*. The name recalls the relation to amperometric titrations discussed in Section 4.3, and the presence of two polarized electrodes.   The method is applicable whenever the titrant forms a reversible couple and the analyte an irreversible couple (or vice versa).[‡]

Stripping Voltammetry

When a solution containing metal ions such as Pb^{2+}, Cu^{2+}, Cd^{2+}, Zn^{2+}, etc. is subjected to polarographic analysis, the reduced metal dissolves in the mercury drop as an amalgam. The process is reversible and if the potential is made anodic, the metal redissolves in the aqueous solution.   In anodic stripping voltammetry (12–14), a mercury electrode is made the cathode of an electrolysis cell. Current is passed for a few minutes through a stirred solution and metals are deposited in the electrode. The potential is then scanned and the anodic current measured, usually using the differential pulse technique. As the potential reaches the dissolution potential of each metal present, a peak is observed in the current–potential curve. The technique achieves very high sensitivity because, in the electrolysis step, metal ions are concentrated in the mercury electrode. In the most common form of anodic stripping voltammetry, a hanging mercury drop electrode is used. If the solution volume is 25 mL and the electrode volume is 5 μL, exhaustive electrolysis would increase the concentrations by a factor of 5000. In practice, electrolysis is far from exhaustive (it would take too long), but a large concentration enhancement is achieved. Since ordinary applications of differential pulse polarography (see Sections 4.5 and 4.8) have sensitivities down to less than 1 μM, a concentration enhancement on the order of 10^3 pushs the sensitivity limit down into the nanomolar range. Clearly the concentration enhancement could be even greater if the volume of the mercury electrode were smaller and electrolysis would be both faster and more complete if the electrode area were larger or if transport were more efficient. A rotating disk electrode, in which in which a thin film of mercury is deposited on a platinum or glassy carbon base, achieves this goal, pushing the sensitivity of the technique down to

---

[‡] One of the earliest applications of the biamperometric method was in the titration of iodine solutions with standard sodium thiosulfate solution from a buret. The $S_2O_3^{2-}/S_4O_6^{2-}$ couple is irreversible, so that a steadily decreasing current is observed up to the endpoint, with essentially zero current beyond. Such a titration was called a *dead–stop titration*, but this name is inappropriate when the titrant furnishes the reversible couple.

concentrations of 10^{-11} M!

Since the controlled potential electrolysis in the concentration step is not exhaustive, and differential pulse peak heights depend on a number of experimental parameters, anodic stripping voltammograms require calibration and careful reproduction of calibration conditions in the analysis. With attention to detail, anodic stripping voltammetry can be the method of choice for trace metal analysis. Sensitivity is usually competitive with atomic absorption spectrophotometry, and, because several metals can be determined in one differential puse voltammetric scan, the analysis can be quicker and easier. Even with care, however, anodic stripping voltammetry is not without some pitfalls. Unanticipated electrode processes sometimes occur during the electrolysis phase which lead to apparently anomalous peaks in the voltammetric scan. Ostapczuk and Kublik *(15)* traced one such peak, observed when solutions containing $AgNO_3$ were subjected to anodic stripping analysis, to hydroxide ion formed in the Ag^+–catalyzed reduction of nitrate ion:

$$NO_3^- + H_2O + 2\ e^- \rightarrow NO_2^- + 2\ OH^-$$

The local increase in pH led to the precipitation of various metal hydroxides on the electrode. On the anodic scan, these gave rise to a mercury oxidation peak:

$$Hg(l) + M(OH)_2 \rightarrow Hg(OH)_2 + M^{2+} + 2\ e^-$$

Another source of trouble is the formation of intermetallic compounds in the mercury phase. Thus reduction of solutions containing gold and cadmium or zinc leads to the intermetallic compounds $AuCd$, Au_3Cd, and $AuZn$. These species oxidize at a different potential than cadmium or zinc amalgam, so that the normal cadmium and zinc peaks in the voltammetric scan are smaller than expected and extra peaks are observed due to oxidation of the intermetallic compounds.

The stripping voltammetry strategy has been most commonly used to concentrate metals in a mercury drop or thin–layer electrode, followed by anodic stripping. The same strategy can often be applied in reverse, depositing the analyte on the electrode surface in an oxidation step; the analysis is then based on a cathodic stripping voltammetric scan.

Example 6.4 Shimizu and Osteryoung *(16)* have used a rotating silver disk electrode to determine traces of sulfide by a cathodic stripping technique. Anodic current is passed, holding the electrode at −0.4 V (*vs.* s.c.e.) for 100–1000 s ($\omega = 260$ rad s^{-1}). A cathodic scan using differential pulse voltammtry then gives a well–defined peak corresponding to the reduction of Ag_2S. The peak current is linear in the initial sulfide concentration over the range 0.01 – 10 μM.

6.3 ELECTROSYNTHESIS

There are an enormous number of applications of electrolysis methods to chemical synthesis, both in inorganic and organic chemistry. Unfortunately, academic synthetic chemists have tended to ignore electrosynthetic methods, and the field has been developed mostly by specialists. This is partly due to a lack of equipment for large–scale electrosynthesis in most synthetic laboratories, but it is more often a result of unfamiliarity with electrochemistry on the part of synthetic chemists. Industrial applications have been relatively more important and it is no accident that research in electrosynthesis has been more common in industrial laboratories than in academia. In this section, we will discuss two of the most important electrosynthetic methods in organic chemistry; in Section 6.4, we will consider some specific examples of industrial electrosynthetic processes. An extensive secondary literature in electrosynthetic methods has developed and the reader is referred to one of the general references *(E1–E7)* for further details.

Reductive Elimination Reactions

Reduction of an organic molecule with a good leaving group generally leads to a neutral radical:

$$RX + e^- \rightarrow R\cdot + X^-$$

Leaving groups include the halide ions (F$^-$, Cl$^-$, Br$^-$, I$^-$) and pseudohalides (CN$^-$, SCN$^-$, etc.) as well as OR$^-$, SR$^-$, NR$_2^-$, RSO$_2^-$, etc. When R is an alkyl group, the ionization step may be concerted with electron transfer, *i.e.*, the C–X bond is somewhat stretched and polarized before electron transfer and ionization is quickly completed after electron transfer. In these cases, the neutral radical is formed at the electrode surface and further reduction to the carbanion nearly always occurs immediately.

$$R \cdot + e^- \rightarrow R^-$$

The strongly basic carbanion usually abstracts a proton from any available source (solvent or supporting electrolyte):

$$R^- + BH \rightarrow RH + B^-$$

Electrolysis thus provides a means for replacement of halogens or other groups by hydrogen. Yields are usually quite good; for example, reduction of benzyl bromide in acetonitrile (Hg cathode, -2.1 V $vs.$ Ag/AgClO$_4$, Et$_4$NBr electrolyte) gives an essentially quantitative yield of toluene.

When the initially formed radical anion is resonance stabilized, the ionization step may be slow enough that the radical anion diffuses away from the electrode. Radical dimerization may then result. For example, reduction of 4–nitrobenzyl bromide in an aprotic solvent gives mostly 4,4'–dinitrobibenzyl:

$$2\ O_2N\!\!-\!\!\bigcirc\!\!-\!\!CH_2Br + 2\ e^- \longrightarrow O_2N\!\!-\!\!\bigcirc\!\!-\!\!CH_2CH_2\!\!-\!\!\bigcirc\!\!-\!\!NO_2 + 2\ Br^-$$

Geminal dihalides, RCHX$_2$, usually reduce in two distinct steps to give RCH$_2$X and then RCH$_3$, presumably via the same mechanism as with alkyl halides. Vicinal dihalides reduce to olefins:

$$RCHX\text{-}CH_2X + 2\ e^- \rightarrow RCH=CH_2 + 2\ X^-$$

The reduction is stereospecific. Thus, for example, reduction of d,l–3,4–dibromobutane in DMF (Hg cathode, -1.1 V $vs.$ s.c.e., Bu$_4$NBF$_4$ electrolyte) gives 100% cis–2–butene. Reduction of the $meso$ isomer under the same conditions gives $trans$–2–butene.

With 1,3– and 1,4–dihalides, cyclization competes with proton abstraction. The detailed mechanism is unclear. It may be that two-electron reduction of one end of the molecule leads to a carbanion,

$$X\text{-}CH_2CH_2CH_2\text{-}X + 2\,e^- \rightarrow X\text{-}CH_2CH_2CH_2^- + X^-$$

which then displaces X^- from the other end, forming a ring. However, there is some evidence to suggest that C–C bond formation is concerted with ionization of the halide ions and electron transfer. The product distribution depends on the reduction potential. Thus, for example, reduction of 1,4–dibromobutane at –1.75 V (vs. s.c.e.) in DMF solution gives 26% cyclobutane and 74% butane; under these conditions, stepwise reduction of the two C–Br bonds seems to occur with protonation competing successfully with cyclization. However, if the reduction is carried out at –2.3 V, the products are 90% cyclobutane and only 10% butane, suggesting that the concerted mechanism may be operative at the more negative potential. Electrochemical reductive elimination processes are particularly useful for formation of strained ring systems. Thus reduction of 1,3–dibromo–2,2–bromomethyl–propane gives first 1,1–di(bromomethyl)cyclopropane and then, on further reduction, spiropentane:

$$Br\text{-}CH_2\text{-}\underset{\underset{CH_2Br}{|}}{\overset{\overset{CH_2Br}{|}}{C}}\text{-}CH_2\text{-}Br \xrightarrow[-2\ Br^-]{+2\ e^-} \bigtriangleup\!\!\!\!<{\overset{CH_2Br}{}}_{CH_2Br} \xrightarrow[-2\ Br^-]{+2\ e^-} \bowtie$$

Kolbe Hydrocarbon Synthesis

The first important organic electrosynthetic reaction was investigated by Kolbe[‡] in 1843 (17). The reaction consists of the oxidation at a Pt anode of partially neutralized carboxylic acids in DMF or methanol solution. One–electron oxidation presumably gives an acyloxy radical intermediate,

$$RCO_2^- \rightarrow RCO_2\cdot + e^-$$

which rapidly loses CO_2 and dimerizes to the hydrocarbon product

---

[‡] Adolf Wilhelm Hermann Kolbe (1818–1884), Professor of Chemistry at Marburg and Leipzig, made very important contributions in organic chemistry and to the development of the concept of radicals. However, he never accepted the idea of structural formulas and wrote some of the most vitriolic polemics in the history of science attacking Kekulé, van't Hoff, and Le Bel.

$$RCO_2\cdot \rightarrow R\cdot + CO_2$$

$$2\,R\cdot \rightarrow R\text{–}R$$

Yields vary from 50 to 90%, quite a bit higher than might be expected considering the reactions open to an alkyl radical; indeed, there is good reason to believe that the radicals are adsorbed on the electrode surface and protected to some extent from the solvent (from which hydrogen abstraction might be expected). Three kinds of side reactions compete with radical dimerization:

(1) In cases where radical rearrangement can occur, a variety of coupling products is obtained. Thus, for example, a β,γ-unsaturated acid leads to an allylic radical intermediate,

$$R\text{–}CH{=}CH\text{–}CH_2\cdot \leftrightarrow R\text{–}\overset{\centerdot}{C}H\text{–}CH{=}CH_2$$

and three coupling products are obtained:

$$R\text{–}CH{=}CH\text{–}CH_2\text{–}CH_2\text{–}CH{=}CH\text{–}R$$

$$CH_2{=}CH\text{–}CHR\text{–}CHR\text{–}CH{=}CH_2$$

$$R\text{–}CH{=}CH_2\text{–}CHR\text{–}CH{=}CH_2$$

(2) When the radical intermediate is secondary or tertiary, or substituted such that the carbonium ion is stabilized, further oxidation takes place:

$$R\cdot \rightarrow R^+ + e^-$$

and the eventual products are derived from carbonium ion reactions.

(3) In cases where the acyloxy radical is stabilized by conjugation (aromatic or α,β-unsaturated acids), Kolbe products are not obtained, most likely because the acyloxy radical survives long enough to escape from the electrode surface. Thus benzoic acid gives benzene under Kolbe conditions, presumably through loss of CO_2 and hydrogen abstraction from the solvent by the resulting phenyl radical.

6.4 INDUSTRIAL ELECTROLYSIS PROCESSES

Electrolysis has found many important applications in the chemical, metallurgical, and metal finishing industries. In this section, we will examine representative examples from four classes of industrial processes:

(1) Electroplating and other metal finishing techniques;

(2) Reduction of ores and purification of metals;

(3) Production of inorganic chemicals; and

(4) Electrosynthesis of organic chemicals.

In an industrial process, the object is to produce a chemical which has an economic value greater than the sum of the costs of raw materials, energy, labor, and capital—in short, to make a profit. An inefficient process might be acceptable if the product is produced a few grams at a time but would be economically ruinous if thousands of tons are needed. Thus several figures of merit are important in industrial electrochemistry.

As in ordinary synthetic chemistry, *percent yield* measures the conversion of the limiting reagent to the desired product:

$$\% \text{ yield} = \frac{\text{moles of starting material converted to product}}{\text{moles of starting material consumed}} \times 100\%$$

When a mixture of products is obtained, it is sometimes more useful to express the yield in terms of the product selectivity:

$$\% \text{ selectivity} = \frac{\text{moles of desired product}}{\text{total moles of all products}} \times 100\%$$

In some processes, the overall yield may be high but only a fraction of starting material is consumed during one pass through the electrochemical cell. The unreacted starting material must then be separated from the product and recycled. The *percent conversion per pass* is an important consideration since separation and recycling add to the cost of the process.

In some electrochemical processes, electrode side reactions such as the production of hydrogen at the cathode or oxygen at the anode may not consume starting material or cause separation problems. Such side reactions draw current, however, and thus consititute a nonproductive use of electrical energy. Thus the *current efficiency*, ϕ, eq (6.1), is also an important figure of merit.

Some industrial electrochemical processes use huge amounts of electrical energy, so that the energy consumption per ton of product is an extremely important consideration. If the potential applied to the cell is $-E$, n Faradays are required to produce one mole of product, and the current efficiency is ϕ, then the energy consumption per mole is

$$\text{Energy consumption} = \frac{nFE}{\phi} \text{ J mol}^{-1}$$

If M is the molecular weight (g mol^{-1}) and we express electrical energy in kilowatt–hours (1 kWh = 3600 kJ), then the energy consumption per ton (1 metric ton = 1000 kg) is

$$\text{Energy consumption} = 0.278 \frac{nFE}{\phi M} \text{ kWh ton}^{-1}$$

Energy consumption may increase either because of low current efficiency (electrode side reactions) or because of energy losses in the electrochemical process (larger applied cell potential). The applied cell potential includes the difference of the reversible half–cell potentials, E_c and E_a, the cathodic and anodic overpotentials, η_c and η_a, and the iR drop in the electrodes, leads, and electrolyte solution:

$$E = E_c - E_a - |\eta_c| - |\eta_a| - i(R_{soln} + R_{circuit})$$

Both the overpotentials and the iR drop increase in magnitude with increasing current. Thus the energy cost per ton of product must increase with the rate of production. Materials costs may be more or less independent of the rate, but the costs of labor and capital per ton of product almost always decrease with increasing rate. Thus the optimum current density in an electrochemical process is determined by a trade–off between energy costs on the one hand and labor and capital costs on the other.

Electroplating

An important application of electrodeposition is the plating of a layer of metal on a substrate to improve the appearance of the object plating or to impart hardness or corrosion resistance. Metals commonly used in electroplating include Cr, Ni, Zn, Cd, Cu, Ag, Au, Sn, and Pb. Some alloys such as brass (Cu/Zn) and bronze (Cu/Sn) also can be electroplated.

The basic theory of electroplating is extremely simple. For example, to plate copper on a steel substrate, the object to be plated is made the cathode in an electrolysis cell where a piece of copper is used as the anode. There are many subtleties, however, which must be considered in practice. Thus, in plating copper on steel, for example, we immediately recognize that the reaction

$$Cu^{2+}(aq) + Fe(s) \rightarrow Cu(s) + Fe^{2+}(aq)$$

is spontaneous. To prevent the dissolution of iron from the substrate, the activity of $Cu^{2+}(aq)$ must be reduced by the addition of a complexing agent which coordinates strongly to Cu(II) but much less so than Fe(II).

In electroplating applications, it is usually desirable to deposit a layer of uniform thickness. This requirement is not difficult to meet if the substrate has a simple geometry. If there are holes or recesses, however, a uniform layer can be quite difficult to achieve. When a potential is applied across an electrolysis cell, the potential drop is the sum of several contributions:

(1) the equilibrium anode–solution and solution–cathode potentials (*i.e.*, the equilibrium cell potential)

(2) the activation and polarization overpotentials at the anode and cathode; and

(3) the solution iR drop.

Since the iR potential drop is proportional to the length of the current path, it will be smallest at that part of the cathode which is closest to the anode. Since the total cell potential is constant, this means that the cathodic overpotential will be greatest at that closest part of the cathode. The current density—and the deposited layer thickness—will be greatest, therefore, at exposed parts of the cathode.

Electroplaters refer to the ability of a plating bath to deposit metal at hard to reach spots in terms of the *throwing power* of the bath.[‡] The throwing power of a plating bath can be controlled (to some extent) in three ways: (1) by adding a large concentration of an inert electrolyte, the solution resistance is lowered and differences in iR drop to various points on the cathode surface are reduced; (2) by operating the bath under conditions where $H_2(g)$ evolution occurs at exposed points where the overpotential is large, the iR drop is increased locally, thus reducing the overpotential and slowing the rate of plating at these exposed points; (3) by adding a complexing agent, the electrode process is converted to

$$ML_x^{n+} + n\,e^- \rightarrow M + x\,L$$

It is generally found that complexing agents decrease the slope of a Tafel plot (log i *vs.* η), so that current variations with overpotential are smaller. In addition, throwing power is usually found to be a function of temperature through the temperature dependences of the various rate processes involved. Plating baths with good throwing power usually have optimized all these parameters, most commonly through trial–and–error investigations.

The nature of an electroplated metal deposit can be modified by the addition of organic additives to perform one or more of the following functions: (1) *Wetting agents* facilitate the release of bubbles of $H_2(g)$ from the surface, preventing the occlusion of hydrogen in the deposit and enhancing current density control by hydrogen evolution. (2) *Levelers* are preferentially adsorbed at surface dislocations and sharp corners and thus inhibit current flow at the points at which the current density would otherwise be highest. (3) *Structure modifiers* and *brighteners* change the nature of the deposited layer, perhaps changing the crystal growth pattern or reducing the crystallite size in the deposited layer.

In most commercial electroplating operations, the current efficiency (*i.e.*, the fraction of the current which produces the desired metal deposit) is high, 90% or better. The most notable exception is in chromium plating, where chromium is added to the bath as CrO_3 and the plating process is based on the six–electron reduction of Cr(VI).

[‡] The concept of throwing power can be made quantitative by defining a standard test cell geometry; see Pletcher *(G6)* for further details.

Here Cr(III) is an unwanted by–product, unreducible under normal experimental conditions.[‡] The detailed electrode process is not well understood, but there are apparently several current–consuming side reactions.

Organic polymer coatings can be applied to metal surfaces by a technique *called* electrophoretic painting, *which is closely related to* electroplating. The polymer to be deposited is solubilized with charged functional groups. In anodic electropainting, negatively charged groups, usually $-CO_2^-$, cause the polymer molecules to move toward the anode, where water is oxidized to O_2 and H^+. The local decrease in pH leads to neutralization of the carboxylate groups and the polymer precipitates. In cathodic electropainting, positively charged polymers with $-NH_3^+$ groups are attracted to the cathode, where water is reduced to H_2 and OH^-, the ammonium groups are neutralized, and the polymer deposited on the surface. Electropainted polymer coatings adhere exceptionally well, apparently because of coordination of surface metal atoms by the polar functional groups. The technique is commonly used to impart corrosion resistance to automobile parts.

For further details on electroplating and related techniques, see Pletcher *(G6)* or Lowenheim *(G3)*.

Anodization

The electrochemical formation of a protective oxide layer on a metal surface is called *anodization*. The process is most commonly applied to aluminum but sometimes also to copper, titanium, and steel. The metal object to be anodized is made the anode of an electrochemical cell and current is passed through an acidic electrolyte solution, so that the electrode process (for aluminum) is

$$2 \text{ Al} + 3 \text{ H}_2\text{O} \rightarrow \text{Al}_2\text{O}_3(\text{s}) + 6 \text{ H}^+(\text{aq}) + 6 \text{ e}^-$$

The process is much like electroplating, although the cell polarity is reversed.

The formation of the oxide layer involves a complicated electrode process which tends to be slow, requiring a large overpotential,

––––––––––––––––––––

[‡] This is obviously a kinetic effect, not a result of thermodynamics.

particularly after a thin oxide layer is formed. For this reason, throwing power is not a problem in anodization; the oxide layer grows most rapidly where it is thinnest.

Aluminum is anodized using sulfuric acid, chromic acid, or oxalic acid baths, depending on the nature of the surface desired. Anodized aluminum surfaces can be colored by adsorbing an organic dye on the surface. Alternatively, transition metals can be deposited in the pores of the oxide layer; colors then arise from interference effects.

Electrolytic Aluminum Production

By far the most important electrometallurgical process is the production of aluminum from bauxite ore—hydrated aluminum oxide with silicates, iron oxides, and other impurities. Worldwide production of aluminum amounts to about 2×10^7 tons per year. Since the energy consumption is about 15,000 kWh ton^{-1} (5.4×10^{10} J ton^{-1}), aluminum production is the largest single industrial consumer of electrical energy.

Since aluminum is quite an active metal, the traditional smelting technique used for iron,

$$\text{Fe}_2\text{O}_3 + 3\,\text{C} \rightarrow 2\,\text{Fe} + 3\,\text{CO}$$

will not work and electrolysis is the only practical method. Reduction of Al(III) from aqueous solution is also impossible since hydrogen would be evolved first even from strongly basic solutions. The solution to these restrictions was discovered in 1886 independently by Hall[‡] (in the United States) and Héroult[#] (in France). The Hall–Héroult process makes use of the solubility of alumina in molten cryolite, Na_3AlF_6, to give a conducting solution from which aluminum metal can be obtained

---

[‡] Charles M. Hall (1863–1914) set to work on development of an aluminum smelting process after graduation from Oberlin College. After his success in 1886, he founded a company which became Aluminum Corporation of America (Alcoa).

[#] Paul L. T. Héroult (1863–1914) went to work on the aluminum smelting problem on completion of his studies at Ecole des Mines in Paris (where he had studied with Henri Le Châtelier). He later developed electric furnace techniques for the production of steel.

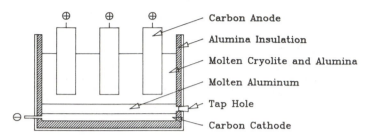

Figure 6.7 Electrochemical cell used in the production of aluminum by the Hall–Héroult process.

at the cathode on electrolysis.

solution (caustic soda to the industrial chemist) under pressure to dissolve the alumina as $NaAl(OH)_4$. After removal of the iron and silicate salts by filtration, hydrated Al_2O_3 is reprecipitated, filtered off, washed, and dried at $1200°C$.

The purified alumina is then added to a bath of molten cryolite at about $1000°C$ up to about 15 weight percent. The electrolysis bath, shown schematically in Figure 6.7, is constructed of steel with alumina insulation and uses carbon anodes. The cathode is molten aluminum (m.p. $660°C$) contacted through steel–reinforced carbon at the bottom of the cell. Aluminum is removed from the cell from time to time. Small amounts of AlF_3 and CaF_2 are normally added to the cryolite to reduce the melting point and increase the conductivity. The exact nature of the species in the cryolite solution is unknown but it is likely that most of the aluminum is in the form of anionic oxyfluorides. The stoichiometry of the cathode process is

$$Al_2O_3 + 6\,e^- \rightarrow 2\,Al(l) + 3\,O^{2-}$$

At the anode, carbon is oxidized to CO_2:

$$C(s) + 2\,O^{2-} \rightarrow CO_2(g) + 4\,e^-$$

Since 0.75 mole of carbon is oxidized per mole of aluminum produced, the preparation of carbon anodes from coal is an important subsidiary activity of an aluminum smelter. Since the anodes are consumed during electrolysis, they are constructed so that they can be gradually lowered into the cryolite melt to maintain an anode–cathode separation of about 5 cm. The oxidation of the carbon anodes provides about half the free energy required for the reduction of Al_2O_3.

The reversible cell potential in the Hall–Héroult process is about -1.2 V and since three Faradays of charge are required per mole of the aluminum, we would expect an electric power consumption of about 350 kJ mol^{-1} (1.3×10^{10} J ton^{-1}). This is only part of the power consumption. The anode process is slow and requires an overpotential of about 0.5 V to achieve a reasonable rate; iR drop in the electrodes and electrolyte adds another 2.5 V, so that the actual working cell potential is nearer -4.2 V. Added to this is the energy cost of heating the cell to 1000°C, purifying and drying the bauxite, and preparing the carbon anodes. As we noted at the outset, aluminum production is a very energy–intensive process.

Other Electrometallurgical Processes

The production of lithium, sodium, and magnesium resembles the Hall–Héroult process in that electrolysis of a molten salt solution is employed. In these cases, chloride salts are electrolyzed and chlorine gas in an important by–product. Sodium, for example, is prepared by electrolysis of a $NaCl/CaCl_2$ eutectic mixture at 600°C. Chlorine is liberated at a graphite anode and sodium at a steel cathode. Sodium is a liquid at 600°C and is considerably less dense that the molten salt mixture, so that it floats to the top of the cell and is taken off into a reservoir. Some calcium is also produced, but this sinks to the bottom of the cell, where it is consumed via the equilibrium

$$Ca + 2\,NaCl \rightleftarrows CaCl_2 + 2\,Na$$

Less active metals can be produced by electrolysis of aqueous solutions. Thus about 10% of world copper production and half the zinc is via hydrometallurgical processes. Small amounts of chromium, manganese, cobalt, nickel, cadmium, silver, gold, gallium, thallium, and indium are also obtained from electrolysis of aqueous solutions. In general, ores are roasted to convert sulfides to oxides,

$$2\,MS + 3\,O_2 \rightarrow 2\,MO + 2\,SO_2$$

and leached with sulfuric acid to separate the desired metal from silicates and aluminosilicates. The resulting metal sulfate solution is then purified, removing traces of less active metals, and then electrolyzed to plate out the desired metal on an aluminum or titanium cathode. The anode is most commonly lead or PbO_2, with a little silver added to catalyze the generation of oxygen.

Some metals obtained by conventional smelting techniques are purified electrolytically. Thus cobalt, nickel, copper, tin, and lead are frequently refined by anodic dissolution and deposition at the cathode of a cell. In the case of copper, which accounts for the largest aqueous electrorefining volume, common impurities are Fe, Co, Ni, Pt, Ag, Au, Zn, Pb, As, Sb, and Bi. The cell potential is controlled such that Pt, Ag, and Au do not dissolve at the anode and these metals eventually fall to the bottom of the cell and are found in the "anode slime". Sn, Pb, Sb, and Bi dissolve at the anode but form insoluble oxides or sulfates and so are also found in the anode slime. Fe, Co, Ni, Zn, and As dissolve and remain in solution but are not reduced at the cathode at the potential provided. The anode slime is collected from time to time and refined to separate the traces of precious metals.

Production of Chlorine and Sodium Hydroxide

Electrolysis of an aqueous sodium chloride solution produces hydrogen at the cathode, chlorine at the anode, and leaves a solution of sodium hydroxide; variations on this process are responsible for most of the world's production of chlorine and sodium hydroxide (caustic soda) and contribute significantly to the production of hydrogen.[‡] In tonnage, this process is the largest electrochemical industry though aluminum production consumes more electric energy.

Although the *chlor–alkali* industry has been in operation since the 1890's, there have been some significant technical modifications to the process. These have consisted both in the development of better electrodes to reduce overpotentials and in changes in the overall strategy of the process. Three distinct approachs to the electrolysis of NaCl brine have been developed and all three are in current use.

The oldest technology divides the overall process into two parts by using a cell with a mercury cathode as shown in Figure 6.8. Since the overpotential for $H_2(g)$ production on Hg is so high, the cathode process is the formation of sodium amalgam,

---

[‡] Hydrogen is also produced from coal or natural gas as "synthesis gas" (a mixture of H_2 and CO) or, when very high purity is required, from the electrolysis of water.

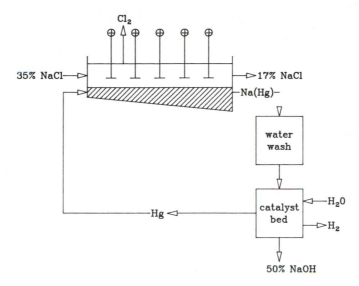

Figure 6.8 Schematic diagram of mercury cell used in the production of chlorine, sodium hydroxide, and hydrogen.

$$Na^+(aq) + e^- \rightarrow Na(Hg)$$

The anode is titanium, coated with RuO_2, Co_2O_3, or other transition metal oxides to catalyze the evolution of chlorine gas,

$$2\ Cl^- \rightarrow Cl_2(g) + 2\ e^-$$

Since O_2 evolution is thermodynamically more favorable than that of $Cl_2(g)$, the anode must have a high overpotential for water oxidation so as to prevent contamination of the Cl_2 product. The reaction of sodium amalgam with water is thermodynamically favorable but is very slow, owing to the large overpotential for H_2 production. Sodium amalgam is withdrawn from the cell at about 0.5% Na content, washed with water to remove NaCl and passed with water through a column packed with graphite impregnated with Fe, Ni, or some other catalyst. Hydrogen is then evolved and collected at the top of the column. Mercury and 50% NaOH solution emerge at the bottom of the column in quite pure form. The NaCl feedstock is typically 50% brine which has been treated with base to precipitate group II and transition metal hydroxides and reacidified to pH 4. The cell normally operates at about 60°C.

The mercury cell technology produces high–purity products with little additional treatment required. However, there is a potentially serious problem of mercury contamination in the plant environment and in the wastewater effluent. Because Na^+ is reduced in the electrolysis step rather than H^+, the cell potential—and thus energy consumption—is significantly higher than should be necessary (in principle) for the electrolysis of NaCl to form $Cl_2(g)$ and $H_2(g)$.

Beginning in the 1950's, environmental concerns prompted the development of another approach, the so–called diaphragm cell. In this method, the cathode material is steel gauze coated with a catalyst to reduce the H_2 overpotential. An asbestos-based diaphragm is used to separate the anode and cathode compartments to prevent mixing of the hydrogen and chlorine gases and to somewhat reduce mixing of the NaCl feedstock on the anode side with the NaOH solution of the cathode side. The arrangement is shown schematically in Figure 6.9.

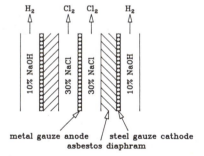

Figure 6.9 Schematic representation of elect– rodes in a diaphragm cell used in the chlor– alkali industry.

While this eliminates the mercury problem, the cell potential is still much larger than theoretical because of an iR drop of about 1 V across the diaphragm. This is not the only problem: (1) OH^- diffusing into the anolyte reacts with Cl_2 to form hypochlorite ion:

$$Cl_2 + 2\ OH^- \rightarrow H_2O + ClO^- + Cl^-$$

reducing the yield of Cl_2; (2) when the anolyte pH goes up, the potential for formation of O_2 at the anode becomes more favorable and contamination of the $Cl_2(g)$ results; (3) to reduce the magnitude of problems (1) and (2), the catholyte cannot be allowed to exceed 10% in NaOH concentration; (4) chloride ion diffusion from the anolyte to the catholyte results in a very considerable contamination of the NaOH solution; (5) to deal with problems (3) and (4), evaporation of the caustic soda solution is required to bring the concentration up to 50% (the usual commercial form of caustic soda), at which point much of the NaCl precipitates; and (6) precipitation of $Ca(OH)_2$ and $Mg(OH)_2$ in the pores of the diaphragm increases the iR drop with time; to minimize this problem, more rigorous brine purification is required than for the mercury cell.

The upshot of all this is that the purity of the NaOH and Cl_2 products from a diaphragm cell is significantly lower and, because of iR drop in the diaphragm and the requirement of an evaporation step, the energy requirement is nearly the same as in a mercury cell. Furthermore, the use of asbestos introduces another potential environmental problem.

The most recent technical development has been the introduction of chlor–alkali cells which incorporate an ion–exchange membrane in place of the asbestos diaphragm. With a cation–permeable membrane, diffusion of OH^- into the anolyte can be drastically reduced, improving the yield and purity of Cl_2. Electrolysis can be continued to produce somewhat more concentrated NaOH solutions, up to about 40%, so that much less evaporation is required to get to 50% commercial caustic soda; because very little Cl^- gets through the membrane, Cl^- contamination of the NaOH is much less serious, and finally, iR drop across the membrane is somewhat less than in a diaphragm cell (but still much greater than in a barrier–free mercury cell).

The major drawback of the membrane cell is that (at present) it is limited in size by the ability to produce and handle large sheets of cation–exchange membrane. To obtain the same product volume, more cells are required and therefore more plumbing and higher capital costs to the manufacturer.

Other Inorganic Electrosyntheses

Although no other electrolytic process approachs the volume of the chlor–alkali process, many other inorganic chemicals are produced electrochemically.

Fluorine. Electrolysis provides the only source of elemental fluorine. An anhydrous mixture of KF and HF (mole ratio 1:2, m.p. 82°C) is used as the electrolyte (K^+ and HF_2^- ions) with carbon anodes and mild steel cathodes. The cell process

$$2 \text{ HF} \rightarrow \text{H}_2(g) + \text{F}_2(g)$$

gives gaseous fluorine and hydrogen. Both gases are contaminated with HF which is removed and recycled by adsorption by solid KF. Fluorine is used in the nuclear industry to produce UF_6 (used in gaseous diffusion isotope separation plants), for the production of SF_6 (a nonflammable, nontoxic gas of low dielectric constant used to prevent voltage breakdown around high–voltage electrical equipment), and for the production of fluorinated hydrocarbons.

Chlorates, bromates, and perchlorates. Sodium chlorate production is based on an undesirable side reaction in the chlor–alkali process. The electrolysis steps are the same,

$$2 \text{ Cl}^- \rightarrow \text{Cl}_2 + 2 \text{ e}^-$$

$$2 \text{ H}^+ + 2 \text{ e}^- \rightarrow \text{H}_2$$

In neutral or basic solution (pH > 6), Cl_2 disproportionates:

$$\text{Cl}_2 + 2 \text{ OH}^- \rightarrow \text{Cl}^- + \text{ClO}^- + \text{H}_2\text{O}$$

Hypochlorite ion also disproportionates:

$$3 \text{ ClO}^- \rightarrow \text{ClO}_3^- + 2 \text{ Cl}^-$$

The last reaction is catalyzed by H^+ and is slow in basic solution. When the pH is in the range 6–7, both disproportionation reactions are relatively slow. However, since the anode process consumes two OH^- ions and the cathode process consumes two H^+ ions, there is a substantial pH gradient across the cell and the rates of the various processes are strongly dependent on stirring and on flow patterns in the electrolyte solution. If hypochlorite ion is formed near the anode, it can be oxidized in a complicated process which generates oxygen with the overall stoichiometry:

$$6 \text{ ClO}^- + 6 \text{ OH}^- \rightarrow 2 \text{ ClO}_3^- + 4 \text{ Cl}^- + 3/2 \text{ O}_2 + 3 \text{ H}_2\text{O} + 6 \text{ e}^-$$

In effect, ClO^- catalyzes the anodic oxidation of water, reducing the current efficiency. This process is thus undesirable and must be suppressed. Efficient stirring of the solution will maintain pH control and thus reduce the rate of hypochlorite formation. In practice, the solution is cycled through a holding tank where the disproportionation reactions can proceed without the complications of the unwanted electrode processes. Other chlorate and bromate salts are produced by similar electrolytic methods.

Electrolysis of sodium chlorate solutions at pH 0–1 results in oxidation to perchlorate:

$$ClO_3^- + H_2O \rightarrow ClO_4^- + 2\,H^+ + 2\,e^-$$

Hydrogen is evolved at the cathode.

Manganese dioxide. Purified pyrolusite provides MnO_2 of adequate quality for most chemical purposes. When MnO_2 of higher purity is required (for Leclanché cells, for example, or for use as an oxidizing agent in industrial organic synthesis), it can be produced by the electrolytic oxidation of Mn(II) in aqueous sulfuric acid solution at a graphite anode. Hydrogen evolution occurs at the cathode.

Potassium permanganate. Pyrolusite (impure MnO_2) is used as a starting material in the production of $KMnO_4$. The ore is ground, slurried in 50% KOH, and air oxidized to manganate ion:

$$2\,MnO_2 + O_2 + 4\,OH^- \rightarrow 2\,H_2O + 2\,MnO_4^{2-}$$

Impurities are filtered off and the manganate ion is oxidized at a nickel anode:

$$MnO_4^{2-} \rightarrow MnO_4^- + e^-$$

The cathode process again is hydrogen evolution, although some MnO_4^- is lost to cathodic reduction.

Potassium dichromate. Potassium dichromate is produced by electrolytic oxidation of Cr(III) in aqueous sulfuric acid solution, using a PbO_2 anode; oxygen evolution is an important competing anodic process. Electrolytic production of Cr(VI) is mostly used in recycling Cr(III) residues from chromate oxidations in industrial organic synthesis or in conditioning chromate plating baths. As such, the scale is usually quite small.

Electrosynthesis of Adiponitrile

The largest volume organic electrosynthetic process presently in use is the production of adiponitrile, a precursor to nylon, from acrylonitrile. The stoichiometric half–cell reactions are

$$2 \text{ CH}_2\text{=CHCN} + 2 \text{ H}_2\text{O} + 2 \text{ e}^- \rightarrow \text{NC-(CH}_2)_6\text{-CN} + 2 \text{ OH}^-$$

$$2 \text{ H}_2\text{O} \rightarrow \text{O}_2(g) + 4 \text{ H}^+ + 4 \text{ e}^-$$

which sum to the overall reaction stoichiometry

$$4 \text{ CH}_2\text{=CHCN} + 2 \text{ H}_2\text{O} \rightarrow \text{O}_2(g) + 2 \text{ NC-(CH}_2)_6\text{-CN}$$

The actual mechanism of the reaction is considerably more complex and probably involves several parallel pathways. Some of the possible

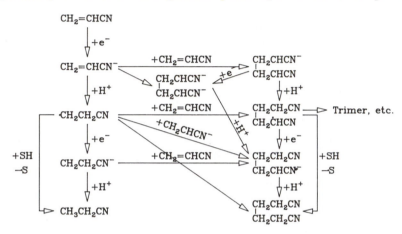

Figure 6.10 Possible reactions occurring in the reduction of acrylonitrile.

reaction pathways are shown in Figure 6.10. The major side reactions which must be minimized are the production of propionitrile (the monomeric reduction product) and trimers and higher polymers.

Monomeric product can be avoided if the initially formed radical anion either dimerizes or reacts with substrate acrylonitrile before it is protonated. This can be achieved if the pH (at least near the electrode surface) is kept quite high. Radical anion dimerization is most desirable since this avoids the neutral radical, which is the most likely route to higher polymers. On the other hand, the pH cannot be allowed to get too high since acrylonitrile reacts with hydroxide ion to form β–hydroxypropionitrile and bis–(2–cyanoethyl)ether. These considerations led to a definite strategy for the process:

(1) The concentration of acrylonitrile should be as high as possible to promote dimerization; in practice an aqueous emulsion is used so

that the aqueous phase is saturated, about 7% by weight. The organic phase is a mixture of acrylonitrile and adiponitrile.

(2) The aqueous phase must be buffered to prevent reaction of acrylonitrile with base; in practice the emulsion contains about 15% Na_2HPO_4.

(3) The electrode surface pH should be high to prevent protonation of the initially formed radical anion. This can be achieved by adsorbing positive ions on the electrode surface (see Section 2.2); the industrial process uses the diquaternary ammonium ion, $Bu_2EtN-(CH_2)_6-NEtBu_2^{2+}$.

(4) Hydrogen evolution at the cathode must be suppressed so that the cathode material should have a large overpotential for H^+ reduction; in practice, cadmium–plated steel is used.

Oxygen is produced at a steel anode, spaced 2 mm from the cathode[‡] to reduce solution iR drop. Corrosion of the anode is a problem since deposition of iron on the cathode would promote hydrogen reduction and a decrease in current efficiency. Additives are used to suppress anode dissolution and EDTA is added to scavenge transition metals before they plate out on the cathode.

The two–phase mixture is cycled rapidly between the electrolysis cell and a reservoir. Solution is continually withdrawn from the reservoir for product isolation, removal of by–products and metal ions, and recycling of unreacted acrylonitrile. The overall process has about 90% selectivity toward adiponitrile. The reversible cell potential is about –2.5 V; the overpotentials and solution iR drop are relatively small at the current density used (about 2000 A m^{-2}) so that the working cell potential is about –3.8 V.

Production of Lead Tetraalkyls

Another major commercial electrosynthetic process is the production of tetraethyl– and tetramethyllead by the oxidation of a Grignard reagent at a lead anode in a mixture of ether solvents.[#] The

---

[‡] About 100 anode–cathode pairs are used in the cell to allow for greater output.

anode reaction stoichiometry is

$$4 \ RMgCl + Pb \rightarrow R_4Pb + 4 \ MgCl^+ + 4 \ e^-$$

At the steel cathode, Mg(II) is reduced:

$$MgCl^+ + 2 \ e^- \rightarrow Mg(s) + Cl^-$$

and the magnesium is used to regenerate the Grignard reagent with excess alkyl chloride,

$$RCl + Mg \rightarrow RMgCl$$

The electrolysis reactor contains many cell assemblies, arranged as concentric vertical tubes as shown in Figure 6.11; lead shot which acts as the anode is fed by gravity down the center tube.

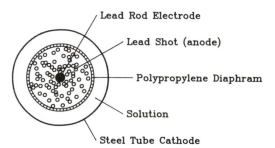

Figure 6.11 Cross section of electrode assembly used in the electrochemical production of lead tetraalkyls.

Careful control of reaction conditions is required since a hydrocarbon is produced if the alkyl chloride concentration is too high:

$$RMgCl + RCl \rightarrow R\text{--}R + MgCl_2$$

If the alkyl chloride concentration is too low, magnesium metal will accumulate on the cathode, eventually bridging the anode–cathode gap and shorting the cell.

\# With the phase–out of leaded gasoline in the United States, this process is decreasing somewhat in importance; tetraethyllead is still used as a motor fuel additive in most of the rest of the world.

Solution flows from a reactor where the Grignard reagent is prepared into the electrolysis cell assemblies; the flow rate must be adjusted to ensure complete reaction of the Grignard reagent by the time the solution exits at the bottom of the cell. Excess alkyl halide is then stripped from the solution and recycled to the Grignard reactor, the $MgCl_2$ is extracted with water, and the ether solvent is separated from the lead tetraalkyl, purified by azeotropic distillation and recycled to the Grignard reactor.

Other Organic Electrosyntheses

Although no other organic processes use electrosynthetic techniques at the scale of the adiponitrile or lead tetraalkyl processes, several other processes are in use. They fall into two categories:

(1) Reduction of aromatic compounds in aqueous acidic solutions at a lead cathode. The initially formed radical anion is rapidly protonated; the resulting neutral radical is then reduced to a carbanion which adds another proton. For example, phthalic anhydride is reduced to 1,2–dihydrophthalic acid, pyridine is reduced to piperidine, and N–methylindole is reduced to N–methyl–2,3–dihydroindole.

(2) Oxidations of organics with inorganic oxidants such as chromic acid, periodate, bromine, or hypobromite. These homogeneous reactions are coupled with a step where the oxidant is regenerated electrolytically. For example, anthroquinone is obtained from the chromic acid oxidation of anthracene and the Cr(III) is cycled through an electrolysis cell to regenerate the oxidant.

6.5 CORROSION

Corrosion is an important problem in all practical uses of metals—in automobiles, bridges, buildings, machinery, pipelines, ships, indeed in most of the works of man. Corrosion consists of the spontaneous oxidation of metals, usually by atmospheric oxygen but also by water or atmospheric pollutants such as SO_2. In simplest terms, we may think of corrosion in terms of the metal oxidation half–cell reaction,

$$M \rightarrow M^{n+} + n\,e^-$$

coupled with the reduction of oxygen

$$O_2 + 2\,H_2O + 4\,e^- \rightarrow 4\,OH^-$$

or the reduction of water

$$2\,H_2O + 2\,e^- \rightarrow H_2 + 2\,OH^-$$

The direction of spontaneous change and the thermodynamically most stable products will depend on the standard reduction potentials of M^{n+}, O_2, and water, and on the pH, O_2 partial pressure, and temperature. The equilibrium thermodynamics of a corrosion problem is most easily visualized in terms of a Pourbaix diagram (see Section 1.3). In the Pourbaix diagram for iron, Figure 6.12, we see that iron is susceptible to oxidation by water or O_2 at all pH values. Thermodynamics thus tells us that corrosion of iron is inevitable in the wet, oxygen–rich environment in which we place our buildings and machinery. The questions then are: (1) How fast does the process occur? and (2) what can we do to make it slower? These questions lead us to a more detailed look at the kinetics and mechanism of corrosion processes.[‡]

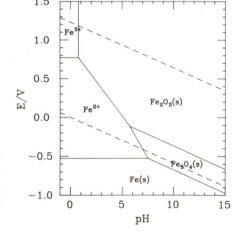

Figure 6.12 Predominance area diagram for iron. The dashed lines correspond to the $H^+(aq)/H_2(g)$ and $O_2(g)/H_2O$ couples.

[‡] For further discussion of corrosion and the prevention thereof, see Uhlig and Revie (G8).

Rate of Dissolution of a Metal in Water

Consider the reaction of a metal with aqueous acid,

$$M(s) + n\,H^+(aq) \rightarrow M^{n+}(aq) + (n/2)\,H_2(g)$$

We could study the rate of this reaction by classical techniques, *e.g.*, by measuring the volume of $H_2(g)$ as a function of time. However, the half–reactions

$$M(s) \rightarrow M^{n+}(aq) + n\,e^-$$

$$2\,H^+(aq) + 2\,e^- \rightarrow H_2(g)$$

can be studied independently using the electrochemical techniques described in Section 5.2. If a piece of the metal were used as an indicator electrode in a well–stirred solution and the current density measured as a function of potential, a current density *vs.* potential curve would be obtained similar to that of Figure 6.13.

Cathodic current at negative potentials corresponds to the reduction of $H^+(aq)$ and, for $E \ll E_e^H$, is given by the Tafel equation, eq (5.21),

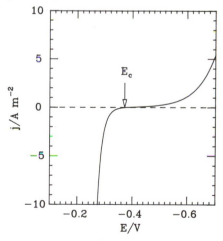
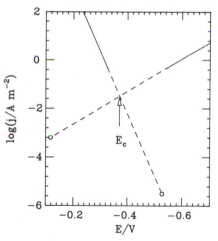

Figure 6.13 Current density *vs.* potential curve for iron in contact with an aqueous acidic solution (pH 2, [Fe^{2+}] = 1 mM) saturated with H_2 (1 bar partial pressure).

Figure 6.14 Tafel plot corresponding to the current density *vs.* potential curve of Figure 6.13.

$$\log j^H = \log j_0^{\;H} - \frac{a^H F(E - E_e^{\;H})}{2.303\ RT} \tag{6.7}$$

where a^H is the apparent cathodic transfer coefficient. Anodic current at more positive potentials ($E \gg E_e^{\;M}$) corresponds to oxidation of the metal[‡] and is given by

$$\log |j^M| = \log |j_0^{\;M}| + \frac{\beta^M F(E - E_e^{\;M})}{2.303\ RT} \tag{6.8}$$

where β^M is the apparent anodic transfer coefficient. Consider now the point on the j–E curve where $j = 0$. Zero current corresponds to the point where the anodic and cathodic components of the current are equal. At the potential of zero current, the rate of reduction of H^+ (aq) is exactly equal to the rate of metal oxidation. This, of course, is just the situation which pertains when we immerse the metal in the acid solution and allow the reaction to proceed. The potential at zero current then corresponds to the metal–solution potential in the absence of electrochemical control circuitry and is called the *corrosion potential*, E_c.

The corrosion potential is easily obtained by equating j_H and $|j^M|$,

$$E_c = \frac{(a^H E_e^{\;H} + \beta^M E_e^{\;M}) + (2.303\ RT/F)(\log j_0^{\;H} - \log |j_0^{\;M}|)}{a^H + \beta^M} \tag{6.9}$$

Substituting eq (6.9) into either eq (6.7) or eq (6.8) gives the corrosion current density at zero net current:

$$\log j_c = \frac{(a^H \log |j_0^{\;M}| + \beta^M \log j_0^{\;H})}{a^H + \beta^M} + \frac{a^H \beta^M F(E_e^{\;H} - E_e^{\;M})}{2.303\ RT(a^H + \beta^M)} \tag{6.10}$$

Thus electrochemical kinetic methods can be applied to a study of the rate of corrosion reactions. Furthermore, given exchange current

---

[‡] We assume that the experiment is done in an oxygen–free solution and that the anion is not electroactive.

densities and standard potentials for the half–cell reactions of interest, we should be able to compute the corrosion potential and the rate of the corrosion reaction.

Example 6.5 Compute the corrosion rate for metallic iron in contact with an aqueous acid solution, pH 2.00, $P(H_2) = 1$ bar, with $[Fe^{2+}] = 0.001$ M using the results of Bockris, Drazic, and Despic (see Example 5.4 and Problem 5.7) for the anodic dissolution of iron and the data of Table 5.1 for the cathodic reduction of $H^+(aq)$.

We first use the Nernst equation to compute the equilibrium half–cell potentials

$$E_e^H = 0.000 + (0.0592) \log[H^+] = -0.118 \text{ V}$$

$$E_e^M = -0.440 + (0.0592/2) \log[Fe^{2+}] = -0.529 \text{ V}$$

According to Figure 5.9, $\log|j_0^M| = -2.2$ at pH 3.1, 0.5 M Fe^{2+}, with $\beta^M = 1.5$. Extrapolation to $[Fe^{2+}] = 0.001$ M at pH 3.1 gives $\log|j_0^M| = -4.4$. Correcting this value to pH 2.0 with the observed pH dependence, we have $\log|j_0^M| = -5.5$.

The reduction of $H^+(aq)$ at an iron electrode has $\log j_0^H = -2.0$, $a_H = 0.4$ for pH 0. Correcting j_0^H to pH 2 using eq (5.25a), we have

$$\log j_0^H = -2.0 + \frac{(1 - a^H)F\Delta E_e}{2.303\ RT} = -3.2$$

Thus the Tafel lines corresponding to the reduction of $H^+(aq)$ and the oxidation of iron under the specified conditions are

$$\log j^H = -3.2 - 6.76(E + 0.118)$$

$$\log|j^M| = -5.5 + 25.4(E + 0.529)$$

These are plotted in Figure 6.14 and the total current density, $j_{tot} = j^H - |j^M|$, is plotted vs. potential in Figure 6.13. The corrosion potential found from eq (6.9) is $E_c = -0.371$ V and the corrosion current density from eq (6.10) is $j_c = 0.033$ A m^{-2}. The rate of the corrosion reaction thus is

$$\text{Rate} = j_c/nF = 1.7 \times 10^{-7} \text{ mol m}^{-2}\text{s}^{-1}$$

Such a rate seems very slow, but corrosion reactions need not be very fast to do their work; this corrosion rate corresponds to 300 g of metal dissolving per square meter per year.

Reaction of a Metal with Air–Saturated Water

In most corrosion problems of practical interest, the aqueous solution (which might be just a thin film) is in contact with air and the reaction of the metal with oxygen is generally much more important that the reaction with H^+ (aq) from the water.

In principle, we should be able to use the model developed above to measure or predict oxygen corrosion rates. Since oxygen is a stronger oxidant than water by 1.23 V, the cathodic current density due to oxygen reduction will almost always be larger than that due to water reduction. The oxygen reduction Tafel line will intersect the line for metal oxidation at a more positive potential and higher current density, thus giving a greater corrosion rate. The actual situation may be considerably less straightforward.

There is relatively little electrochemical kinetic data available for the reduction of oxygen and the mechanism of the process is, for the most part, not understood. Even when reliable data are available for the electrochemical reduction of oxygen, they may not be relvant to the actual corrosion process. For example, in low pH solutions, the reaction of dissolved oxygen with Fe(II) is very much faster than the reaction with solid iron. The Fe(III) product, however, undergoes electron transfer at the metal surface quite rapidly. The overall reaction then is the sum of the homogeneous and heterogeneous steps:

$$4\,Fe^{2+} + O_2 + 4\,H^+ \rightarrow 4\,Fe^{3+} + 2\,H_2O$$

$$2\,Fe^{3+} + Fe(s) \rightarrow 3\,Fe^{2+}$$

$$\overline{2\,Fe(s) + O_2 + 4\,H^+ \rightarrow 2\,Fe^{2+} + 4\,H_2O}$$

Thus the stoichiometry is identical to that of the direct reaction of oxygen with solid iron, although the mechanism is more complicated.

Passivation

Some metals—most notably aluminum, chromium, and nickel—are essentially unreactive in air–saturated water despite a large thermodynamic driving force. Aluminum, for example, can be used in cookware despite an expected free energy of oxidation of -717 kJ mol^{-1}. Metals are said to be *passive* when they corrode anomalously slowly. Many other metals, including iron, show passive behavior under some circumstances and "normal behavior" under others. It has been known since the eighteenth century that iron is attacked rapidly by dilute nitric acid but is essentially immune to attack by concentrated nitric acid. Treatment of iron with concentrated nitric acid imparts temporary corrosion resistance when the iron is immersed in dilute acid. Iron can also be temporarily passivated by electrochemical anodization. Whereas iron loses passivity in a matter of minutes, chromium and nickel show long–term passive behavior.

The passivation of a metal can be demonstrated by a current–potential curve such as that shown in Figure 6.15. The experiment

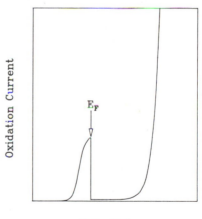

Figure 6.15 Anodic current–potential curve showing the onset of passivation at the Fladé potential, E_F.

starts with a clean metal surface in contact with a stirred solution saturated in oxygen with the potential of the metal held sufficiently negative that no metal oxidation occurs. A positive–going potential scan then gives an anodic current as the metal begins to oxidize. Initially, the current increases exponentially as expected. Instead of a continued increase, however, the current levels off and then drops precipitously, perhaps by several powers of 10; the current remains very small until the potential becomes sufficiently positive to oxidize water. The potential of onset of passivation is called the *Fladé potential*. The details of the mechanism of passivation—in particular, the sudden onset of the passive region at the Fladé potential—are imperfectly understood. The Fladé potential depends on the pH, usually according to

$$E_F = E_F° - 0.0592 \text{ pH}$$

suggesting the formation of an oxide film by the process

$$M + H_2O \rightarrow M \cdot O + 2 H^+ + 2 e^-$$

The Fladé potentials (at pH 0) for iron, nickel, and chromium are, respectively, $+0.6$, $+0.2$, and -0.2 V, and these values correlate well with the apparent kinetic stability of the protective oxide film. The Fladé potential of iron–chromium alloys decreases rapidly with increasing chromium content; $E_F°$ is about 0 V at 20% chromium. Thus the corrosion resistance of "stainless" steels is brought about through stabilization of the passivating oxide film.

In the case of main–group metals such as magnesium, aluminum, or zinc, the oxide film apparently grows into a layer of the metal oxide and can be regarded as a separate phase. The passivity of such metals then depends on the slow rate of diffusion of water or oxygen through the oxide layer. Transition metals, on the other hand, show no evidence of a surface oxide phase and there is reason to believe that the passivating film is a layer of oxygen atoms chemisorbed to the metal surface, $M \cdot O$. Coulometric experiments suggest that the passivating layer on iron is 4–5 oxygen atoms thick, but, because of surface roughness, this is an upper limit to the film thickness. There is some reason to suspect that the film contains both oxygen atoms and oxygen molecules. Uhlig *(18)* has used other thermodynamic data to estimate a potential of 0.57 V for the process

$$Fe \cdot O \cdot O_2 + 6 H^+ + 6 e^- \rightarrow Fe + 3 H_2O$$

This is sufficiently close to the experimental value, $E_F° = 0.63$ V to encourage belief in the model. If the passivating film is a layer of chemisorbed oxygen atoms and molecules, its success in protecting the metal from further oxidation must mean that the oxide film is lower in free energy than either the bare metal or a *thin* layer of the metal oxide phase. Thus, although thermodynamics says that complete conversion of the metal to metal oxide is spontaneous, the chemisorbed oxide layer represents a pronounced local free energy minimum on the pathway from metal to oxide.

Differential Aeration

Although ionic conduction across a passivating metal oxide film is very slow, electronic conduction may be much faster. Thus even at a passivated metal surface, reduction of oxygen may occur provided that a source of electrons is available. Consider the experimental arrangement shown in Figure 6.16. Two initially identical pieces of metal are dipped into identical electrolyte solutions. Nitrogen is bubbled over one electrode and oxygen over the other and we suppose that the metal exposed to oxygen is passivated. If we connect the two electrodes by closing the switch, we have set up an electrochemical cell in which, at the left–hand electrode, metal is oxidized,

$$M(s) \rightarrow M^{n+} + n\,e^-$$

and at the right–hand electrode, oxygen is reduced,

$$O_2(g) + 4\,H^+ + 4\,e^- \rightarrow 2\,H_2O$$

This overall cell reaction will almost always be spontaneous and the rate may be quite significant, despite the fact that the right–hand electrode is passivated.

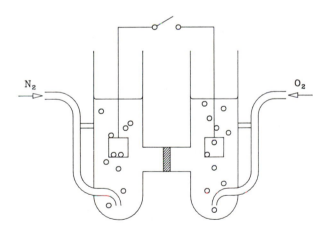

Figure 6.16 Experiment to demonstrate corrosion through differential aeration.

This experiment gives an insight into one of the most troublesome sources of corrosion, *differential aeration*. If a metal object is exposed to well–aerated water, it will often passivate, so that the rate of corrosion is negligible. If, however, part of the object is exposed to oxygen–deficient water, an electrochemical cell exactly like that of Figure 6.16 is set up. One can think of many situations where this might occur. For example, a metal piling in a river is exposed to well–aerated water near the surface but anaerobic mud at the bottom. The stagnant water trapped in a crevice in a metal object may have an oxygen concentration below the point required for passivation. Notice that in both these examples, the surface area of metal exposed to aerated water is large, so that oxygen reduction may be quite rapid overall; the rate of oxygen reduction must be balanced by metal oxidation, often in a relatively small area. Thus corrosion by differential aeration is usually characterized by rapid dissolution of metal in very localized areas. Cracks in a metal structure are thus enlarged, pits deepened, imperfect welds destroyed, etc.

Passive metal surfaces remain susceptible to corrosion in other ways as well. Although a passivating oxide layer behaves as if it were impermeable to oxygen or metal ions, it generally is either permeable to or is disrupted by halide ions. Thus metal passivity is usually lost when exposed to a solution containing chloride ions. The action of chloride ions tends to be localized, resulting in the formation of corrosion pits. Apparently chloride attacks thin or otherwise weak spots in the oxide layer, creating small anodic areas of active metal surrounded by large cathodic areas of passive metal. The mechanism thus is analogous to different aeration. Once the process starts, corrosion occurs primarily at the anodic sites and a corrosion pit forms and grows.

Prevention of Corrosion

Perhaps the most common approach to the prevention of corrosion is to cover the metal surface with a protective coating, *e.g.,* paint, plastic, ceramic, or an electroplated layer of a passive metal such as chromium, nickel, or tin. A mechanical barrier is effective, of course, only as long as it remains a barrier. Bare metal exposed by a scratch is susceptible to corrosion. Once corrosion begins at a breach in the protective layer, it often continues under the layer through the differential aeration mechanism, lifting the coating and eventually leading to massive damage. Differential aeration is particularly effective with damaged metal–plated surfaces since oxygen reduction can occur over the entire undamaged surface to balance the base metal oxidation at the point of damage.

Coatings can be made more effective by the addition of *corrosion inhibitors*. A corrosion inhibitor works by decreasing the rate of metal oxidation, by decreasing the rate of oxygen (or water) reduction, or by shifting the surface potential into the passive region. Thus rust–inhibiting paints often contain chromate salts which passivate the metal surface by making the surface potential positive. Other common paint additives are aliphatic or aromatic amines which appear to adsorb on the metal, slowing the metal oxidation half–reaction. Since inhibitors impart a resistance to corrosion over and above the effect of the physical barrier, small scratches do not usually lead to massive damage.

Another way to protect a metal object from corrosion is to attach a *sacrificial anode* made of zinc, magnesium, or an aluminum alloy. In effect a short–circuited galvanic cell is produced where the protected object is the cathode (oxygen is reduced) and the more active metal is oxidized. Consider, for example, the corrosion potential of zinc. Since the reduction potential of zinc(II) is –0.76 V, then (assuming similar kinetics) its corrosion potential will be about 0.3 V more negative than that of iron. Thus if the two metals are in electrical contact, the surface potential of iron will be more negative. According to Figure 6.14, a 0.3–V negative shift should reduce the rate of metal oxidation by about five powers of 10.

As the name implies, sacrificial anodes do corrode and may have to be replaced from time to time. The ideal anode material has some passivity so that the rate of corrosion is low, but a completely passive anode would impart no protection at all; thus aluminum must be alloyed (with magnesium, for example) to increase its corrosion rate. Sacrificial anodes are commonly used to protect such objects as ship hulls, pipelines, and oil–field drilling rigs from environmental corrosion.

Surface treatment with chromate imparts corrosion protection by making the metal surface anodic and hopefully passive. Contact with a sacrificial anode makes the protected surface cathodic and thus slows the rate of metal dissolution. Either of these approaches could be used if we add an inert electrode and control the surface potential with a potentiostat. This sounds like a strictly laboratory approach to the problem, but in fact potentiostatic control of corrosion is used in certain kinds of applications, *e.g.,* in the protection of ship hulls.

REFERENCES

(Reference numbers preceded by a letter, *e.g.*, *(F6)*, refer to a book listed in the Bibliography.)

1. W. Cruikshank, *Ann. Phys.* **1801**, *7*, 105.
2. M. Faraday, Experimental Researches in Electricity, London: Bernard Quaritch, 1839 [facsimile reprint, London: Taylor & Francis, 1882; included in Vol. 45 of *Great Books of the Western World*, R. M. Hutchins, ed, Chicago: Encyclopedia Brittanica, 1952], Vol. 1, Series V.
3. O. W. Gibbs, *Z. anal. Chem.* **1864**, *3*, 327.
4. N. Tanaka in *Treatise on Analytical Chemistry*, I. M. Kolthoff and P. J. Elving, eds, Part I, Vol. 4, New York: Interscience, 1963, p 2417.
5. L. Szebelledy and Z. Somogyi, *Z. anal. Chem.* **1938**, *112*, 313.
6. D. D. DeFord and J. W. Miller in *Treatise on Analytical Chemistry*, I. M. Kolthoff and P. J. Elving, eds, Part I, Vol. 4, New York: Interscience, 1963, p 2475.
7. J. E. Harrar, *Electroanalytical Chemistry* **1975**, *8*, 1.
8. J. J. Lingane, *J. Am. Chem. Soc.* **1945**, *67*, 1916.
9. L. Meites and J. Meites, *Anal. Chem.* **1956**, *28*, 103.
10. D. J. Curran in *Laboratory Techniques in Electroanalytical Chemistry*, P. T. Kissinger and W. R. Heineman, eds, New York: Marcel Dekker, 1984, p 539.
11. J. C. Verhoef and E. Barendrecht, *J. Electroanal. Chem.* **1976**, *71*, 305; **1977**, *75*, 705.
12. I. Shain in *Treatise on Analytical Chemistry*, I. M. Kolthoff and P. J. Elving, eds, Part I, Vol. 4, New York: Interscience, 1963, p 2533.
13. E. Barendrecht, *Electroanalytical Chemistry* **1967**, *2*, 53.
14. W. R. Heineman, H. B. Mark, Jr., J. A. Wise, and D. A. Roston, *Laboratory Techniques in Electroanalytical Chemistry*, P. T. Kissinger and W. A. Heineman, eds, New York: Marcel Dekker, 1984, p 499.
15. P. Ostapczuk and Z. Kublik, *J. Electroanal. Chem.* **1976**, *68*, 193; **1977**, *83*, 1.
16. K. Shimizu and R. A. Osteryoung, *Anal. Chem.* **1981**, *53*, 584.
17. H. Kolbe, *Ann.* **1843**, *69*, 257.
18. H. H. Uhlig, *Z. Elektrochem.* **1958**, *62*, 626.

PROBLEMS

6.1 Show that the charge passed through an electrolysis cell of volume V is

$$Q(t) = nFVC_O{}^*(0) \left[1 - \exp(-kt)\right]$$

where k is given by eq (6.2), $C_O{}^*(0)$ is the initial bulk concentration of O and the mass transport rate constant k_D is assumed constant.

6.2 Repeat the calculation of Example 6.1 assuming that the solution is unstirred and that the current remains diffusion–controlled indefinitely. The diffusion coefficient of Cu^{2+} is about 7 x 10^{-10} $m^2 s^{-1}$.

6.3 If the electrode surface area to solution volume ratio is large, density gradients usually lead to convective mixing so that the mass transport rate is often much greater than might be expected from diffusion alone and may remain more or less constant during an electrolysis. For example, radical anions are prepared for electron spin resonance study by electrolysis in a 0.2 mL cell equipped with a cylindrical working electrode 5 mm long and 1 mm in diameter. With a substrate concentration of 1 mM, a convection–limited initial current of about 10 μA is obtained.
(a) How long will it take to reduce half the substrate?
(b) Suppose that the radical anions produced decay by first–order kinetics with a half–life of 5 minutes. What is the radical anion concentration after electrolysis for 5 minutes? 10 minutes? 25 minutes?

6.4 Show that the optimum conditions for an electroseparation process are achieved when the electrode potential is halfway between the standard potentials of the two couples (assuming that $n_1 = n_2$).

6.5 Derive eq (6.6) given eqs (6.4) and (6.5).

6.6 What is the minimum charge passed through a silver coulometer if the charge is to be measured with a precision of $\pm 0.1\%$ if the cathode is weighed before and after electrolysis to ± 0.1 mg?

6.7 An electronic coulometer is used in the controlled–potential coulometric determination of trichloroacetic acid.
(a) With $R_1 = 1$ kΩ, $R_2 = 1$ MΩ, $C = 50$ μF, the final output potential of the coulometer was 4.679 V. What charge was passed through the electrolysis cell?
(b) If the volume of the Cl_3CCOOH solution was 25 mL, what was the molar concentration?

6.8 The operational amplifier galvanostat circuit of Figure 6.6 is used to supply a constant current of 1 mA through an electrolysis cell.
(a) If the control battery supplies a potential of 1.00 V, what size resistor should be used in the circuit?
(b) If the maximum output voltage of the operational amplifier is ± 10 V, what is the maximum allowable cell resistance?

6.9 A constant current of 13.64 mA was used in a coulometric Karl Fischer titration of water in 5 mL of acetonitrile. The endpoint was obtained after 248 s. What was the water concentration?

6.10 Suppose that you have a contant current source which can deliver 1, 2, 5, 10, 20, 50, or 100 mA ($\pm 0.25\%$), an endpoint detection system and timer which is reliable to ± 1 s, and a collection of pipets (1, 2, 5, 10, 25, and 50 mL), each of which is accurate to ± 0.02 mL. You wish to do a coulometric titration of a solution where the analyte concentration is approximately 1 mM. What volume and current should be chosen if the analytical uncertainty must be less than 1.0% and it is desirable to use minimum sample size and do the titration in minimum time? Assume a one–electron oxidation is involved.

6.11 Unsaturated organic compounds can be determined by coulometric titration with bromine

$$RCH=CH_2 + Br_2 \rightarrow RCHBr-CH_2Br$$

where bromine is generated by oxidation of bromide ion and the endpoint is determined biamperometrically with two small platinum electrodes polarized by 100 mV. 10 mL of a solution of an unsaturated acid was placed in a coulometric cell, sulfuric acid and potassium bromide was added and a constant current of 18.5 mA passed. The current between the detector electrodes was measured as a function of time with the following results:

t/s	250	260	270	280	290	300
$i/\mu A$	0	6	15	24	34	43

What was the molar concentration of the unsaturated acid? Assume mono–unsaturation.

6.12 Suppose that copper is deposited on a platinum electrode at a cathode potential of -0.30 V (vs. s.c.e.) and an initial current density of $j = 100$ A m^{-2}. Early in the experiment, the electrode surface is mostly platinum. If the exchange current density for the reduction of H$^+$ (aq) on Pt is 10 A m^{-2}, $a_{app} = 1.5$, at pH 0, what is the initial current density contribution from hydrogen reduction if the pH is 1.0? What is the current efficiency of the copper reduction process? When the current density has dropped to 10 A m^{-2}, the cathode is covered with copper. If the exchange current density for H$^+$ reduction on copper is 2 mA m^{-3}, $a_{app} = 0.5$, at pH 1, what is the hydrogen reduction current? What is the current efficiency for copper reduction?

6.13 An industrial electrolysis process has raw material costs of $1.00 per kilogram of product and capital and labor costs of $5.00 per electrolysis cell per day (independent of whether any product is made). The cell potential (for $i > 1$ A) is

$$E = E_0 + \beta \log i + iR$$

where $\beta = 1.0$ V, $R = 0.10$ Ω, and $E_0 = 1.50$ V. The cost of electric power is $0.10 per kilowatt–hour. What is the optimum current per cell for minimum cost per kilogram of product? Assume that 50 g of product is produced per Faraday of charge passed, that the current efficiency is 100%, and that the plant runs 24 hours a day. What is the production per cell per day? What are the costs per kilogram for raw materials, capital and labor, and electrical energy? If the cost per kilowatt–hour doubled, what would be the optimum current? What would be the total cost of product be?

6.14 Two common ways of protecting steel from corrosion are tin plating and zinc coating. Typical examples of this approach are "tin cans" and galvanized buckets. Discuss qualitatively the chemical reactions which would occur if pieces of tin–plated steel and zinc–coated steel were scratched to expose the steel and placed in a wet oxygen–rich environment. Indicate clearly where (relative to the scratch) each reaction would take place.

6.15 Consider the reaction of zinc with 1 M hydrochloric acid if $[Zn^{2+}] = 1$ M. The exchange current densities are

$$j_0^{Zn} = 0.2 \text{ A m}^{-2}, \beta^{Zn} = 1.5$$

$$j_0^{H} = 1 \times 10^{-6} \text{ A m}^{-2}, a^{H} = 0.5$$

for $[Zn^{2+}] = 1$ M, pH 0, and 25°C. Assume that

$$\partial \log|j_0^{Zn}| / \partial \log[Zn^{2+}] = 0.75.$$

(a) Compute the corrosion potential, E_c, the zinc corrosion rate in g m^{-2}s^{-1} and the rate of hydrogen gas evolution in mL m^{-2}s^{-1}
(b) Corrosion chemists often express corrosion rates in units of mm year^{-1}. Given that the density of zinc is 7.14 g cm^{-3}, convert the corrosion rates to units of mm year^{-1}.

APPENDIX 1. BIBLIOGRAPHY

A. Introductory Electrochemistry Texts

A1. C. N. Reilley and R. W. Murray, *Electroanalytical Principles*, New York: Interscience, 1963.

A2. E. H. Lyons, Jr., *Introduction to Electrochemistry*, Boston: D. C. Heath, 1967.

A3. B. B. Damaskin, *The Principles of Current Methods for the Study of Electrochemical Reactions*, New York: McGraw–Hill, 1967.

A4. J. B. Headridge, *Electrochemical Techniques for Inorganic Chemists*, London: Academic Press, 1969.

A5. J. Robbins, *Ions in Solution (2): An Introduction to Electrochemistry*, London: Oxford University Press, 1972.

A6. W. J. Albery, *Electrode Kinetics*, London: Oxford University Press, 1975.

A7. N. J. Selley, *Experimental Approach to Electrochemistry*, New York: John Wiley, 1976.

A8. D. R. Crow, *Principles and Applications of Electrochemistry*, 2nd ed, London: Chapman and Hall, 1979.

A9. J. Koryta, *Ions, Electrodes and Membranes*, New York: John Wiley, 1982.

439

A10. D. Hibbert and A. M. James, *Dictionary of Electrochemistry*, 2nd ed, New York: John Wiley, 1985.

B. More Advanced Texts

B1. P. Delahay, *New Instrumental Methods in Electrochemistry*, New York: Interscience, 1954.

B2. H. S. Harned and B. B. Owen, *Physical Chemistry of Electrolytic Solutions*, New York: Reinhold, 1958.

B3. G. Charlot, J. Badoz–Lambling, and B. Tremillon, *Electrochemical Reactions*, Amsterdam: Elsevier, 1962.

B4. D. A. MacInnes, *The Principles of Electrochemistry*, New York: Dover, 1966 (corrected reprint of 1947 edition).

B5. G. Kortüm, *Treatise on Electrochemistry*, 2nd ed, Amsterdam: Elsevier, 1965.

B6. B. E. Conway, *Theory and Principles of Electrode Processes*, New York: Ronald Press, 1965.

B7. K. J. Vetter, *Electrochemical Kinetics*, New York: Academic Press, 1967.

B8. J. O'M. Bockris and A. Reddy, *Modern Electrochemistry*, New York: Plenum Press, 1970.

B9. R. A. Robinson and R. H. Stokes, *Electrolyte Solutions*, 2nd ed (rev), London: Butterworths, 1970.

B10. J. S. Newman, *Electrochemical Systems*, Englewood Cliffs, NJ: Prentice–Hall, 1972.

B11. L. I. Antropov, *Theoretical Electrochemistry*, Moscow: Mir, 1972.

B12. A. J. Bard and L. R. Faulkner, *Electrochemical Methods*, New York: John Wiley, 1980.

B13. R. Greef, R. Peat, L. M. Peter, D. Pletcher, and J. Robinson, *Instrumental Methods in Electrochemistry*, Chicester: Ellis Horwood, 1985.

C. Specialized Books and Monographs

C1. W. Ostwald, *Electrochemistry, History and Theory*, Leipzig: Veit, 1896. Republished in English translation for the Smithsonian Institution, New Delhi: Amerind Publishing, 1980.

C2. V. G. Levich, *Physiochemical Hydrodynamics*, Englewood Cliffs, NJ: Prentice–Hall, 1962.

C3. R. N. Adams, *Electrochemistry at Solid Electrodes*, New York: Marcel Dekker, 1969.

C4. C. K. Mann and K. K. Barnes, *Electrochemical Reactions in Nonaqueous Systems*, New York: Marcel Dekker, 1970.

C5. A. Weissberger and B. W. Rossiter, eds, *Physical Methods of*

Chemistry, Vol. 1 (*Techniques of Chemistry*), Parts IIA and IIB (*Electrochemical Methods*), New York: John Wiley, 1971.

C6. J. S. Mattson, H. B. Mark, Jr., and H. C. MacDonald, Jr., *Electrochemistry: Calculations, Simulation, and Instrumentation (Computers in Chemistry and Instrumentation, Vol. 2)*, New York: Marcel Dekker, 1972.

C7. Yu. V. Pleskov and V. Yu. Filinovskii, *The Rotating Disc Electrode*, New York: Consultants Bureau, 1976.

C8. J. O'M. Bockris and S. U. M. Khan, *Quantum Electrochemistry*, New York: Plenum Press, 1979.

D. Electroanalytical Methods

D1. I. M. Kolthoff and J. J. Lingane, *Polarography*, 2nd ed, New York: Interscience, 1952.

D2. J. J. Lingane, *Electroanalytical Chemistry*, 2nd ed, New York: Interscience, 1958.

D3. W. C. Purdy, *Electroanalytical Methods in Biochemistry*, New York: McGraw–Hill, 1965.

D4. L. Meites, *Polarographic Techniques*, 2nd ed, New York: John Wiley, 1965.

D5. J. Heyrovský and J. Kuta, *Principles of Polarography*, New York: Academic Press, 1966.

D6. H. Rossotti, *Chemical Applications of Potentiometry*, Princeton, NJ: Van Nostrand, 1969.

D7. R. G. Bates, *Determination of pH: Theory and Practice*, 2nd ed, New York: John Wiley, 1973.

D8. Z. Galus, *Fundamentals of Electrochemical Analysis*, Chichester: Ellis Harwood, 1976.

D9. G. Dryhurst, *Electrochemistry of Biological Molecules*, New York: Academic Press, 1977.

D10. C. C. Westcott, *pH Measurements*, New York: Academic Press, 1978.

D11. J. Vesely, D. Weiss, and K. Stulik, *Analysis with Ion–Selective Electrodes*, Chichester: Ellis Horwood, 1978.

D12. A. M. Bond, *Modern Polarographic Methods in Analytical Chemistry*, New York: Marcel Dekker, 1980.

D13. J. A. Plambeck, *Electroanalytical Chemistry*, New York: John Wiley, 1982.

D14. J. Koryta and K. Stulik, *Ion–Selective Electrodes*, 2nd ed, London: Cambridge University Press, 1983.

E. Organic Electrosynthesis

E1. A. J. Fry, *Synthetic Organic Electrochemistry*, New York: Harper and Row, 1972.

E2. M. R. Rifi and F. H. Covitz, *Introduction to Organic Electrochemistry*, New York: Marcel Dekker, 1974.

E3. N. L. Weinberg, ed, *Technique of Electroorganic Synthesis (Technique of Chemistry, Vol. V)*, New York: John Wiley, 1974 (Part I), 1975 (Part II).

E4. D. K. Kyriacou, *Basics of Electroorganic Synthesis*, New York: John Wiley, 1981.

E5. M. M. Baizer and H. Lund, eds, *Organic Electrochemistry*, New York: Marcel Dekker, 1983.

E6. K. Yoshida, *Electrooxidation in Organic Chemistry*, New York: John Wiley, 1984.

E7. T. Shono, *Electroorganic Chemistry as a New Tool in Organic Synthesis*, Berlin: Springer–Verlag, 1984.

F. Experimental Methods

F1. D. J. G. Ives and G. J. Janz, eds, *Reference Electrodes, Theory and Practice*, New York: Academic Press, 1961.

F2. W. J. Albery and M. L. Hitchman, *Ring–Disc Electrodes*, Oxford: Clarendon Press, 1971.

F3. D. T. Sawyer and J. L. Roberts, Jr., *Experimental Electrochemistry for Chemists*, New York: John Wiley, 1974.

F4. E. Gileadi, E. Kirowa–Eisner, and J. Penciner, *Interfacial Electrochemistry—An Experimental Approach*, Reading, MA: Addison–Wesley, 1975.

F5. D. D. MacDonald, *Transient Techniques in Electrochemistry*, New York: Plenum Press, 1977.

F6. P. T. Kissinger and W. R. Heineman, eds, *Laboratory Techniques in Electroanalytical Chemistry*, New York: Marcel Dekker, 1984.

G. Technological Applications of Electrochemistry

G1. J. O'M. Bockris and S. Srinivasan, *Fuel Cells: Their Electrochemistry*, New York: McGraw–Hill, 1969.

G2. C. L. Mantell, *Batteries and Energy Systems*, New York: McGraw–Hill, 1970.

G3. F. A. Lowenheim, ed, *Modern Electroplating*, New York: John Wiley, 1974.

G4. S. W. Angrist, *Direct Energy Conversion*, Boston: Allyn and Bacon, 1976.

G5. V. S. Bagotzky and A. M. Skundin, *Chemical Power Sources*, New York: Academic Press, 1980.

G6. D. Pletcher, *Industrial Electrochemistry*, London: Chapman and Hall, 1982.

G7. N. L. Weinberg and B. V. Tilak, eds, *Technique of Electroorganic Synthesis (Technique of Chemistry, Vol. V), Part III*, New York: John Wiley, 1982.

G8. H. H. Uhlig and R. W. Revie, *Corrosion and Corrosion Control*, New York: John Wiley, 1984.

G9. H. V. Ventatasetty, ed, *Lithium Battery Technology*, New York: John Wiley, 1984.

G10. R. E. White, ed, *Electrochemical Cell Design*, New York: Plenum Press, 1984.

G11. Z. Nagy, *Electrochemical Synthesis of Inorganic Compounds*, New York: Plenum Press, 1985.

H. Electrochemical Data

H1. W. M. Latimer, *Oxidation Potentials*, 2nd ed, Englewood Cliffs, NJ: Prentice–Hall, 1952.

H2. B. E. Conway, *Electrochemical Data*, Amsterdam: Elsevier, 1952.

H3. R. Parsons, *Handbook of Electrochemical Data*, London: Butterworths, 1959.

H4. A. J. de Bethune and N. A. S. Loud, *Standard Aqueous Electrode Potentials and Temperature Coefficients at 25°C* Skokie, IL: Hampel, 1964.

H5. M. Pourbaix, *Atlas of Electrochemical Equilibria*, New York: Pergamon Press, 1966.

H6. G. J. Janz and R. P. T. Tomkins, *Nonaqueous Electrolytes Handbook*, New York: Academic Press, 1972.

H7. A. J. Bard and H. Lund, eds, *The Encyclopedia of the Electrochemistry of the Elements*, New York: Marcel Dekker, 1973.

H8. L. Meites and P. Zuman, *Electrochemical Data. Part I. Organic, Organometallic, and Biochemical Systems*, New York: John Wiley, 1974.

H9. D. Dobos, *Electrochemical Data*, Amsterdam: Elsevier, 1975.

H10. G. Milazzo and S. Caroli, *Tables of Standard Electrode Potentials*, New York: John Wiley, 1977.

H11. M. S. Antelman, *The Encyclopedia of Chemical Electrode Potentials*, New York: Plenum Press, 1982.

H12. A. J. Bard, R. Parsons, and J. Jordan, eds, *Standard Potentials in Aqueous Solution*, New York: Marcel Dekker, 1985.

I. Review Series

I1. J. O'M. Bockris and B. E. Conway, eds, *Modern Aspects of Electrochemistry*, New York: Plenum Press, from 1954.

I2. P. Delahay (Vols. 1-9), C. W. Tobias (Vols. 1-), and H. Gerischer (Vols. 10-), *Advances in Electrochemistry and Electrochemical Engineering*, New York: John Wiley, from 1961.

I3. A. J. Bard, ed, *Electroanalytical Chemistry*, New York: Marcel Dekker, from 1966.

I4. E. B. Yeager and A. J. Salkind, eds, *Techniques of Electrochemistry*, New York: John Wiley, from 1972.

I5. G. J. Hills (Vols. 1-3) and H. R. Thirsk (Vol. 4-), Senior Reporters, *Electrochemistry*, A Specialist Periodical Report, London: Royal Society of Chemistry, from 1971.

I6. *Analytical Chemistry*, Fundamental Annual Reviews (April issue of even-numbered years), Washington: American Chemical Society.

APPENDIX 2. SYMBOLS AND UNITS

TABLE A.1 Values of Physical Constants

Constant	Symbol	Value
permittivity of free space	ϵ_0	$8.8541878 \times 10^{-12}$ $C^2J^{-1}m^{-1}$
electronic charge	e	1.602189×10^{-19} C
Avogadro's number	N_A	6.02204×10^{23} mol^{-1}
Faraday constant	F	96484.6 C mol^{-1}
gas constant	R	8.3144 J $mol^{-1}K^{-1}$
Boltzmann constant	k	1.38066×10^{-23} J K^{-1}
Planck constant	h	6.62618×10^{-34} J s
gravitational acceleration	g	9.80665 m s^{-2}

TABLE A.2 The International System (SI) of Units

Physical Quantity	Name of Unit	Symbol
Fundamental units:		
Length	meter	m
mass	kilogram	kg
time	second	s
electric current	ampere	A
temperature	kelvin	K
amount of substance	mole	mol
luminous intensity	candela	cd
Derived units:		
force	newton	N (kg m s^{-2})
energy	joule	J (N m)
power	watt	W (J s^{-1})
pressure	pascal	Pa (N m^{-2})
electric charge	coulomb	C (A s)
electric potential	volt	V (J C^{-1})
electric resistance	ohm	Ω (V A^{-1})
electric conductance	siemens	S (A V^{-1})
electric capacitance	farad	F (C V^{-1})
frequency	hertz	Hz (s^{-1})

TABLE A.3 List of Symbols

Symbol	Name	Units
a	activity	none
a	radius	m
A	area	m^2
C_i	molar concentration of species i	$M = mol\ L^{-1}$ $mM = mol\ m^{-3}$
C	differential capacity	$F\ m^{-2}$
d	density	$kg\ m^{-3}$
D_i	diffusion coefficient of species i	$m^2 s^{-1}$
E	energy	J
E	cell potential	V
$E_{\frac{1}{2}}$	half–wave potential	V
E	electric field strength	$V\ m^{-1}$
f_i	frictional coefficient of species i	$kg\ s^{-1}$
F	force	N
G	Gibbs free energy	$J\ mol^{-1}$
H	enthalpy	$J\ mol^{-1}$
i	electric current	A
i_D	diffusion–limited current	A
i_L	limiting current	A
I	ionic strength	$mol\ L^{-1}$
I	a.c. current amplitude	A

Symbol	Name	Units
I_D	diffusion current constant	$\mu A\ mM^{-1}(mg\ s^{-1})^{-\frac{2}{3}}s^{-\frac{1}{6}}$
j	volume flux	$m^3 s^{-1}$
j	current density	$A\ m^{-2}$
J	molar flux density	$mol\ m^{-2}s^{-1}$
k	rate constant	variable
k_{ij}	potentiometric selectivity coefficient	none
K	equilibrium constant	none
L	length	m
m	mass	kg
m_i	molal concentration of species i	$mol\ kg^{-1}$
M	molecular weight	$g\ mol^{-1}$
n	number of moles	mol
n_i	kinetic order in species i	none
N	number of molecules	none
P	pressure	$bar = 10^5\ Pa$
q	heat	J
Q	electric charge	C
r, R	radial distance	m
R	resistance	Ω
S	entropy	$J\ mol^{-1}K^{-1}$

Symbol	Name	Units
t	time	s
t_i	transference number of species i	none
T	temperature	K
u_i	mobility of species i	$m^2 V^{-1} s^{-1}$
u	mass flow rate	$kg\ s^{-1}$
U	internal energy	$J\ mol^{-1}$
$\mathbf{v}$	velocity	$m\ s^{-1}$
V	volume	m^3
υ	potential scan rate	$V\ s^{-1}$
w	work	J
x	distance	m
x_A	ion atmosphere thickness (Debye length)	m
x_D	diffusion layer thickness	m
x_H	hydrodynamic distance parameter	m
x_R	reaction layer thickness	m
X_i	mole fraction of species i	none
z_i	charge on species i	none
Z	impedance	Ω
α	degree of dissociation	none
α	cathodic transfer coefficient	none

Symbol	Name	Units
a_i	electrokinetic coefficient	variable
a, β, γ, δ	stoichiometric coefficients	none
a, β	phase labels	none
β	anodic transfer coefficient	none
γ	surface tension	$N\ m^{-1}$
γ_i	activity coefficient of species i (molar scale)	none
γ_i^m	activity coefficient of species i (molal scale)	none
γ_i^x	activity coefficient of species i (mole fraction scale)	none
$\gamma_\pm$	mean ionic activity coefficient	none
ϵ	dielectric constant	none
η	coefficient of viscosity	$kg\ m^{-1}s^{-1} = Pa\text{-}s$
η	overpotential	V
θ	$\exp[F(E - E^\circ)/RT]$	none
κ	conductivity	$S\ m^{-1}$
λ	reaction zone parameter	none
Λ	molar conductivity	$S\ m^2 mol^{-1}$
μ	chemical potential	$J\ mol^{-1}$
ν	kinematic viscosity	$m^2 s^{-1}$
ν_i	moles of ion i per mole of salt	none
ν_i	stoichiometric coefficient of species i	none

Symbol	Name	Units
ξ	$(D_O/D_R)^{1/2}$	none
ρ	resistivity	Ω m
ρ	space charge density	$C\ m^{-3}$
σ	surface charge density	$C\ m^{-2}$
τ	characteristic time in an experiment	s
ϕ	current efficiency	none
ϕ	phase angle	none
Φ	electric potential	V
ω	angular frequency	$rad\ s^{-1}$
ζ	zeta potential	V

APPENDIX 3. ELECTROCHEMICAL DATA

TABLE A.4 Standard Reduction Potentials

Half–Cell Reaction	$E°/V$ (25°C)

Main group elements:

Half–Cell Reaction	$E°/V$ (25°C)
$2\ H^+ + 2\ e^- \rightarrow H_2(g)$	0.0000
$Li^+ + e^- \rightarrow Li(s)$	-3.045
$Na^+ + e^- \rightarrow Na(s)$	-2.714
$K^+ + e^- \rightarrow K(s)$	-2.925
$Rb^+ + e^- \rightarrow Rb(s)$	-2.925
$Cs^+ + e^- \rightarrow Cs(s)$	-2.923
$Be^{2+} + 2\ e^- \rightarrow Be(s)$	-1.97
$Mg^{2+} + 2\ e^- \rightarrow Mg(s)$	-2.356
$Ca^{2+} + 2\ e^- \rightarrow Ca(s)$	-2.84
$Ba^{2+} + 2\ e^- \rightarrow Ba(s)$	-2.92
$Al^{3+} + 3\ e^- \rightarrow Al(s)$	-1.67
$CO_2(g) + 2\ H^+ + 2\ e^- \rightarrow CO(g) + H_2O$	-0.106
$CO_2(g) + 2\ H^+ + 2\ e^- \rightarrow HCOOH$	-0.199
$2\ CO_2(g) + 2\ H^+ + 2\ e^- \rightarrow H_2C_2O_4$	-0.475
$Pb^{2+} + 2\ e^- \rightarrow Pb(s)$	-0.1251
$PbO_2(s) + 4\ H^+ + 2\ e^- \rightarrow Pb^{2+} + 2\ H_2O$	1.468

Half–Cell Reaction	$E°$/V (25°C)
$NO_3^- + 3\,H^+ + 2\,e^- \rightarrow HNO_2 + H_2O$	0.94
$NO_3^- + 4\,H^+ + 3\,e^- \rightarrow NO(g) + 2\,H_2O$	0.96
$NO_3^- + 10\,H^+ + 8\,e^- \rightarrow NH_4^+ + 3\,H_2O$	0.875
$PO(OH)_3 + 2\,H^+ + 2\,e^- \rightarrow HPO(OH)_2 + H_2O$	−0.276
$HPO(OH)_2 + 2\,H^+ + 2\,e^- \rightarrow H_2PO(OH) + H_2O$	−0.499
$HPO(OH)_2 + 3\,H^+ + 3\,e^- \rightarrow P(s) + 3\,H_2O$	−0.454
$P(s) + 3\,H^+ + 3\,e^- \rightarrow PH_3$	−0.111
$AsO(OH)_3 + 2\,H^+ + 2\,e^- \rightarrow As(OH)_3 + H_2O$	0.560
$As(OH)_3 + 3\,H^+ + 3\,e^- \rightarrow As(s) + 3\,H_2O$	0.240
$As(s) + 3\,H^+ + 3\,e^- \rightarrow AsH_3(g)$	−0.225
$O_2(g) + H^+ + e^- \rightarrow HO_2$	−0.125
$O_2(g) + 2\,H^+ + 2\,e^- \rightarrow H_2O_2$	0.695
$H_2O_2 + H^+ + e^- \rightarrow HO\cdot + H_2O$	0.714
$H_2O_2 + 2\,H^+ + 2\,e^- \rightarrow 2\,H_2O$	1.763
$S_2O_8^{2-} + 2\,e^- \rightarrow 2\,SO_4^{2-}$	1.96
$SO_4^{2-} + H_2O + 2\,e^- \rightarrow SO_3^{2-} + 2\,OH^-$	−0.94
$2\,SO_4^{2-} + 4\,H^+ + 2\,e^- \rightarrow S_2O_6^{2-} + 2\,H_2O$	−0.25
$2\,SO_2(aq) + 2\,H^+ + 4\,e^- \rightarrow S_2O_3^{2-} + H_2O$	0.40
$SO_2(aq) + 4\,H^+ + 4\,e^- \rightarrow S(s) + 2\,H_2O$	0.50
$S_4O_6^{2-} + 2\,e^- \rightarrow 2\,S_2O_3^{2-}$	0.08

Half–Cell Reaction	$E°/V$ (25°C)
$S(s) + 2 H^+ + 2 e^- \rightarrow H_2S(aq)$	0.14
$F_2(g) + 2 e^- \rightarrow 2 F^-$	2.866
$ClO_4^- + 2 H^+ + 2 e^- \rightarrow ClO_3^- + H_2O$	1.201
$ClO_3^- + 3 H^+ + 2 e^- \rightarrow HClO_2 + H_2O$	1.181
$ClO_3^- + 2 H^+ + e^- \rightarrow ClO_2 + H_2O$	1.175
$HClO_2 + 2 H^+ + 2 e^- \rightarrow HOCl + H_2O$	1.701
$2 HOCl + 2 H^+ + 2 e^- \rightarrow Cl_2(g) + 2 H_2O$	1.630
$Cl_2(g) + 2 e^- \rightarrow 2 Cl^-$	1.35828
$Cl_2(aq) + 2 e^- \rightarrow 2 Cl^-$	1.396
$BrO_4^- + 2 H^+ + 2 e^- \rightarrow BrO_3^- + H_2O$	1.853
$2 BrO_3^- + 12 H^+ + 10 e^- \rightarrow Br_2(l) + 6 H_2O$	1.478
$2 HOBr + 2 H^+ + 2 e^- \rightarrow Br_2(l) + 2 H_2O$	1.604
$Br_2(l) + 2 e^- \rightarrow 2 Br^-$	1.0652
$Br_2(aq) + 2 e^- \rightarrow 2 Br^-$	1.0874
$IO(OH)_5 + H^+ + e^- \rightarrow IO_3^- + 3 H_2O$	1.60
$2 IO_3^- + 12 H^+ + 10 e^- \rightarrow I_2(s) + 6 H_2O$	1.20
$2 HOI + 2 H^+ + 2 e^- \rightarrow I_2(s) + 2 H_2O$	1.44
$I_2(s) + 2 e^- \rightarrow 2 I^-$	0.5355
$I_3^- + 2 e^- \rightarrow 3 I^-$	0.536
$I_2(aq) + 2 e^- \rightarrow 2 I^-$	0.621

Half–Cell Reaction	E°/V (25°C)

Transition and post–transition elements:

Half–Cell Reaction	E°/V (25°C)
$VO_2^+ + 2\,H^+ + e^- \rightarrow VO^{2+} + H_2O$	1.000
$VO^{2+} + 2\,H^+ + e^- \rightarrow V^{3+} + H_2O$	0.337
$V^{3+} + e^- \rightarrow V^{2+}$	−0.255
$V^{2+} + 2\,e^- \rightarrow V(s)$	−1.13
$Cr_2O_7^{2-} + 14\,H^+ + 6\,e^- \rightarrow 2\,Cr^{3+} + 7\,H_2O$	1.38
$Cr^{3+} + e^- \rightarrow Cr^{2+}$	−0.424
$Cr^{2+} + 2\,e^- \rightarrow Cr(s)$	−0.90
$MnO_4^- + e^- \rightarrow MnO_4^{2-}$	0.56
$MnO_4^- + 8\,H^+ + 5\,e^- \rightarrow Mn^{2+} + 4\,H_2O$	1.51
$MnO_2(s) + 4\,H^+ + 2\,e^- \rightarrow Mn^{2+} + 2\,H_2O$	1.23
$Mn^{3+} + e^- \rightarrow Mn^{2+}$	1.5
$Mn^{2+} + 2\,e^- \rightarrow Mn(s)$	−1.18
$Fe^{3+} + e^- \rightarrow Fe^{2+}$	0.771
$Fe(phen)_3^{3+} + e^- \rightarrow Fe(phen)_3^{2+}$	1.13
$Fe(CN)_6^{3-} + e^- \rightarrow Fe(CN)_6^{4-}$	0.361
$Fe(CN)_6^{4-} + 2\,e^- \rightarrow Fe(s) + 6\,CN^-$	−1.16
$Fe^{2+} + 2e^- \rightarrow Fe(s)$	−0.44
$Co^{3+} + e^- \rightarrow Co^{2+}$	1.92
$Co(NH_3)_6^{3+} + e^- \rightarrow Co(NH_3)_6^{2+}$	0.058

Half–Cell Reaction	$E°/V$ (25°C)
$Co(phen)_3^{3+} + e^- \rightarrow Co(phen)_3^{2+}$	0.327
$Co(C_2O_4)_3^{3-} + e^- \rightarrow Co(C_2O_4)_3^{4-}$	0.57
$Co^{2+} + 2\,e^- \rightarrow Co(s)$	−0.277
$NiO_2(s) + 4\,H^+ + 2\,e^- \rightarrow Ni^{2+} + 2\,H_2O$	1.593
$Ni^{2+} + 2\,e^- \rightarrow Ni(s)$	−0.257
$Ni(OH)_2(s) + 2\,e^- \rightarrow Ni(s) + 2\,OH^-$	−0.72
$Cu^{2+} + e^- \rightarrow Cu^+$	0.159
$CuCl(s) + e^- \rightarrow Cu(s) + Cl^-$	0.121
$Cu^{2+} + 2\,e^- \rightarrow Cu(s)$	0.340
$Cu(NH_3)_4^{2+} + 2\,e^- \rightarrow Cu(s) + 4\,NH_3$	−0.00
$Ag^{2+} + e^- \rightarrow Ag^+$	1.980
$Ag^+ + e^- \rightarrow Ag(s)$	0.7991
$AgCl(s) + e^- \rightarrow Ag(s) + Cl^-$	0.2223
$Zn^{2+} + 2\,e^- \rightarrow Zn(s)$	−0.7626
$Zn(OH)_4^{2-} + 2\,e^- \rightarrow Zn(s) + 4\,OH^-$	−1.285
$Cd^{2+} + 2\,e^- \rightarrow Cd(s)$	−0.4025
$2\,Hg^{2+} + 2\,e^- \rightarrow Hg_2^{2+}$	0.9110
$Hg_2^{2+} + 2\,e^- \rightarrow 2\,Hg(l)$	0.7960
$Hg_2Cl_2(s) + 2\,e^- \rightarrow 2\,Hg(l) + 2\,Cl^-$	0.26816

Data from Bard, Parsons, and Jordan (H12).

TABLE A.5 Some Formal Reduction Potentials

| Couple | $E^{\circ\prime}$/V (25°C) | | |
	1 M $HClO_4$	1 M HCl	1 M H_2SO_4
Ag(I)/Ag(0)	0.792	0.228	0.77
As(V)/As(III)	0.577	0.577	
Ce(IV)/Ce(III)	1.70	1.28	1.44
Fe(III)/Fe(II)	0.732	0.700	0.68
H(I)/H(0)	−0.005	−0.005	
Hg(II)/Hg(I)	0.907		
Hg(I)/Hg(0)	0.776	0.274	0.674
Mn(IV)/Mn(II)	1.24		
Pb(II)/Pb(0)	−0.14		−0.29
Sn(IV)/Sn(II)		0.14	
Sn(II)/Sn(0)	−0.16		

Data from E. H. Swift and E. A. Butler, *Quantitative Measurements and Chemical Equilibria*, San Francisco: Freeman, 1972.

Reference Electrode Potentials:

Calomel (0.1 M KCl)	$E^{\circ\prime} = 0.336$ V	$dE/dT = -0.08$ V k^{-1}
Calomel (1.0 M KCl)	$E^{\circ\prime} = 0.283$ V	$dE/dT = -0.29$ V k^{-1}
Calomel (satd. KCl)	$E^{\circ\prime} = 0.244$ V	$dE/dT = -0.67$ V k^{-1}
Ag/AgCl (3.5 M KCl)	$E^{\circ\prime} = 0.205$ V	$dE/dT = -0.73$ V k^{-1}
Ag/AgCl (satd. KCl)	$E^{\circ\prime} = 0.199$ V	$dE/dT = -1.01$ V k^{-1}

TABLE A.6 Biochemical Reduction Potentials

Half–Cell Reaction	E'/V

Reduction of a carboxyl group to an aldehyde:

1,3–diphosphoglycerate + 2 e⁻

$\qquad\qquad \rightarrow$ 3–phosphoglyceraldehyde + $HPO_4{}^{2-}$ –0.286

acetyl–CoA + 2 H^+ + 2 e⁻ → acetaldehyde + coenzyme A –0.412

oxalate + 3 H^+ + 2 e⁻ → glyoxalate   –0.462

gluconate + 3 H^+ + 2 e⁻ → glucose   –0.47

acetate + 3 H^+ + 2 e⁻ → acetaldehyde   –0.598

Reduction of a carbonyl group to an alcohol:

dehydroascorbic acid + H^+ + 2 e⁻ → ascorbate   0.077

glyoxylate + 2 H^+ + 2 e⁻ → glycolate   –0.090

hydroxypyruvate + 2 H^+ + 2 e⁻ → glycerate   –0.158

oxaloacetate + 2 H^+ + 2 e⁻ → malate   –0.166

pyruvate + 2 H^+ + 2 e⁻ → lactate   –0.190

acetaldehyde + 2 H^+ + 2 e⁻ → ethanol   –0.197

acetoacetate + 2 H^+ 2 e⁻ → β–hydroxybutyrate   –0.349

Carboxylation:

pyruvate + CO_2(g) + H^+ + 2 e⁻ → malate   –0.330

a–ketoglutarate + CO_2(g) + H^+ + 2 e⁻ → iso–citrate   –0.363

succinate + CO_2(g) + 2 H^+ + 2 e⁻

$\qquad\qquad \rightarrow$ a–ketoglutarate + H_2O –0.673

acetate + CO_2(g) + 2 H^+ + 2 e⁻ → pyruvate + H_2O –0.699

Half–Cell Reaction	E'/V

Reduction of a carbonyl group with formation of an amino group:

oxaloacetate $+ NH_4^+ + 2 H^+ + 2 e^- \rightarrow$ aspartate $+ H_2O$ $\quad$ −0.107

pyruvate $+ NH_4^+ + 2 H^+ + 2 e^- \rightarrow$ alanine $+ H_2O$ $\quad$ −0.132

α–ketoglutarate $+ NH_4^+ + 2 H^+ + 2 e^-$
$\rightarrow$ glutamate $+ H_2O$ $\quad$ −0.133

Reduction of a carbon–carbon double bond:

crotonyl–CoA $+ 2 H^+ + 2 e^- \rightarrow$ butyryl–CoA $\quad$ 0.187

fumarate $+ 2 H^+ + 2 e^- \rightarrow$ succinate $\quad$ 0.031

Reduction of disulfide:

cystine $+ 2 H^+ + 2 e^- \rightarrow$ 2 cysteine $\quad$ −0.340

glutathione dimer $+ 2 H^+ + 2 e^- \rightarrow$ 2 glutathione $\quad$ −0.340

Other reductions of biochemical interest:

$O_2 + 4 H^+ + 4 e^- \rightarrow 2 H_2O$ $\quad$ 0.816

cytochrome c $(Fe^{3+}) + e^- \rightarrow$ cytochrome c (Fe^{2+}) $\quad$ 0.25

$FAD^+ + H^+ + 2 e^- \rightarrow$ FADH $\quad$ −0.20

$NAD^+ + H^+ + 2 e^- \rightarrow$ NADH $\quad$ −0.320

$2 H^+ + 2 e^- \rightarrow H_2$ $\quad$ −0.414

Reduction potentials at 25°C, pH 7 standard state; data from H. A. Krebs, H. L. Kornberg, and K. Burton, *Erg. Physiol.* **1957,** *49,* 212.

TABLE A.7 Molar Ionic Conductivities

Ion	$\Lambda°$	Ion	$\Lambda°$	Ion	$\Lambda°$
H^+	349.8	Pb^{2+}	139.0	IO_3^-	40.5
Li^+	38.7	Mn^{2+}	107.	IO_4^-	54.6
Na^+	50.1	Fe^{2+}	107.	MnO_4^-	62.8
K^+	73.5	Co^{2+}	110.	HCO_3^-	44.5
Cs^+	77.3	Ni^{2+}	108.	$H_2PO_4^-$	36.
NH_4^+	73.6	Al^{3+}	189.	HCO_2^-	54.6
$(CH_3)_4N^+$	44.9	Cr^{3+}	201.	$CH_3CO_2^-$	40.9
$(C_2H_5)_4N^+$	32.7	Fe^{3+}	204.	$C_2H_5CO_2^-$	35.8
$(C_3H_7)_4N^+$	23.4	OH^-	199.2	$C_6H_5CO_2^-$	32.4
Ag^+	61.9	F^-	55.4	CO_3^{2-}	138.6
Mg^{2+}	106.1	Cl^-	76.3	SO_4^{2-}	160.0
Ca^{2+}	119.0	Br^-	78.1	$S_2O_3^{2-}$	174.8
Sr^{2+}	118.9	I^-	76.8	CrO_4^{2-}	170.
Ba^{2+}	127.3	CN^-	78.	HPO_4^{2-}	114.
Cu^{2+}	107.2	NO_2^-	72.	$C_2O_4^{2-}$	148.3
Zn^{2+}	105.6	NO_3^-	71.5	$P_3O_9^{3-}$	250.8
Cd^{2+}	108.0	ClO_3^-	64.6	$Fe(CN)_6^{3-}$	302.7
Hg^{2+}	127.2	ClO_4^-	67.4	$Fe(CN)_6^{4-}$	442.

Conductivities from Robinson and Stokes *(B9)* and Dobos *(H9)* in units of 10^{-4} S $m^2 mol^{-1}$, aqueous solutions at infinite dilution, 25°C.

TABLE A.8 Solvent Properties

Solvent[a]	Liquid Range/°C	Vapor Pressure[b]	Dielectric Constant	Viscosity[c]
Water	0 to 100	3.2	78.4	0.89
Propylene carbonate (PC)	−49 to 242	0.0	64.4	2.5
Dimethylsulfoxide (DMSO)	19 to 189	0.1	46.7	2.00
N,N–Dimethylformamide (DMF)	−60 to 153	0.5	36.7	0.80
Acetonitrile	−44 to 82	11.8	37.5	0.34
Nitromethane	−29 to 101	4.9	35.9	0.61
Methanol	−98 to 65	16.7	32.7	0.54
Hexamethylphosphoramide (HMPA)	7 to 233	0.01	30.[d]	3.47[d]
Ethanol	−114 to 78	8.0	24.6	1.08
Acetone	−95 to 56	24.2	20.7	0.30
Dichloromethane	−95 to 40	58.1	8.9	0.41
Trifluoroacetic acid	−15 to 72	14.4	8.6	0.86
Tetrahydrofuran (THF)	−108 to 66	26.3	7.6	0.46
1,2–Dimethoxyethane (glyme, DME)	−58 to 93	10.0	7.2	0.46
Acetic acid	17 to 118	2.0	6.2[d]	1.13
p–Dioxane	12 to 101	4.9	2.2	1.2

[a] Data at 25°C from J. A. Riddick and W. B. Bunger, *Organic Solvents*, 3rd ed, New York: Wiley–Interscience, 1970.
[b] Vapor pressure in units of kPa.
[c] Viscosity in units of 10^{-3} kg m^{-1}s^{-1}.
[d] 20°C.

TABLE A.9 Potential Range for Some Solutions

Solvent	Electrolyte	Potential Range/V (vs. s.c.e.)	
		Pt	Hg
Propylene carbonate	Et_4NClO_4	1.7 to −1.9	0.5 to −2.5
Dimethylsulfoxide	$NaClO_4$	0.7 to −1.8	0.6 to −2.9
	Et_4NClO_4	0.7 to −1.8	0.2 to −2.8
	Bu_4NI		−0.4 to −2.8
Dimethylformamide	$NaClO_4$	1.6 to −1.6	0.5 to −2.0
	Et_4NCl_4	1.6 to −2.1	0.5 to −3.0
	Et_4NBF_4		— to −2.7
	Bu_4ClO_4	1.5 to −2.5	0.5 to −3.0
	Bu_4I		−0.4 to −3.0
Acetonitrile	$NaClO_4$	1.8 to −1.5	0.6 to −1.7
	Et_4NClO_4		0.6 to −2.8
	Et_4NBF_4	2.3 to —	— to −2.7
	Bu_4NI		−0.6 to −2.8
	Bu_4PF_6	3.4 to −2.9	
Acetone	$NaClO_4$	1.6 to —	
	Et_4NClO_4		— to −2.4
Dichloromethane	Bu_4NClO_4	1.8 to −1.7	0.8 to −1.9
	Bu_4NI	0.2 to −1.7	−0.5 to −1.7
1,2–Dimethoxyethane	Bu_4NClO_4		0.6 to −2.9

Data from C. K. Mann, *Electroanal. Chem.* **1969,** *3,* 57.

APPENDIX 4. LAPLACE TRANSFORM METHODS

The method of Laplace transforms provides a powerful aid to the solution of differential equations.[‡] The method is particularly useful in solving the coupled partial differential equations which are encountered in electrochemical diffusion problems. Here we will briefly introduce the technique and demonstrate the method by deriving a few of the results quoted in the text.

Laplace Transformations

The Laplace transform of a function $F(t)$ is defined by

$$f(s) = \int_0^\infty e^{-st} F(t)\, dt \qquad (A.1)$$

Not all functions possess a Laplace transform. Clearly, $F(t)$ must be finite for finite t and $F(t) \exp(-st)$ must go to zero as $t \to \infty$. The Laplace transformation can be thought of as an operation in linear algebra:

$$f(s) = \mathbf{L}[F(t)]$$

which is reversible by the inversion operation

$$F(t) = \mathbf{L}^{-1}[f(s)]$$

The Laplace transformation is a linear operation, that is, sums or differences of functions are transformed as

$$\mathbf{L}[F(t) + G(t)] = f(s) + g(s) \qquad (A.2)$$

[‡] F. E. Nixon, *Handbook of Laplace Transformations: Tables and Examples*, Englewood Cliffs, NJ: Prentice–Hall, 1960; R. V. Churchill, *Modern Operational Mathematics in Engineering*, 2nd ed, New York: McGraw–Hill, 1963; P. A. McCollum and B. F. Brown, *Laplace Transform Tables and Theorems*, New York: Holt, Rinehart, and Winston, 1965; M. G. Smith, *Laplace Transform Theory*, London, D. Van Nostrand, 1966.

Multiplicative constants are unaffected by Laplace transformation:

$$L[a\ F(t)] = a\ f(s) \tag{A.3}$$

The utility of Lapace transforms in the solution of differential equations is that the transform of a derivative is a simple function

$$L[dF(t)/dt] = s\ f(s) - F(0) \tag{A.4a}$$

$$L[d^2F(t)/dt^2] = s^2 f(s) - s\ F(0) - (dF/dt)_0 \tag{A.4b}$$

The Laplace tranform of a constant is

$$L(a) = a/s$$

Functions of variables other than t behave as constants in the transformation

$$L[H(x)] = H(x)/s$$

A short selection of Laplace transforms are found in Table A.10.

Two properties of the Laplace transformation are sometimes useful in finding the inverse transform. The "shift theorem" allows the zero of s to be displaced by a constant:

$$L^{-1}[f(s + a)] = F(t)\ \exp(-at) \tag{A.5}$$

The "convolution theorem" is useful when the inverse transformation of f(s) cannot be found, but f(s) can be written as the product of two functions, f(s) = g(s)h(s), the inverse transforms of which can be found. If

$$G(t) = L^{-1}[g(s)]$$

$$H(t) = L^{-1}[h(s)]$$

then

$$L^{-1}[g(s)h(s)] = \int_0^t G(t - \tau)H(\tau)\ d\tau \tag{A.6}$$

TABLE A.10 Some Laplace Transforms

F(t)	f(s)	F(t)	f(s)
a (a constant)	a/s	$\sin at$	$\dfrac{a}{s^2 + a^2}$
t	s^{-2}	$\cos at$	$\dfrac{s}{s^2 + a^2}$
$\dfrac{t^{n-1}}{(n-1)!}$	s^{-n}	$\sinh at$	$\dfrac{a}{s^2 - a^2}$
$(\pi t)^{-\frac{1}{2}}$	$s^{-\frac{1}{2}}$	$\cosh at$	$\dfrac{s}{s^2 - a^2}$
$2(t/\pi)^{\frac{1}{2}}$	$s^{-\frac{3}{2}}$	$\exp at$	$\dfrac{1}{s-a}$

F(t)	f(s)
$\dfrac{\exp at - \exp bt}{a - b}$	$\dfrac{1}{(s-a)(s-b)}$
$\exp(at)\,\mathrm{erf}(at)^{\frac{1}{2}}$	$\dfrac{a^{\frac{1}{2}}}{s^{\frac{1}{2}}(s-a)}$
$\exp(at)[1 - \mathrm{erf}(at)^{\frac{1}{2}}]$	$\dfrac{1}{s^{\frac{1}{2}}(s^{\frac{1}{2}} + a^{\frac{1}{2}})}$
$1 - \exp(at)[1 - \mathrm{erf}(at)^{\frac{1}{2}}]$	$\dfrac{a^{\frac{1}{2}}}{s(s^{\frac{1}{2}} + a^{\frac{1}{2}})}$
$[1/2t(\pi at)^{\frac{1}{2}}]\exp(-1/4at)$	$\exp[-(s/a)^{\frac{1}{2}}]$
$(\pi t)^{-\frac{1}{2}}\exp(-1/4at)$	$(s^{-\frac{1}{2}})\exp[-(s/a)^{\frac{1}{2}}]$
$1 - \mathrm{erf}[1/2(at)^{\frac{1}{2}}]$	$(s^{-1})\exp[-(s/a)^{\frac{1}{2}}]$

The specific solution to a differential equation depends on the initial and boundary conditions on the problem. The solution to a differential equation using Laplace transform methods in general follows the steps:

(1) Transform the differential equation to remove derivatives with respect to one of the variables. An ordinary differential equation will then be an algebraic equation and a partial differential equation with two independent variables will become an ordinary differential equation.

(2) Transform the initial and boundary conditions.

(3) Solve the resulting system of algebraic equations or ordinary differential equations, using the transformed boundary conditions to evaluate constants.

(4) Take the inverse transform to obtain the solution to the original differential equation.

Consider as an example the ordinary differential equation

$$\frac{d^2 F(t)}{dt^2} + a^2 F(t) = 0$$

with boundary conditions $F(0) = 0$, $dF(0)/dx = a$. Taking the Laplace transform, using eqs (A.2)–(A.4), we have

$$s^2 f(s) - a + a^2 f(s) = 0$$

from which we obtain

$$f(s) = \frac{a}{a^2 + s^2}$$

The inverse transform from Table A.10 gives

$$F(t) = \sin at$$

We really didn't need a fancy method to solve this problem, but other cases arise which are not quite so simple.

Solutions of the Diffusion Equation

Now let us apply Laplace transform methods to the solution of the one–dimensional diffusion equation

$$\frac{\partial C(x,t)}{\partial t} = D \frac{\partial^2 C(x,t)}{\partial x^2} \tag{A.7}$$

We can do step (1) of the solution procedure in general. Writing the Laplace transform of $C(x,t)$ as $c(x,s)$, the transformed diffusion equation is

$$s\, c(x,s) - C(x,0) = D \frac{\partial^2 c(x,s)}{\partial x^2} \tag{A.8}$$

We need initial and boundary conditions to solve eq (A.8), and these differ from one problem to another.

Derivation of eq (3.22). Let us start with the problem posed in Section 3.3. We considered a solution layered on pure solvent so that the initial condition was $C = C^*$ for $x < 0$, $C = 0$ for $x > 0$. The boundary condition is $C \to C^*$ as $x \to -\infty$, $C \to 0$ as $x \to +\infty$. We will divide the problem into two regimes, $-\infty < x < 0$ and $0 < x < +\infty$, with the requirement that $C(x,t)$ and $J(x,t)$ be continuous at $x = 0$ for $t > 0$. Thus for $x > 0$, we have

$$D \frac{\partial^2 c(x,s)}{\partial x^2} - s\, c(x,s) = 0$$

the general solution to which is

$$c(x,s) = A(s)\, \exp[-(s/D)^{\frac{1}{2}}x] + B(s)\, \exp[+(s/D)^{\frac{1}{2}}x] \tag{A.9a}$$

where $A(s)$ and $B(s)$ are to be determined from the boundary conditions. One of the boundary conditions requires $c(x,s) \to 0$ as $x \to \infty$ so that $B(s) = 0$. For $x < 0$, eq (A.8) gives

$$D \frac{\partial^2 c'(x,s)}{\partial x^2} - s\, c'(x,s) + C^* = 0$$

The general solution to this differential equation is

$$c'(x,s) = C^*/s + A'(s) \exp[-(s/D)^{\frac{1}{2}}x] + B'(s) \exp[+(s/D)^{\frac{1}{2}}x] \quad (A.9b)$$

The boundary condition requires $c'(x,s) \to C^*/s$ as $x \to -\infty$ so that $A'(s)$ = 0. We now apply the continuity restraints to determine $A(s)$ and $B'(s)$. If $C(0,t) = C'(0,t)$, then $c(0,s) = c'(0,s)$. Thus we have

$$c(0,s) = A(s)$$

$$c'(0,s) = C^*/s + B'(s)$$

so that

$$A(s) - B'(s) = C^*/s$$

The equal fluxes at $x = 0$ means that

$$\frac{\partial c(0,s)}{\partial x} = \frac{\partial c'(0,s)}{\partial x}$$

Differentiating eqs (A.9) and setting $x = 0$, we have

$$-(s/D)^{\frac{1}{2}}A(s) = +(s/D)^{\frac{1}{2}}B'(s)$$

Thus

$$A(s) = -B'(s) = C^*/2s$$

Equations (A.9) then become

$$c(x,s) = (C^*/2s) \exp[-(s/D)^{\frac{1}{2}}x] \qquad x > 0$$

$$c'(x,s) = C^*/s - (C^*/2s) \exp[-(s/D)^{\frac{1}{2}}|x|] \qquad x < 0$$

Taking the inverse transforms, we have

$$C(x,t) = \tfrac{1}{2} C^* [1 - \mathrm{erf}(x/2D^{\frac{1}{2}}t^{\frac{1}{2}}] \qquad x > 0$$

$$C'(x,t) = C^* - \tfrac{1}{2} C^* [1 - \mathrm{erf}(|x|/2D^{\frac{1}{2}}t^{\frac{1}{2}})] \qquad x < 0$$

Since the error function is an odd function of the argument, *i.e.*, $\mathrm{erf}(-\psi) = -\mathrm{erf}(\psi)$, we see that these two functions are in fact identical:

$$C(x,t) = \tfrac{1}{2} C^* [1 - erf(x/2D^{\frac{1}{2}}t^{\frac{1}{2}})] \qquad (3.22)$$

Derivation of eqs (4.2). In typical electrochemical applications of the diffusion equation, the concentrations of the diffusing species are uniform at the beginning of the experiment, $C(x,0) = C^*$ and, at later times, approach the initial concentration at sufficient distance from the electrode, $C(x,t) \to C^*$ as $x \to \infty$. With the electrode as $x = 0$, we need not consider negative values for x. Thus eq (A.8) is

$$D \frac{\partial^2 c(x,s)}{\partial x^2} - s\, c(x,s) + C^* = 0$$

The solution consistent with the boundary condition (at $x \to \infty$) is

$$c(x,s) = C^*/s + A(s) \exp[-(s/D)^{\frac{1}{2}}x] \qquad (A.10)$$

where $A(s)$ must be determined by the boundary condition at $x = 0$. If we have two species, O and R, which are involved in an electrode process, each transformed concentration will have the form of eq (A.10). If the initial concentrations are $C_O(x,0) = C_O{}^*$, $C_R(0,t) = 0$, then eq (A.10) gives

$$c_O(x,s) = C_O{}^*/s + A(s) \exp[-(s/D_O)^{\frac{1}{2}}x] \qquad (A.11a)$$

$$c_R(x,s) = B(s) \exp[-(s/D)^{\frac{1}{2}}x] \qquad (A.11b)$$

For a reversible electrode process, the surface boundary conditions are: (1) the concentration ratio at $x = 0$, governed by the Nernst equation

$$\frac{C_O(0,t)}{C_R(0,t)} = \theta = \exp \frac{nF(E - E^\circ)}{RT} \qquad (A.12a)$$

and (2) the continuity restriction

$$J_O(0,t) = -J_R(0,t)$$

or

$$-D_O \frac{\partial C_O(0,t)}{\partial x} = D_R \frac{\partial C_R(0,t)}{\partial x} \qquad (A.12b)$$

The boundary conditions transform to

$$c_O(0,s) = \theta \; c_R(0,s) \qquad\qquad \text{(A.13a)}$$

and

$$- D_O \frac{\partial c_O(0,s)}{\partial x} = D_R \frac{\partial c_R(0,s)}{\partial x} \qquad\qquad \text{(A.13b)}$$

Differentiating $c_O(x,s)$ and $c_R(x,s)$ with respect to x and setting x = 0, we have on substitution in eq (A.13b),

$$(D_O s)^{\frac{1}{2}} A(s) = - (D_R s)^{\frac{1}{2}} B(s)$$

or

$$B(s) = - \xi \; A(s)$$

where

$$\xi = (D_O/D_R)^{\frac{1}{2}}$$

Setting x = 0 in eqs (A.11) and substituting in eq (A.13a) gives

$$C_O{}^*/s + A(s) = \theta \, B(s) = - \xi \theta \; A(s)$$

so that

$$A(s) = - \frac{C_O{}^*}{s(1 + \xi \theta)}$$

Thus the transformed concentrations are

$$c_O(x,s) = \frac{C_O{}^*}{s} \left[1 - \frac{\exp[-(s/D_O)^{\frac{1}{2}}x]}{1 + \xi \theta} \right]$$

$$c_R(x,s) = \frac{\xi \, C_O{}^*}{s(1 + \xi \theta)} \exp[-(s/D_R)^{\frac{1}{2}}x]$$

Taking the inverse transform, we have

$$C_O(x,t) = C_O{}^* \frac{\xi\theta + \text{erf}(x/2D_O{}^{\frac{1}{2}}t^{\frac{1}{2}})}{1 + \xi\theta} \qquad (4.2a)$$

$$C_R(x,t) = C_O{}^* \frac{\xi[1 - \text{erf}(x/2D_O{}^{\frac{1}{2}}t^{\frac{1}{2}})]}{1 + \xi\theta} \qquad (4.2b)$$

Derivation of eq (4.20b). In double potential step chronoamperometry, the electrode is polarized for a time τ at a sufficiently negative potential that $C_O(0,t) = 0$ ($\theta = 0$); the potential is then stepped to a positive potential so that $C_R(0, t - \tau) = 0$. This problem is easily solved using Laplace transforms by noting that eq (4.2b) gives the "initial" concentration distribution of R for the second potential step. Thus substituting eq (4.2b) with $t = \tau$ and $\theta = 0$ into eq (A.10), we have

$$c_R(x,s) = (\xi C_O{}^*/s)[1 - \text{erf}(x/2D_R{}^{\frac{1}{2}}\tau^{\frac{1}{2}})] + A(s)\exp[-(s/D_R)^{\frac{1}{2}}x]$$

where the transform variable s corresponds to $(t - \tau)$. The boundary condition $C_R(0,t) = 0$ for $t > \tau$ implies that $c_R(0,s) = 0$. Thus

$$(\xi C_O{}^*/s) + A(s) = 0$$

so that

$$A(s) = - \xi C_O{}^*/s$$

We need the flux of R at the electrode in order to calculate the current. Thus we compute the derivative with respect to x

$$\frac{\partial c_R(0,s)}{\partial x} = - \frac{\xi C_O{}^*}{s(\pi D_R\tau)^{\frac{1}{2}}} + \frac{\xi C_O{}^*}{(D_R s)^{\frac{1}{2}}}$$

and take the inverse transform to obtain

$$\frac{\partial C_R(0,t)}{\partial x} = - \frac{\xi C_O{}^*}{(\pi D_R\tau)^{\frac{1}{2}}} + \frac{\xi C_O{}^*}{[\pi D_R(t - \tau)]^{\frac{1}{2}}}$$

Since the current is

$$i = -nFAD_R \frac{\partial C_R(0,t)}{\partial x}$$

we get

$$i = -nFA (D_O/\pi)^{\frac{1}{2}} C_O^* [(t - \tau)^{-\frac{1}{2}} - t^{-\frac{1}{2}}] \qquad t > \tau \qquad (4.20b)$$

Derivation of eq (4.21a). In a constant current experiment such as chronopotentiometry, the flux of O at the electrode surface is constant up to the transition time when $C_O(0,t) \to 0$ and the potential swings negative. Thus the boundary condition is

$$D_O \frac{\partial C_O(0,t)}{\partial x} = \frac{i}{nFA}$$

Taking the Laplace transform, we have

$$D_O \frac{\partial c_O(0,s)}{\partial x} = \frac{i}{nFAs}$$

Differentiating eq (A.10) and setting $x = 0$, we have

$$-D_O(s/D_O)^{\frac{1}{2}} A(s) = i/nFAs$$

Solving for A(s) and substituting in eq (A.10) with $x = 0$:

$$c_O(0,s) = \frac{C_O^*}{s} - \frac{i}{nFAD_O^{\frac{1}{2}} s^{\frac{3}{2}}}$$

Taking the inverse transform, we get

$$C_O(0,t) = C_O^* - \frac{2i t^{\frac{1}{2}}}{nFA(\pi D_O)^{\frac{1}{2}}}$$

Apparently, $C_O(0,t)$ goes to zero when

$$nFAC_O^*(\pi D_O)^{\frac{1}{2}} = 2i t^{\frac{1}{2}}$$

so that the transition time is given by

$$\tau^{\frac{1}{2}} = nFA(\pi D_O)^{\frac{1}{2}} C_O^* / 2i \tag{4.21a}$$

Derivation of eqs (5.36). Allowing for a finite electron transfer rate, the surface boundary conditions are

$$D_O \frac{\partial C_O(0,t)}{\partial x} = -D_R \frac{\partial C_R(0,t)}{\partial x} = k_c C_O(0,t) - k_a C_R(0,t)$$

Transforming the boundary conditions, we again have eq (A.13b), but instead of eq (A.13a), we get

$$D_O \frac{\partial c_O(0,s)}{\partial x} = k_c c_O(0,s) - k_a c_R(0,s)$$

Thus, with $c_O(x,s)$ and $c_R(x,s)$ given by eqs (A.11), and $B(s) = -\xi A(s)$, we have

$$-(D_O s)^{\frac{1}{2}} A(s) = k_c [C_O^*/s + A(s)] + k_a \xi A(s)$$

or

$$A(s) = -\frac{k_c D_O^{-\frac{1}{2}} C_O^*}{s(\lambda + s^{\frac{1}{2}})}$$

where

$$\lambda = k_c / D_O^{\frac{1}{2}} + k_a / D_R^{\frac{1}{2}}$$

Substituting $A(s)$ into eqs (A.11) with $x = 0$, we have

$$c_O(0,s) = \frac{C_O^*}{s} \left[1 - \frac{k_c / D_O^{\frac{1}{2}}}{\lambda + s^{\frac{1}{2}}} \right]$$

$$c_R(0,s) = \frac{C_O^* k_c / D_R^{\frac{1}{2}}}{s(\lambda + s^{\frac{1}{2}})}$$

Taking the inverse transforms, we have

$$C_O(0,t) = C_O^* - \frac{k_c C_O^*/D_O^{\frac{1}{2}}}{\lambda} \{1 - \exp(\lambda^2 t)[1 - \text{erf}(\lambda t^{\frac{1}{2}})]\}$$

$$C_R(0,t) = \frac{k_c C_O^*/D_R^{\frac{1}{2}}}{\lambda} \{1 - \exp(\lambda^2 t)[1 - \text{erf}(\lambda t^{\frac{1}{2}})]\}$$

Defining

$$F(\lambda t^{\frac{1}{2}}) = \lambda(\pi t)^{\frac{1}{2}} \exp(\lambda^2 t)[1 - \text{erf}(\lambda t^{\frac{1}{2}})]$$

we have

$$C_O(0,t) = C_O^* - \frac{C_O^*}{1 + \xi\theta} \left[1 - \frac{F(\lambda t^{\frac{1}{2}})}{\lambda(\pi t)^{\frac{1}{2}}} \right] \qquad \text{(A.14a)}$$

$$C_R(0,t) = \frac{\xi C_O^*}{1 + \xi\theta} \left[1 - \frac{F(\lambda t^{\frac{1}{2}})}{\lambda(\pi t)^{\frac{1}{2}}} \right] \qquad \text{(A.14b)}$$

where we have used the relation $k_a/k_c = \theta$. Equations (A.14) are identical to eqs (5.36) when $D_O = D_R$ so that $\xi = 1$. Equations (A.14) reduce to eqs (4.2) in the limit $F(\lambda t^{\frac{1}{2}})/\lambda(\pi t^{\frac{1}{2}}) = 0$.

Derivation of eqs (5.46). When the boundary condition is determined by a sinusoidal current

$$I(t) = I_0 \sin\omega t$$

the flux of O at the electrode surface is

$$-D_O \frac{\partial C_O(0,t)}{\partial x} = -(I_0/FA) \sin\omega t$$

Taking the Laplace transform, we have

$$D_O \frac{\partial c_O(0,s)}{\partial x} = \frac{I_0}{FA} \frac{\omega}{s^2 + \omega^2}$$

Substituting the first derivative of eq (A.11), evaluated at $x = 0$, we get

$$-(sD_O)^{\frac{1}{2}}A(s) = \frac{I_0}{FA} \frac{\omega}{s^2 + \omega^2}$$

Solving for A(s) and substituting in eq (A.11) with x = 0, we have

$$c_O(0,s) = \frac{C_O^*}{s} - \frac{I_0}{FAD_O^{\frac{1}{2}}} \frac{\omega s^{-\frac{1}{2}}}{s^2 + \omega^2}$$

The inverse transform of this function cannot be found in tables, so we have recourse to the convolution theorem, eq (A.6), taking

$$g(s) = \omega/(s^2 + \omega^2) \rightarrow G(t) = \sin \omega t$$

$$h(s) = s^{-\frac{1}{2}} \rightarrow H(t) = (\pi t)^{-\frac{1}{2}}$$

Thus

$$L^{-1}\left[\frac{\omega s^{-\frac{1}{2}}}{s^2 + \omega^2}\right] = \int_0^t (\pi \tau)^{-\frac{1}{2}} \sin \omega (t - \tau) \, d\tau$$

Using the trigonometric identity

$$\sin \omega (t - \tau) = \sin \omega t \cos \omega \tau - \cos \omega t \sin \omega \tau$$

the integral becomes

$$\frac{\sin \omega t}{\sqrt{\pi}} \int_0^t \tau^{-\frac{1}{2}} \cos \omega \tau \, d\tau - \frac{\cos \omega t}{\sqrt{\pi}} \int_0^t \tau^{-\frac{1}{2}} \sin \omega \tau \, d\tau$$

The factor of $1/\sqrt{\tau}$ in the integrand represents a transient response to the application of the sinusoidal current which dies off to give a steady–state sinusoidal variation in the concentrations. Since we are interested only in the steady state, the limits on the integrals can be extended to infinity, obtaining

$$\int_0^\infty \tau^{-\frac{1}{2}} \cos \omega\tau \ d\tau \ = \ \int_0^\infty \tau^{-\frac{1}{2}} \sin \omega\tau \ d\tau \ = \ (\pi/2\omega)^{\frac{1}{2}}$$

Thus we have

$$C_O(0,t) = C_O{}^* - \frac{I_0(\sin \omega t - \cos \omega t)}{FA(2\omega D_O)^{\frac{1}{2}}} \qquad (5.46a)$$

Equation (5.46b) for $C_R(0,t)$ results from a similar development.

APPENDIX 5. DIGITAL SIMULATION METHODS

The theoretical description of an electrochemical experiment usually requires the solution of a set of coupled partial differential equations. based on the diffusion equation. In experiments which include forced convection, such as r.d.e. voltammetry, a driving term is added to each equation as in eq (4.29). When a diffusing species is involved in a chemical reaction, reaction rate terms must be added to the equation describing its concentration. The set of equations often has time–dependent boundary conditions and can be devilishly difficult to solve. In some fortunate cases, such as those treated in Appendix 4, the use of Laplace transforms leads to closed–form analytical solutions, but more often solutions are obtained in terms of infinite series or intractable integrals which must be evaluated numerically. In the 1950's and 1960's, a great deal of effort was expended by theorists in obtaining mathematical descriptions of electrochemical experiments. While the results provide an invaluable aid to understanding (we have quoted many of these results), one often finds that the theoretical results available in the literature do not quite cover the experimental case at hand. If the problem seems to be of sufficient generality and interest, it may be worthwhile attempting an analytical approach. More often, however, electrochemists have turned to the digital computer to simulate experiments. The details of digital simulation are beyond the scope of this text, but a brief outline of the strategy is in order. For further details see reviews by Feldberg,[‡] Maloy,[#] or Britz.[§]

In general the problem to be solved involves equations of the form

$$\frac{\partial C}{\partial t} = D \frac{\partial^2 C}{\partial x^2} + \text{kinetic and/or driving terms}$$

[‡]   S. W. Feldberg, *Electroanalytical Chemistry* **1969**, *3*, 199.

[#]   J. T. Maloy in *Laboratory Techniques in Electroanalytical Chemistry*, P. T. Kissinger and W. R. Heineman, eds, New York: Marcel Dekker, 1984.

[§]   D. Britz, *Digital Simulation in Electrochemistry, Lecture Notes in Chemistry*, Vol. 23, Heidelberg: Springer–Verlag, 1981.

together with a set of initial and boundary conditions. The concentrations are functions of time and distance from the electrode and the general strategy in a digital simulation is to divide the time and distance axes into discrete elements of size δt and δx, respectively. Writing Fick's first law as

$$J(x,t) = -D\,\frac{\partial C(x,t)}{\partial x} \cong -D\,\frac{C(x+\delta x/2,t) - C(x-\delta x/2,t)}{\delta x}$$

the second law,

$$\frac{\partial C(x,t)}{\partial t} = -\frac{\partial J(x,t)}{\partial x}$$

can be written

$$\frac{C(x,t+\delta t) - C(x,t)}{\delta t} = -\frac{J(x+\delta x/2,t) - J(x-\delta x/2,t)}{\delta x}$$

or

$$C(x,t+\delta t) = C(x,t) + \frac{D\delta t}{(\delta x)^2}\,[C(x+\delta x,t) - 2C(x,t) + C(x-\delta x,t)] \quad (A.15)$$

If the spatial boxes, of equal width δx, are labeled 1,2,3,...j... and the time boxes of width δt, are labeled 1,2,3,...k... we can rewrite eq (A.15) as

$$C(j,k+1) = C(j,k) + \mathbf{D}[C(j+1,k) - 2C(j,k) + C(j-1,k)] \quad (A.16)$$

where $\mathbf{D} = D\delta t/(\delta x)^2$ is a dimensionless diffusion coefficient.[‡] The simulation program is structured with an outer loop over k (the time

---

[‡] A requirement for the stability of the simulation is $\mathbf{D} < 0.5$. This is a consequence of the model which only considers diffusion between adjacent boxes within one cycle of iteration. A value of $\mathbf{D} > 0.5$ would require the simultaneous consideration of more than three boxes.

index) and an inner loop over j (the distance index). Starting with an initial concentration distribution, the effect of diffusion on the concentrations is modeled using eq (A.16).

The net flux at the electrode (related to the current) is determined by the changes in concentrations in the j = 1 box (adjacent to the electrode) which are required to satisfy the surface boundary condition. For example, for a reversible electrode process, the boundary condition corresponds to the surface concentration ratio $C_O/C_R = \theta$, specified by the Nernst equation. The equilibrium surface concentration of O can be written as

$$C_O(eq) = \frac{\theta}{1 + \theta} \, [C_O(1,k) + C_R(1,k)]$$

The net flux then corresponds to the difference between $C_O(1,k)$ and the equilibrium surface concentration

$$\delta C_O(k) = \frac{C_O(1,k) - \theta \, C_R(1,k)}{1 + \theta}$$

The surface concentrations are thus corrected to

$$C_O(1,k + 1) = C_O(1,k) - \delta C_O(k) \tag{A.17a}$$

$$C_R(1,k + 1) = C_R(1,k) + \delta C_O(k) \tag{A.17b}$$

In an experiment where the electrode potential is time dependent (*e.g.*, cyclic voltammetry), the potential (and thus the concentration ratio θ) will be different for each time increment.

If the species of interest is consumed in a first–order chemical reaction, the contribution to the rate equation is

$$\frac{\partial \, C(x,t)}{\partial t} = - \, kC(x,t)$$

In terms of the finite differences, this expression is

$$\frac{C(x,t+\delta t) - C(x,t)}{\delta t} = - k\, C(x,t)$$

or, using the indices j and k,

$$C(j,k+1) = C(j,k) - \mathbf{k}C(j,k) \tag{A.18}$$

where $\mathbf{k} = k\delta t$ is the dimensionless rate constant. In practice, $\mathbf{k}$ must be small (0.1 or less) in order to accurately model the system; for a given value of k, this places a restriction on δt. This restriction can be somewhat relaxed ($\mathbf{k} \approx 1$) by using an analytical solution for the extent of reaction during a time increment, $e.g.$, for a first–order process

$$C(j,k+1) = C(j,k)\, e^{-\mathbf{k}} \tag{A.19}$$

Since the size of the time increment δt is determined by the rate constant, the number of time increments to be used in the simulation is determined by

$$n_t = t/\delta t \tag{A.20}$$

where t is the total time of the experiment to be modeled. The size of the spatial increment is determined by

$$\delta x = (D\delta t/\mathbf{D})^{\frac{1}{2}} \tag{A.21}$$

During the time of the experiment, the diffusion layer grows to a thickness of about

$$6(Dt)^{\frac{1}{2}} = j_{max}\delta x$$

Thus the number of spatial boxes required in the simulation is

$$n_x = j_{max} = 6[Dn_t\delta t/(\delta x)^2]^{\frac{1}{2}}$$

or

$$n_x = 6(n_t\mathbf{D})^{\frac{1}{2}} \tag{A.22}$$

or $n_x \cong 4\sqrt{n_t}$ if $\mathbf{D} = 0.45$.

Even with the use of analytical expressions such as eq (A.19) to model the effects of chemical reactions, n_t can be very large for schemes with fast reactions. Since computer execution time increases as the product of n_t and n_x, there is a practical upper limit to rate constants. One solution to this problem is to use variable–width increments in the simulation. For very fast reactions, the reaction layer, x_R, is thin and, in the simulation, most of the action takes place in the first few spatial boxes adjacent to the electrode. Thus some saving in execution time can be realized if the size of the spatial boxes is allowed to increase with increasing distance from the electrode.[‡] (This approach leads to a different **D** for each box.) A still more efficient approach is to expand the time grid for boxes far from the electrode.[#] Thus boxes in which not much is happening are sampled less frequently.

The overall structure of the simulation program then is as follows:

(1) Set concentrations to initial values.

(2) Correct each concentration in each box for the results of the chemical reactions using eqs (A.18) or (A.19).

(3) Correct each concentration in each box for the results of diffusion using eq (A.16).

(4) Change the concentrations in box 1 to satisfy the surface boundary condition, *e.g.*, using eqs (A.17) for a nernstian process the changes correspond to the flux and thus to the current.

(5) Go to the next time increment and adjust time–dependent parameters such as the electrode potential.

(6) Repeat steps (2)–(5) for the required number of time increments.

[‡] T. Joslin and D. Pletcher, *J. Electroanal. Chem.* **1974**, *49*, 171.

[#] R. Seeber and S. Stefani, *Anal. Chem.* **1981**, *53*, 1011.

APPENDIX 6. R.D.E. STEADY–STATE METHODS

One of the advantages of rotating disk voltammetry is that it is relatively easy to derive current–potential curves for electrode processes perturbed by coupled chemical reactions. In Section 5.3, we used the steady–state approximation to compute voltammetric curves for irreversible processes. An equivalent procedure is to take advantage of the nearly linear concentration gradients across the diffusion layer at an r.d.e. (see Section 4.6). Here we will examine the problem in general and derive some expressions which will be useful in discussions of r.d.e. voltammetry for the CE, EC, EC', and ECE mechanisms.

The flux of a species X diffusing to the electrode across a linear concentration gradient can be expressed by

$$J_X(0) = -(D/x_D)[C_X^* - C_X(0)] \tag{A.23}$$

where x_D is the diffusion layer thickness, given by eq (4.31). This result is equivalent to neglecting the forced convection term in eq (4.29) and assuming steady–state conditions so that

$$\frac{\partial C_X}{\partial t} = D\,\frac{\partial^2 C_X}{\partial x^2} = 0$$

If $\partial^2 C_X/\partial x^2 = 0$, then the concentration gradient, $\partial C_X/\partial x$, is indeed constant and eq (A.23) results. This approximation greatly simplifies calculation of current–potential curves for r.d.e. voltammetry, particularly compared with the time–dependent problems encountered in the analysis of other voltammetric techniques.

In our discussions of r.d.e. voltammetry, we will assume identical diffusion coefficients for all diffusing species. This assumption is for convenience rather than necessity. It is easy enough to derive expressions where each species has its own diffusion coefficient, but the algebra becomes significantly more complex.

In each of the kinetic schemes, two species are coupled by a chemical step,

$$Y \quad \overset{k_1}{\underset{k_{-1}}{\rightleftarrows}} \quad Z$$

The steady–state diffusion equations for Y and Z are then modified to

$$D \frac{\partial^2 C_Y}{\partial x^2} - k_1 C_Y + k_{-1} C_Z = 0$$

$$D \frac{\partial^2 C_Z}{\partial x^2} + k_1 C_Y - k_{-1} C_Z = 0$$

These equations can be solved by the introduction of a new variable,

$$C' = K C_Y - C_Z$$

where $K = k_1/k_{-1}$ and C' measures the departure of the Y–Z reaction from equilibrium. Combining the differential equations in C_Y and C_Z, we obtain an equation in C':

$$D \frac{d^2 C'}{dx^2} = (k_1 + k_{-1}) C'$$

The general solution to this differential equation is

$$C'(x) = A \exp(-x/x_R) + B \exp(x/x_R)$$

where x_R, the *reaction layer thickness*, is given by

$$x_R^2 = \frac{D}{k_1 + k_{-1}} \tag{A.24}$$

We assume that for $x > x_D$, the solution is homogeneous and at equilibrium so that

$$C'(x_D) = 0$$

With this boundary condition, the parameters A and B can be evaluated to give

$$C'(x) = C'(0) \frac{\exp[(x_D - x)/x_R] - \exp[-(x_D - x)/x_R]}{\exp(x_D/x_R) - \exp(-x_D/x_R)}$$

where

$$C'(0) = KC_Y(0) - C_Z(0)$$

Since we know $C'(x)$, we can compute its first derivative,

$$\frac{dC'(x)}{dx} = - \frac{C'(0)}{x_R} \frac{\exp[(x_D - x)/x_R] + \exp[-(x_D - x)/x_R]}{\exp(x_D/x_R) - \exp(-x_D/x_R)}$$

and then obtain an expression involving the fluxes of Y and Z:

$$KJ_Y(0) - J_Z(0) = (D/x_R')[KC_Y(0) - C_Z(0)] \qquad (A.25)$$

where

$$x_R' = x_R \tanh(x_D/x_R) = x_R \frac{\exp(x_D/x_R) - \exp(-x_D/x_R)}{\exp(x_D/x_R) + \exp(-x_D/x_R)} \qquad (A.26)$$

The parameter x_R'/x_D is important in the kinetic schemes discussed in Sections 5.6–5.9. A plot of x_R'/x_D vs. x_R/x_D is shown in Figure A.1a. Notice that $x_R' \cong x_R$ when $x_R/x_D < 0.5$. Combining eqs (4.31) and (A.24), we have

$$x_D/x_R = 1.61 \, (\nu/D)^{\frac{1}{6}} [(k_1 + k_{-1})/\omega]^{\frac{1}{2}} \qquad (A.27)$$

Figure A.1b shows x_R'/x_D as a function of $\omega^{\frac{1}{2}}$ for several values of $(k_1 + k_{-1})$, assuming that $\nu/D = 10^3$.

If Y and Z are involved only in one chemical reaction which interconverts them, then the sum of the two concentrations, $C_Y(x) + C_Z(x)$, must have a linear gradient. In other words, we can write

$$J_Y(0) + J_Z(0) = - (D/x_D)[C_Y^* + C_Z^* - C_Y(0) - C_Z(0)] \qquad (A.28)$$

If O and R are interconverted by electron transfer, then their fluxes at the electrode must be equal and opposite in sign,

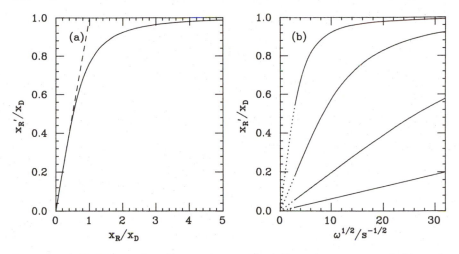

Figure A.1 Plots of x_R'/x_D (a) as a function of x_R/x_D, and (b) as a function of $\omega^{\frac{1}{2}}$ for $(k_1 + k_{-1}) = 1$, 10, 100, and 1000 s^{-1} (top to bottom).

$$J_O(0) = - J_R(0) \qquad (A.29)$$

Finally, we will assume nernstian behavior for the electron transfer steps with

$$\frac{C_O(0)}{C_R(0)} = \theta = \exp \frac{nF(E - E^\circ)}{RT} \qquad (A.30)$$

Thus in general, eqs (A.23), (A.25), and (A.28)–(A.30) interrelate the fluxes and surface concentrations. Derivation of the r.d.e. current for the various kinetic schemes then can make use of these equations to obtain an expression for the flux of O at the electrode surface, which determines the current.[‡]

Derivation of eq (5.65). We can use the results obtained above to write expressions for the fluxes of the species involved in the CE mechanism,

---

[‡] For further discussion of kinetic studies using the r.d.e. electrode, see Albery *(A6)*.

$$A \quad \overset{k_1}{\underset{k_{-1}}{\rightleftarrows}} \quad O$$

$$O + ne^- \rightleftarrows R$$

From eqs (A.23) and (A.29), we have

$$J_R(0) = (D/x_D)C_R(0) = -J_O(0)$$

where we have assumed that $C_R{}^* = 0$. Equation (A.30) can be used to eliminate $C_R(0)$:

$$J_O(0) = -(D/x_D)C_O(0)/\theta$$

From eq (A.25), we have

$$J_O(0) = -(D/x_R')[KC_A(0) - C_O(0)]$$

where $K = k_1/k_{-1}$ and we have assumed A to be electro–inactive so that $J_A(0) = 0$. Finally, eq (A.28) gives

$$J_O(0) = -(D/x_D)[C^* - C_A(0) - C_O(0)]$$

where $C^* = C_A{}^* + C_O{}^*$. We next combine the three expressions for $J_O(0)$ to eliminate $C_A(0)$ and $C_O(0)$, obtaining

$$J_O(0) = -\frac{DKC^*}{x_R' + Kx_D + \theta(1 + K)x_D}$$

Multiplying by $(-nFA)$, we obtain the current

$$i = \frac{nFADKC^*}{x_R' + Kx_D + \theta(1 + K)x_D} \tag{5.65}$$

Derivation of eq (5.84). For the EC' mechanism,

$$O + ne^- \rightleftarrows R$$

$$R + A \overset{k}{\rightarrow} O + B$$

the modified diffusion equation for C_O is

$$\frac{\partial C_O}{\partial t} = D \frac{\partial^2 C_O}{\partial x^2} + k'C_R$$

where $k' = kC_A$ and we assume that $C_A^* \gg C_O^*$. The sum of the O and R concentrations should be constant,

$$C_O(x) + C_R(x) = C_O^*$$

so that

$$C_R(x) = C_O^* - C_O(x)$$

and application of the steady–state approximation gives

$$D \frac{d^2 C_O}{dx^2} + k'[C_O^* - C_O(x)] = 0$$

Defining the reaction layer thickness

$$x_R^2 = D/k'$$

the solution to the differential equation is

$$C_O(x) = C_O^* - [C_O^* - C_O(0)] \frac{\exp[(x_D - x)/x_R] - \exp[-(x_D - x)/x_R]}{\exp(x_D/x_R) - \exp(-x_D/x_R)}$$

The flux of O at the electrode surface then is

$$J_O(0) = - (D/x_R')[C_O^* - C_O(0)]$$

where x_R' is given by eq (A.26). When the electrode potential is sufficiently negative that $C_O(0) = 0$, we obtain the limiting current

$$i_L = nFAJ_O(0) = nFAC_O^*(D/x_R') \tag{5.84}$$

Derivation of eqs (5.89) and (5.91). The ECE mechanisms are

$$O_1 + e^- \rightleftarrows R_1 \qquad E_1^\circ$$

$$O_2 + e^- \rightleftarrows R_2 \qquad E_2^\circ$$

$$R_1 \overset{k}{\underset{\rightarrow}{\rightarrow}} O_2 \qquad \overset{\rightarrow}{} \overset{\rightarrow}{} \text{(ECE)}$$

$$R_1 \overset{k}{\underset{\rightarrow}{\rightarrow}} R_2 \qquad \overset{\rightarrow}{} \overset{\leftarrow}{} \text{(ECE)}$$

We assume equal diffusion coefficients for all species and $C_{O2}^* = C_{R1}^* = C_{R2}^* = 0$. Following the general procedure outlined above, eq (A.23) gives the flux of the substrate

$$J_{O1}(0) = - (D/x_D)[C_{O1}^* - C_{O1}(0)]$$

The flux of R_1 is given by eq (A.25) with $K = 0$ since the chemical step is assumed to be irreversible:

$$J_{R1}(0) = (D/x_R')C_{R1}(0)$$

Since $J_{R1}(0) = - J_{O1}(0)$, we can combine the expressions for these fluxes to obtain

$$C_{O1}^* - C_{O1}(0) = (x_D/x_R')C_{R1}(0)$$

and defining

$$\frac{C_{O1}(0)}{C_{R1}(0)} = \theta_1 = \exp \frac{F(E - E_1^\circ)}{RT}$$

we can solve for $C_{O1}(0)$,

$$C_{O1}(0) = \frac{\theta_1 C_{O1}^*}{\theta_1 + x_D/x_R'} = \frac{\theta_1 x_R' C_{O1}^*}{x_D + \theta_1 x_R'} \tag{A.31}$$

and compute $J_{O1}(0)$:

$$J_{O1}(0) = - \frac{DC_{O1}^*}{x_D + \theta_1 x_R'} \tag{A.32}$$

For the $\overrightarrow{E}\overrightarrow{C}E$ mechanism, we can write the flux of R_2, using eq (A.23):

$$J_{R2}(0) = (D/x_D)C_{R2}(0)$$

and the total flux of R_1 and O_2, using eq (A.28):

$$J_{R1}(0) + J_{O2}(0) = (D/x_D)[C_{R1}(0) + C_{O2}(0)]$$

The corresponding flux expressions for the $\overrightarrow{E}\overleftarrow{C}E$ mechanism are

$$J_{O2}(0) = (D/x_D)C_{O2}(0)$$

and

$$J_{R1}(0) + J_{R2}(0) = (D/x_D)[C_{R1}(0) + C_{R2}(0)]$$

In either case, $J_{R1}(0) = -J_{O1}(0)$, $J_{R2}(0) = -J_{O2}(0)$, $C_{R1}(0) = C_{O1}(0)/\theta_1$, and

$$\frac{C_{O2}(0)}{C_{R2}(0)} = \theta_2 = \exp\frac{F(E - E_2°)}{RT}$$

Either set of flux expressions can be combined with eqs (A.31) and (A.32) to obtain $C_{O2}(0)$:

$$C_{O2}(0) = \frac{\theta_2}{1 + \theta_2}\frac{C_{O1}{}^*(x_D - x_R')}{x_D + \theta_1 x_R'} \qquad (A.33)$$

The current is proportional to the sum of the O1 and O2 fluxes:

$$i = -FA[J_{O1}(0) + J_{O2}(0)]$$

For the $\overrightarrow{E}\overrightarrow{C}E$ mechanism, this becomes

$$i = FA(D/x_D)[C_{O1}{}^* - C_{O1}(0) + C_{O2}(0)/\theta_2]$$

Substituting eqs (A.31) and (A.33) for the surface concentrations, we get

$$i = FA \left[\frac{DC_{O1}^*}{x_D + \theta_1 x_R'} \right] \left[1 + \frac{1 - x_R'/x_D}{1 + \theta_2} \right] \qquad (5.89)$$

For the $\overrightarrow{E}\overleftarrow{C}E$ mechanism, the current is somewhat different:

$$i = FA(D/x_D)[C_{O1}^* - C_{O1}(0) - C_{O2}(0)]$$

On substitution of eqs (A.31) and (A.33) for the surface concentrations, we have

$$i = FA \left[\frac{DC_{O1}^*}{x_D + \theta_1 x_R'} \right] \left[1 - \frac{\theta_2(1 - x_R'/x_D)}{1 + \theta_2} \right] \qquad (5.91)$$

APPENDIX 7.   ANSWERS TO SELECTED PROBLEMS

1.1   (a) $Pt | Fe^{2+}, Fe^{3+}) \| MnO_4^-, Mn^{2+}, H^+ | Pt$

(b) $Fe^{3+} + e^- \rightarrow Fe^{2+}$
$MnO_4^- + 8 H^+ + 5 e^- \rightarrow Mn^{2+} + 4 H_2O$

(c) $5 Fe^{2+} + MnO_4^- + 8 H^+ \rightarrow 5 Fe^{3+} + Mn^{2+} + 4 H_2O$

(d) $E^\circ = 1.51 - 0.771 = 0.74$ V

(e) $E = 0.56$ V

(f) $\Delta G^\circ = -357$ kJ mol^{-1}, $K = 3 \times 10^{62}$

1.2   (a) 1.0662 V
(b) –0.607 V
(c) –0.76 V

1.4 $K_{sp} = 1.6 \times 10^{-8}$

1.5 $K_{sp} = 4.5 \times 10^{-18}$, pH = 7.00

1.7 $E = 0.077, 0.085, 0.107, 0.145, 0.302, 0.458, 0.491$ V

1.8 1.27×10^{-4} M

1.9 $K = 9.8 \times 10^{18}$

1.10 (b) pH 5.0

1.11 flux $= 2.2 \times 10^{-3}$ mole Na$^+$ s^{-1}; firing rate $= 0.022$ s^{-1}

1.12 (a) $E' = +0.043, +0.353, -0.154$ V
(b) $\Delta G' = -8.3, -68.1, +29.7$ kJ mol^{-1}
(c) $K' = 28, 8.6 \times 10^{11}, 6.2 \times 10^{-6}$

1.13 (a) $\Delta G^\circ = -130$ kJ mol^{-1}
(b) $\Delta G^\circ = -76$ kJ mol^{-1}
(c) $\Delta G^\circ = -206$ kJ mol^{-1}

1.14 (a) $\Delta G^\circ = -35$ kJ mol^{-1}
(b) $\Delta G^\circ = +234$ kJ mol^{-1}

(c) $\Delta G^\circ = -458$ kJ mol^{-1}

1.15 pH $= 10.39 \pm 0.08$

1.16 $[Na^+] = (1.12 \pm 0.03) \times 10^{-4}$ M

1.17 (a) $k_{H,Na} = 7.8 \times 10^{-12}$
(b) pH 11.19 solution would give apparent pH 11.14

1.18 (a) $Zn(s) + 2 OH^- \rightarrow ZnO(s) + H_2O + 2 e^-$
$Ag_2O(s) + H_2O + 2 e^- \rightarrow 2 Ag(s) + 2 OH^-$
$Ag_2O(s) + Zn(s) \rightarrow 2 Ag(s) + ZnO(s)$
(b) 1100 J g^{-1}

2.4 $\sigma = \epsilon \epsilon_0 \Phi_a (1/a + 1/x_A)$

2.8 (a) $Q_\infty = 2.0 \ \mu C$
(b) $i_0 = 1.0$ mA
(c) $t = 0.092$ s

2.9 (a) $\gamma_\pm (NaCl)$ (b) $\gamma_\pm (NaF)/\gamma_\pm (NaCl)$ (c) $\gamma_\pm (NaF)$

2.11 $\gamma_\pm (exptl) = 0.905, 0.875, 0.854, 0.826, 0.807, 0.786$
$\gamma_\pm (57) = 0.889, 0.847, 0.816, 0.769, 0.733, 0.690$
$\gamma_\pm (56) = 0.903, 0.872, 0.851, 0.821, 0.800, 0.776$

2.12 $[KOH] = 0.988$ M, $\gamma_\pm = 0.742$

2.14 $E = 0.767$ V

3.1 (a) $L/A = 29.05$ m^{-1}
(b) $\Lambda = 14.66 \times 10^{-4}$ S m^{2}mol^{-1}
(c) $\Lambda^\circ = 390.7 \times 10^{-4}$ S m^{2}mol^{-1}
(d) $a = 0.0375$, $K = 1.5 \times 10^{-5}$

3.2 $\Lambda^\circ = 133.4 \times 10^{-4}$ S m^{2}mol^{-1}

3.3 $s(calc) = 2.88 \times 10^{-4}$ S m$^{\frac{1}{2}}$mol$^{-\frac{3}{2}}$
$s(expt) = 3.03 \times 10^{-4}$

3.4 $s = 1.94 \times 10^{-3}$ S m$^{\frac{1}{2}}$mol$^{-\frac{3}{2}}$

3.5 0.203

3.6 1.4×10^{-4} M

3.7 $K = 1.6 \times 10^{-4}$

3.8 $u = 1.046 \times 10^{-7}$, 1.145×10^{-7} m^{2}V^{-1}s^{-1}
$r = 274, 334$ pm
$f = 4.60 \times 10^{-12}$, 5.60×10^{-12} kg s^{-1}
$D = 8.96 \times 10^{-10}$, 7.36×10^{-10} m^{2}s^{-1}

3.12 $D = 2.45 \times 10^{-11}$ m^{2}s^{-1}, $t \cong 6.5$ years

3.13 $D = 1.59 \times 10^{-9}$ m^{2}s^{-1}

3.14 $t_{Cu} = 0.366$

3.15 $[Ba(OH)_2] = 0.0422$ M, $\Lambda = 411 \times 10^{-4}$ S m^{2}mol^{-1}

3.17 $[Na^+]_a = 0.0319$ M, $[Na^+]_\beta = 0.0281$ M, $\Delta\Phi = 3.3$ mV

4.4 $E_{\frac{1}{2}} = 0.764$ V

4.5 A square wave signal is required.

4.7 (a) Gain $> 10^4$
(b) Gain $> 4 \times 10^4$
(c) Scan rate is 2 ppm smaller when output voltage is –1 V.
(d) $i = 100$ μA, $R_{cell} < 10$ kΩ, nominal current correct to 0.001%

4.8 $C_O = 0.14$ mM

4.12 4.75 μA

4.13 Diameter greater than 0.5 mm

4.14 Electrode radius less than about 2.3 μm

4.15 (a) $E_{\frac{1}{2}} = -0.693$ V
(b) $R = 1480$ Ω

4.17 $\beta \cong 2.4$ eV (Figure 32a), $\beta \cong 2.1$ eV (Figure 32b)

4.20 (a) $p = 2$, $E_{\frac{1}{4}} = -0.258$ V
(b) $\beta \cong 10^{35}$

4.23 $\beta \cong 4 \times 10^{12}$

5.2

$$i = nFA \left[\frac{k_c C_O^* - k_a C_R^*}{1 + k_a/k_D + k_c/k_D} \right]$$

5.5 $j_0 = 0.0079$ A m^{-2}, $a_{app} = 0.58$, $k_0 = 2.2 \times 10^{-10}$ m s^{-1}

5.9 (a) $E_{\frac{1}{2}} - E^\circ = -24.7$ mV
(b) $k_0^{\frac{2}{3}}/k_D = 0.141$

5.10 (b) $k_0 \cong 2.1 \times 10^{-4}$ m s^{-1}

5.12 $k_0 = 5.8 \times 10^{-5}$ m s^{-1}

5.13 Resistive component: $R_s = |Z_f| \cos \phi$
Capacitive component: $1/\omega C_s = |Z_f| \sin \phi$

5.14 Width at half height $= 90.6$ mV.

5.15 Width between extrema $= 67.6$ mV.

5.16 (a) $k_0 \leq 10^{-5}$ m s^{-1}
(b) $k_0 \leq 0.0002$ m s^{-1}
(c) $k_0 \leq 0.002$ m s^{-1}

5.17 $k_0 \cong 5 \times 10^{-4}$ m s^{-1}

5.18 $k_1 K = 0.008$ s^{-1}

5.19 (a) $C_A > 10^{-6}$ M
(b) $A = 1$ cm^2

5.25 (a) $(i_L - i)/i = \theta_2(1 + 2\theta_1)/(2 + \theta_2)$
(b) $E_{\frac{1}{2}} = \frac{1}{2}(E_1^\circ + E_2^\circ)$
(c) The Heyrovský–Ilkovič equation is obtained when $E_2^\circ \gg E_1^\circ$
(d) $E_{\frac{3}{4}} - E_{\frac{1}{4}} = 42.9, 33.8, 30.3, 28.5$ mV for $E_2^\circ - E_1^\circ = 0$, 50, 100, and 200 mV. The Tomeš criterion for a reversible two-electron wave gives 28.2 mV.

5.26 k = 0.8 s^{-1}

5.28 (b) k = 1.0 s^{-1}

6.2 t = 27.5 days

6.3 (a) 22 min
(b) C_B(t) = 0.10, 0.14, and 0.12 mM at t = 5, 10, and 25 min

6.6 126 C

6.7 (a) 0.234 C
(b) 48.5 μM

6.8 (a) R = 1000 Ω
(b) R_{cell} < 9 kΩ

6.9 3.51 mM

6.10 5–mL sample, 2 ˙mA current give t = 241 s, uncertainty is ±0.6%

6.12 Hydrogen reduction current: j = 0.27 A m^{-2} on Pt (ϕ = 0.997), j = 1.9 mA m^{-2} on Cu (ϕ = 0.9992)

6.13 Optimum current = 133 A
Cost = (raw materials + capital and labor + energy)
Cost = (1.00 + 0.84 + 0.91) = $2.75 per kilogram
With doubled energy costs, optimum current = 91 A
Cost = $3.57 per kilogram

6.15 (a) E_c = –0.73 V
Corrosion rate = 0.5 mg Zn m^{-2}s^{-1}
Hydrogen evolution rate = 0.18 mL m^{-2}s^{-1}
(b) Corrosion rate = 2.2 mm year^{-1}

INDEX OF NAMES

SUBJECT INDEX